Green Power: Perspectives on Sustainable Energy Generation

Green Power: Perspectives on Sustainable Energy Generation

Editor: Marrianne Fox

CALLISTO REFERENCE

www.callistoreference.com

Callisto Reference,
118-35 Queens Blvd., Suite 400,
Forest Hills, NY 11375, USA

Visit us on the World Wide Web at:
www.callistoreference.com

ISBN: 978-1-63239-873-4 (Hardback)

The publisher's policy is to use permanent paper from mills that operate a sustainable forestry policy. Furthermore, the publisher ensures that the text paper and cover boards used have met acceptable environmental accreditation standards.

Trademark Notice: Registered trademark of products or corporate names are used only for explanation and identification without intent to infringe.

Printed in the United States of America.

Cataloging-in-Publication Data

Green power : perspectives on sustainable energy generation / edited by Marrianne Fox.
    p. cm.
Includes bibliographical references and index.
ISBN 978-1-63239-873-4
1. Renewable energy sources. 2. Clean energy. 3. Biomass energy. 4. Agriculture and energy.
5. Energy development--Technological innovations. I. Fox, Marrianne.
TJ808 .G74 2017
621.042--dc23

Table of Contents

Permissions

List of Contributors

Index

Preface

The purpose of the book is to provide a glimpse into the dynamics and to present opinions and studies of some of the scientists engaged in the development of new ideas in the field from very different standpoints. This book will prove useful to students and researchers owing to its high content quality.

Green power or sustainable energy is defined as the harvesting of electric power from sources that are replenishable. This book on green power deals with the various technologies that are used in the creation of sustainable energy. Sustainable energy is closely related to sustainable development and green power. The aim of this book is to present researches that have transformed this discipline and aided its advancement. The topics introduced here are of utmost significance and are bound to provide detailed knowledge to readers about the various theories and techniques of sustainable energy. This book is a complete source of knowledge on the present status of this important field. With state-of-the-art inputs by acclaimed experts of this field, this book targets students and professionals.

At the end, I would like to appreciate all the efforts made by the authors in completing their chapters professionally. I express my deepest gratitude to all of them for contributing to this book by sharing their valuable works. A special thanks to my family and friends for their constant support in this journey.

Editor

Determination of the performance characteristics of a modified solar water heating system

Orovwode Evwiroghene Hope[1], Wara Tita Samuel[1], Agbetuyi Ayoade Felix[1],
Adediran Olalekan Gideon[2], Adoghe Uwakhonye Anthony[1]

[1]Electrical and Information Engineering Department, School of Applied Engineering, College Engineering, Covenant University, Ota, Nigeria
[2]Department of Electrical / Electronics Engineering, College of Engineering, Federal University of Agriculture, Abeokuta, Ogun State, Nigeria

Email address:
hope.orovwode@covenantuniversity.edu.ng (O. E. Hope), samuel.wara@covenantuniversity.edu.ng (W. T. Samuel)
ayo.agbetuyi@covenantuniversity.edu.ng (A. A. Felix), Anthony.adoghe@covenantuniversity.edu.ng (A. O. Gideon),
Walesco4real2002@yahoo.com (A. U. Anthony)

Abstract: Heated water is very essential for domestic, agricultural, commercial as well as industrial operations. Obtaining hot water usually comes with a cost. Though conventional water heating systems exist, however, the operational cost and environmental impact of this method is contributing seriously to the already worsened energy situation in Nigeria. Solar water heating system (SWHS) is a practical application to replace the conventional electrical water heater. The low efficiency of solar water heaters has been the main reason why many people still prefer the conventional methods. To improve on the performance characteristics of the existing solar water heater, a modified version was developed by re-shaping the solar collector and making it adjustable so that it could attract maximum solar radiation. The pipe was replaced with another pipe of smaller diameter to allow for more heating surface area using the same collector frame. The tank and the sides of the collector frame were lagged with a special polysterine material to reduce heat loss due to radiation and a unidirectional control valve was introduced to the tank so that the already heated water that can be preserved in the tank. Tests were carried out on the existing as well as the modified heaters. The results obtained show a significant improvement on the performance of the SWHS. The inlet average temperature increased form $24.64\,^{\circ}C$ to $35.14^{\circ}C$ while the outlet temperature increased from $51.53^{\circ}C$ to $60.73^{\circ}C$ which corresponds to an improvement in the performance of the system's inlet water temperature of 15.65% and outlet water temperature of 15.14%.

Keywords: Solar Energy, Heating System, Modified System, Collector Design, Solar Insulator

1. Introduction

Warm water is a necessity in most homes for various chores including but not limited to bathing, cooking, washing and sterilization of babies' accessories. In agriculture, warm water is needed for pen cleaning in piggeries, dairy operations like cleaning or sterilizing equipment and to warm and stimulate cow's udder [1]. Obtaining hot water usually comes with a cost. Water is heated using conventional energy sources such as wood and coal in the rural areas while electricity, kerosene and liquefied petroleum gas (LPG) are used in the urban areas. It has been reported that energy consumed in the residential sector is about one-third of the total delivered energy [2] and more than a quarter of this energy is used for heating applications.

Various water heating schemes evolved over time to satisfy human needs, the choice of the type and method depends on the resources available and the derivable benefits. Some of the water heating methods are highlighted below:

1.1. Storage Water Heaters

Storage water heaters remain the most popular type for residential heating needs. The systems involves the heating of water stored in a tank using conventional fuel options such as electricity, natural gas, oil, and propane [3]. It is cheap to install and very efficient. However, the water is constantly heated in the tank hence energy can be wasted even when heated water is needed at that time. This makes the storage water heaters method very expensive to operate. This is called standby heat loss. Also, burning of these fuels can contribute to environmental pollution.

1.2. Heat Pump Water Heating

Heat pump water heaters use electricity to move heat from air and transfers it to heat water instead of generating heat directly using a compressor and evaporator are built into the water heater unit [4]. The benefit of electric heat pump water heater is that though it runs on electricity yet is roughly three times more efficient than a conventional electric and that makes it highly cost effective advantage over time. When used in the right environment, it saves energy, money and reduces greenhouse gas emissions. A major drawback of electric heat pump water heaters is that the initial purchase price and installation cost may be higher than for a traditional electric water heater.

1.3. Tankless Water Heaters

Tankless water heaters use conventional method (gas, electricity, wood or coal) to heat the water on an as-needed basis without storing any heated water in a tank of any kind [5]. This method is preferred to storage waters in that money is not being spent heating a tank full of water and keeping it hot. Hence savings on energy costs as the boiler does not need to operate frequently. One of the disadvantages of the tankless water heater is that it is somewhat expensive to install.

2. Alternative Energy Sources for Water Heating

The challenges with the use of conventional energy sources for water heating include its availability and the high cost of obtaining these fuels [6]. Also, environmental concerns and the need to de-carbonize water heating processes has made it imperative to source for alternative heating techniques using non- conventional means like the solar energy.

A solar hot water system uses energy from the sun to heat water. Through specially designed solar collectors, sun's rays can be harnessed to heat the water, which then flows to a storage tank.

Two important components of the solar water heating system are the collector and storage tank. Various system designs can be classified as passive or active and as direct (also called open loop) or indirect (also called closed loop).Passive water heater systems operate without pumps and controls and can be more reliable, more durable, easier to maintain, longer lasting, and less expensive to operate than active systems. Active solar water heaters incorporate pumps and controls to move heat-transfer fluids from the collectors

to the storage tanks. Both active and passive solar water-heating systems often require conventional water heaters as backups, or the solar systems function as pre-heaters for the conventional units because of their low efficiency. The main advantages of this heating system are the fact that it is cheap to run and it does not cause pollution, as conventional heating systems do. A disadvantage of solar water heaters, however, is the fact that the cost of installation is very high compared to conventional water heaters but this cost can be off-set after a period [6]

3. Materials and Methods

In designing a solar water heating system, the following parameters determine the efficiency of the system.

3.1. Solar Collector

The heat benefit of a solar heating system depends on the amount of solar energy that can be captured. Hence, the solar collector is the most important part of the solar heating system. The collector design is a function of the amount of solar insolation (which is the amount of electromagnetic energy i.e. solar radiation incident on the surface of the earth) [7] available in a particular place at a time. The size of a solar collector that is required in any application depends on the level of solar insolation of that region.

The solar insolation of any place can be gotten by measuring the radiation value generally expressed in $kWh/m^2/day$. This is defined as the amount of solar energy that strikes a square meter of the earth's surface in a single day. This value is averaged to account for differences in the days' length.) Using a pyrometer, recording and analyzing the data using a Computer, the monthly or yearly values can be gotten for any particular locality.

3.2. Efficient Heat Transfer

The solar heat energy absorbed by the metallic tube has to be transferred to the water to be heated. Therefore, thermal conductivity is of high priority in a solar Water Heating System. The tube has to be made of a material that has high thermal conductivity, cheap and non corrosive as the tube will always be in contact with water. Copper materials (pipes and fins) meet the highlighted characteristics.

The diameter of the pipe to be used should be such that it allows more heating per surface area. Smaller diameter pipes should be preferred to larger diameter pipes.

3.3. Glaze Material

All solar radiation passing through the glazing material must be transmitted into the copper pipe/tube, so as to heat the water to a higher temperature. Transmittance varies not only for different materials but also varies with the wavelength of the radiation. The glazing should be very transparent to incoming shortwave radiation but opaque to outgoing long wave (thermal) radiation, because radiant losses account for over 70 per cent of collector heat loss [8]

3.4. Absorptance and Emissivity

The percentage of incoming radiation that is absorbed by a material is referred to as its absorptance and is a measure of the ease with which a material or surface collects energy. [9]

The best materials are those with high absorptance and very low emittance. Black paint has absorptance rate of 0.95 and emissivity of 0.875 This makes it a good choice for the SWHS [10]

3.5. Collector Casing

The purpose of the casing is to provide thermal insulation for the collector to prevent heat loss into the environment. Insulating materials like wood or ceramics with proper lagging could be ideal.

3.6. Area of Collector

The surface area of the collector is another important design factor to be considered. The efficiency of a flat plate collector is given as the heat gained by water with respect to the actual solar energy received by the flat plate collector expressed as

$$mcp\Delta T = Ac\, I\eta \quad [11] \qquad (3.1)$$

where m = mass of water in Kilograms (kg)
cp = Specific heat of water(J/g°C)
ΔT = Temperature difference (Outlet Temperature – Inlet Temperature °C)
A_c = Area of Collector
I = intensity of solar radiation, (W/m^2)
η = Collector efficiency
Therefore,

$$A_c = \frac{mcp\Delta T}{I\eta} \qquad (3.2)$$

but efficiency , η is given as [12]

$$\eta = F_R\tau\alpha - F_R U_L\left[\frac{Ti - Ta}{I}\right] \qquad (3.3)$$

where F_R = Collector heat removal factor
I = Intensity of solar radiation, W/m^2 27
T_i = Inlet fluid temperature, °C
T_a = Ambient temperature, °C
U_L = Collector overall heat loss coefficient, W/m^2 K
τ = Transmission coefficient of glazing
α = Absorption coefficient of plate

The equation (3.3) shows that the efficiency of a flat plate collector is dependent on the Transmission coefficient of glazing (τ), the Absorption coefficient of plate, collector heat removal factor (F_R), intensity of solar radiation (I), ambient temperature (T_a), inlet fluid temperature (T_i) and the collector overall heat loss coefficient (U_L).

For the SWHS from the efficiency equation, F_R, τ, α and U_L are fixed by the collector design. T_i, T_a and I are variables determined by the application. This simply implies that the solar collector efficiency can change when T_i, T_a and I are changed.

4. The Existing Solar Water Heating System

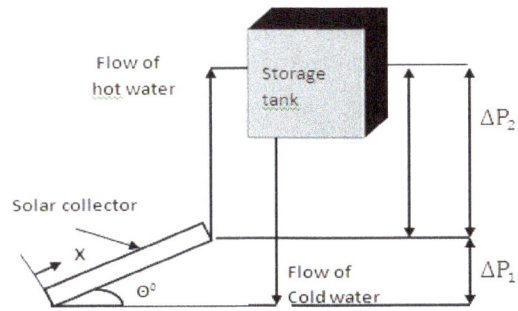

Figure 1. *The schematic diagram of a natural circulation solar water heater*

The existing SWHS is a natural circulation solar water heating system whose schematic diagram is shown in Figure1. The system consists of a flat-plate solar collector, storage tank and connecting pipes. The plate of the solar collector has the water pipes and headers in the grooves to maintain good contacts. The Pipe is 9m long and has a diameter of ¾ inches with a single glass cover of 4mm and an Aluminum foil of 0.8m x 0.45m. The concept on which the SWHS operates is based on the thermosyhon principle. A thermosyphon solar water heater relies on warm water rising, a phenomenon known as natural convection, to circulate water through the solar collector and to the storage water tank [14]. Temperature in the storage water tank is a function of the buoyancy-induced flow of heated water from the water heater to the tank while cooler water in the tank flows down pipes to the bottom of the solar collector, causing circulation throughout the system. The water storage tank is attached to the top of the solar collector. The connection between flat-plate collector and the storage tank are in two parts; the return pipe and the flow pipe. The return pipe connects the outlet of the storage tank and the inlet of the collector together; while the flow pipe connects the outlet of the collector and the inlet of the storage tank together. The flat-plate collector is oriented in such a way that it receives maximum solar radiation during the desired season of use. The absorbing surfaces were painted with black paint. Being a natural convection system, the water flows through the pipes by the thermosyphonic force and enters the storage tank. The existing system is as shown in Figure 2.

Figure 2. *The existing solar water heating system*

5. Modifications Made to the Existing System

In other to improve on the efficiency of the existing system, the following modifications were made to produce the modified version shown in Figure 3.

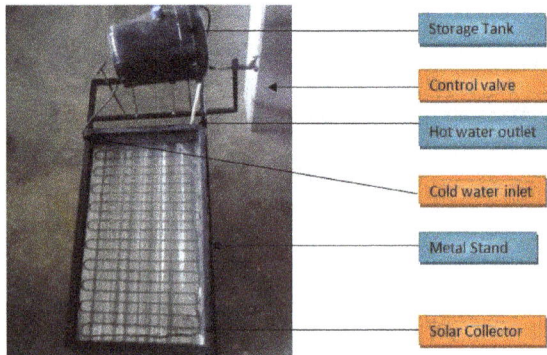

Figure 3. *The modified solar water heating system*

- The solar collector was re-shaped and also made adjustable so that it could attract maximum solar radiation. Aluminum foil was placed at the back of the wooden case so that any solar radiation not absorbed by the collector on the first pass will bounce back over the collector for secondary absorption.
- The pipe was replaced with another pipe of smaller diameter to allow for more heating surface area using the same collector frame.
- The tank and the sides of the collector frame were lagged with a special polysterine material to reduce heat loss due to radiation.
- A unidirectional control valve was introduced to the tank so that the already heated water that can be preserved in the tank.

6. Tests, Results and Discussion

The two systems (existing and modified) were placed side by side in the open atmosphere for a period of seven days between the hours of 8:00 and 18:00. The ambient temperature (T_{am}) was read as well as inlet water temperature (T_{in}) and the outlet temperature (T_{out}) for both systems. The obtained readings are as tabulated in Table 1 for the existing and Table 2 for the modified system. The hourly average for both systems is also shown in Tables 1 and 2 while the graphical presentation of the hourly average data is as shown in Figure 4.

Table 1. *performance record of the existing system.*

Time	Parameters	Day 1	Day 2	Day 3	Day 4	Day 5	Day 6	Day 7	Week Average (Existing)
8.00	T_{am}(°C)	25.50	25.00	24.00	25.00	25.70	24.00	25.50	24.96
	T_{in} (°C)	22.00	28.00	23.00	23.50	27.00	25.00	25.70	24.89
	T_{out} (°C)	37.50	31.00	30.00	35.00	35.00	33.00	38.00	34.21
9.00	T_{am}(°C)	26.00	26.00	25.00	26.00	26.00	25.50	27.00	25.93
	T_{in} (°C)	26.20	27.50	24.00	24.50	25.00	26.60	27.00	25.83
	T_{out} (°C)	50.00	35.00	37.50	39.00	45.00	40.00	39.00	40.79
10.00	T_{am}(°C)	30.00	26.50	28.00	26.00	27.00	28.00	27.00	27.50
	T_{in} (°C)	28.00	28.00	28.40	26.00	28.00	27.00	28.00	27.63
	T_{out} (°C)	52.00	50.20	63.50	44.00	55.00	50.50	60.60	53.69
11.00	T_{am}(°C)	29.80	28.00	28.00	27.00	30.00	28.00	29.00	28.54
	T_{in} (°C)	31.00	30.20	31.00	27.50	30.50	29.00	31.00	30.03
	T_{out} (°C)	70.00	63.00	68.00	57.00	65.00	63.00	64.50	64.36
12.00	T_{am}(°C)	30.00	30.00	28.50	29.00	31.00	28.00	30.50	29.57
	T_{in} (°C)	33.00	32.00	30.00	32.00	33.00	32.00	31.00	31.86
	T_{out} (°C)	73.00	52.00	76.00	48.00	72.00	67.00	69.00	65.29
13.00	T_{am}(°C)	31.00	29.00	28.00	31.00	32.00	29.00	30.00	30.00
	T_{in} (°C)	35.50	32.50	31.50	33.00	33.00	32.00	31.00	32.64
	T_{out} (°C)	70.00	68.00	49.50	61.00	67.00	65.00	67.00	63.93
14.00	T_{am}(°C)	28.20	32.00	28.00	29.50	29.70	26.00	28.00	28.77
	T_{in} (°C)	33.00	33.20	32.00	30.00	33.00	31.00	32.00	32.03
	T_{out} (°C)	52.00	68.00	48.00	67.00	65.00	64.00	66.50	61.50
15.00	T_{am}(°C)	28.00	31.00	28.00	31.00	29.00	28.00	31.00	29.43
	T_{in} (°C)	29.90	34.00	32.00	32.90	32.00	31.00	33.00	32.11
	T_{out} (°C)	50.00	63.00	48.50	72.00	62.00	61.00	63.00	59.93
16.00	T_{am}(°C)	29.00	26.00	28.50	29.50	29.00	28.00	29.00	28.43
	T_{in} (°C)	30.00	30.20	32.00	33.00	32.00	29.00	31.00	31.03
	T_{out} (°C)	46.00	38.50	44.00	52.00	47.00	45.00	51.00	46.21
17.00	T_{am}(°C)	26.50	26.50	28.00	28.50	27.00	26.00	28.00	27.21
	T_{in} (°C)	28.00	29.00	29.00	32.00	29.00	28.50	29.00	29.21
	T_{out} (°C)	31.00	33.20	55.50	43.00	45.00	43.00	44.00	42.10
18.00	T_{am}(°C)	25.00	25.00	28.00	27.00	26.00	24.00	26.00	25.86
	T_{in} (°C)	28.00	28.50	29.00	30.00	28.00	28.00	30.00	28.79
	T_{out} (°C)	29.00	31.00	34.40	31.00	41.00	38.00	39.00	34.77

T_{am}(°C)= ambient temperature

T_{in} (°C)= inlet water temperature

T_{out} (°C)= outlet water temperature

Table 2. performance record of the modified system.

Time	Parameters	Day 1	Day 2	Day 3	Day 4	Day 5	Day 6	Day 7	Week Average
8.00	T_{am}(°C)	25.50	25.00	24.00	25.00	25.70	24.00	25.50	24.96
	T_{in} (°C)	23.00	27.00	25.00	26.00	27.00	24.00	26.50	25.50
	T_{out} (°C)	39.00	37.00	33.00	37.00	48.00	36.00	39.50	38.50
9.00	T_{am}(°C)	26.00	26.00	25.00	26.00	26.00	25.50	27.00	25.93
	T_{in} (°C)	27.00	29.00	26.00	39.00	38.00	35.00	37.00	33.00
	T_{out} (°C)	53.00	52.00	42.00	47.00	59.00	60.00	63.00	53.71
10.00	T_{am}(°C)	30.00	26.50	28.00	26.00	27.00	28.00	27.00	27.50
	T_{in} (°C)	29.00	31.00	31.50	41.00	37.00	42.00	41.00	36.07
	T_{out} (°C)	54.00	56.00	56.00	46.00	68.00	69.00	72.00	60.14
11.00	T_{am}(°C)	29.80	28.00	28.00	27.00	30.00	28.00	29.00	28.54
	T_{in} (°C)	33.00	32.00	34.00	42.00	44.00	39.00	38.00	37.43
	T_{out} (°C)	76.00	64.00	74.00	59.00	71.00	69.00	71.00	69.14
12.00	T_{am}(°C)	30.00	30.00	28.50	29.00	31.00	28.00	30.50	29.57
	T_{in} (°C)	35.00	33.00	35.00	43.00	45.00	38.00	39.00	38.29
	T_{out} (°C)	78.00	62.00	77.00	57.00	74.00	68.00	69.50	69.36
13.00	T_{am}(°C)	31.00	29.00	28.00	31.00	32.00	29.00	30.00	30.00
	T_{in} (°C)	38.00	34.00	36.00	42.00	46.00	35.00	38.00	38.43
	T_{out} (°C)	75.00	72.00	67.00	73.00	71.00	72.00	68.00	71.14
14.00	T_{am}(°C)	28.20	32.00	28.00	29.50	29.70	26.00	28.00	28.77
	T_{in} (°C)	35.00	35.00	38.00	44.00	45.00	36.00	37.00	38.57
	T_{out} (°C)	56.00	70.00	65.00	79.00	67.00	68.00	69.00	67.71
15.00	T_{am}(°C)	28.00	31.00	28.00	31.00	29.00	28.00	31.00	29.43
	T_{in} (°C)	32.00	38.00	39.00	39.00	44.00	35.00	36.00	37.57
	T_{out} (°C)	54.00	68.00	63.00	75.00	65.00	65.00	67.00	65.29
16.00	T_{am}(°C)	29.00	26.00	28.50	29.50	29.00	28.00	29.00	28.43
	T_{in} (°C)	30.00	36.00	40.00	37.00	45.00	31.00	34.00	36.14
	T_{out} (°C)	50.00	60.00	62.00	73.00	51.00	62.00	65.00	60.43
17.00	T_{am}(°C)	26.50	26.50	28.00	28.50	27.00	26.00	28.00	27.21
	T_{in} (°C)	29.00	34.00	37.00	36.00	33.00	30.00	32.00	33.00
	T_{out} (°C)	47.00	58.00	60.00	71.00	48.00	61.00	63.00	58.29
18.00	T_{am}(°C)	25.00	25.00	28.00	27.00	26.00	24.00	26.00	25.86
	T_{in} (°C)	29.00	31.00	35.00	30.00	35.00	40.00	28.00	32.57
	T_{out} (°C)	43.00	52.00	56.00	63.00	46.00	58.00	62.00	54.29

T_{am}(°C)= ambient temperature

T_{in} (°C)= inlet water temperature

T_{out} (°C)= outlet water temperature

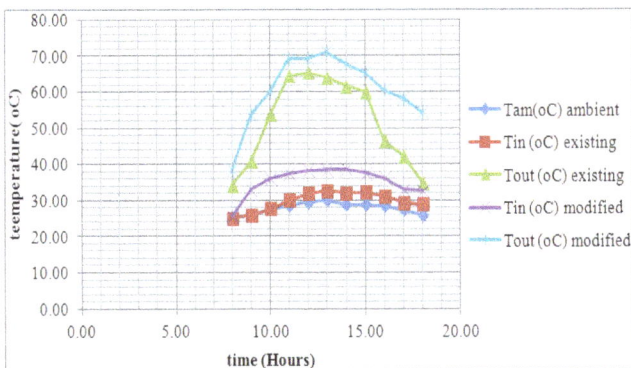

Figure 4. The graphical presentation of the data of Table 3

The results of Tables 1 and 2 and the graphical presentation of Figure 4 show a significant improvement on the performance of the SWHS. The inlet average temperature increased form 24.64 °C to 35.14°C while the outlet temperature increased from 51.53°C to 60.73°C which corresponds to an improvement in the performance of the system's inlet temperature of 15.65% and outlet temperature of 15.14%.

7. Conclusion

The performance characteristic of an existing solar water heater was improved upon by modifying some of the parameters. The modified version was developed by re-shaping the solar collector and also made adjustable so that it could attract maximum solar radiation. The pipe was replaced with another pipe of smaller diameter to allow for more heating surface area using the same collector frame. The tank and the sides of the collector frame were lagged with a special polysterine material to reduce heat loss due to radiation and a unidirectional control valve was introduced to the tank so that the already heated water that can be preserved in the tank. Tests were carried out on the existing as well as the modified heaters. The results obtained show a significant improvement on the performance of the SWHS. The inlet average temperature increased form 24.64 °C to 35.14°C while the outlet temperature increased from 51.53°C to 60.73°C which corresponds to an improvement in the performance of the system's inlet temperature of 15.65% and outlet temperature of 15.14%.

References

[1] Donald B. Roark,S G. H. Beck and H. C. Fryer 'Differences In Milking Response Under Prescribed Variations In Methods Employed To Stimulate Milk Let-Down' Online, available At Www.Journalofdairyscience.Org/Article/S0022-0302(52)93789.../Pdf Accessed July, 2014

[2] Sambo A. S. 'Renewable For Rural Development: The Nigeria Perspective' Proceedings of The ISESCO Science and Technology Vision Vol 1 (12-22)

[3] UK Department of Energy and Climate Change'The Future of Heating: A strategic framework for low carbon heat in the UK' (2012) Online, available At http://www.decc.gov.uk/en/content/cms/consultations/cons_s mip/cons_smip.aspx *Accessed August, 2014*

[4] Ozalla solar energy ' heat pumps' Online, available At www.ozalla.com/index.php?option=com_content&task=view. *Accessed August, 2014*

[5] U.S. Department of Energy's Office "Conventional Water Heating Efficiency" by U.S. Department of Energy's Office of Energy Efficiency and Renewable Energy, Building Technologies Program, Online, available At www.eere.energy.gov/buildings/info/components/waterheating /conventional.html. *Accessed August, 2014*

[6] Lee, D.W and Sharma, A (2007) 'Thermal Performance of the Active and Passive Heating Systems Based on Annual Operation' Solar Energy, Vol 81 No2, 207-215

[7] J. A. Duffie and W. A. Beckman, *Solar Engineering of Thermal Processes*, John Wiley and Sons, New York, NY, USA, 2nd edition, 1991.

[8] S. T. Wara and S. E. Abe (2013) 'Mitigating Climate Change by the Development and Deployment of Solar Water Heating Systems' Journal of Energy Hindawi Publishing Corporation Volume 2013, Article ID 679035

[9] Kalogirou S (2009). Solar Energy Engineering: Processes and Systems, 1st edition, Elsevier Publications, London, UK,

[10] Andrea Ambrosini, Timothy N. Lambert, Chad L. Staiger, Aaron C. Hall, Marlene Bencomo, Ellen B. Stechel (2010) 'Improved High Temperature Solar Absorbers for use in Concentrating Solar Power Central Receiver Applications' SANDIA REPORT SAND2010-7080.

[11] Yi-Mei Liu, Kung-Ming Chung *, Keh-Chin Chang and Tsong-Sheng Lee (2012), 'Performance of Thermosyphon Solar Water Heaters in Series' *Energies 5*, 3266-3278; ISSN 1996-1073

[12] Duffie, J.A.; Beckman, W.A. *Solar Engineering of Thermal Processes*; John Wiley Sons:

[13] Hoboken, NJ, USA, 1980; Chapter 12, pp. 487–497.

[14] Belessiotis, V.; Mathioulakis, E. (2002) 'Analytical approach of thermosyphon solar domestic hot water system performance'. *Sol. Energy* 2002, *72*, 307–315.

[15] Bansal, N.K. and K. Uhlemann, 1986. Thermosypon water heating system. Reviews of Renewable Energy Resources. Wiley Eastern Limited, New Delhi, India, 3: 189-196

2

Biodiesel production from unrefined palm oil on pilot plant scale

Jean Baptiste Nduwayezu, Theoneste Ishimwe, Ananie Niyibizi, Alexis Munyentwali[*]

Institute of Scientific and Technological Research (IRST), P.O. Box 227 Butare, Rwanda

Email address:

jbuwayezu@yahoo.co.uk (J. B. Nduwayezu), istheos@yahoo.fr (T. Ishimwe), niyibizan@yahoo.co.uk (A. Niyibizi),
alexmune@yahoo.fr (A. Munyentwali)

Abstract: As global warming and climate change issues are defying modern society sustainable development; biofuels, biodiesel included, are among promising solutions. Biodiesel is generally produced from renewable vegetable oils and animal fats via acid or base catalyzed transesterification. Depending on regional availability, biodiesel production feedstocks vary from vegetable oils such as rapeseed oil, soya oil, palm oil, and jatropha oil, to used cooking oil and animal fats, with each type of feedstock presenting its own process challenges rooting from its chemical composition. This paper reports about biodiesel production from crude palm oil on a pilot plant scale, subsequent to a laboratory scale investigation of biodiesel synthesis from various vegetable oil feedstocks. Prior to transesterification, pretreatment processes have been applied due to the fact that crude palm oil as a biodiesel feedstock possesses a high free fatty acid (FFA) content, water, solid impurities and waxes, all of which hinder an efficient transesterification if not dealt with accordingly. Those processes are mainly filtering, water evaporation, and FFA esterification which is done with 99.9% methanol and 96% sulfuric acid as a catalyst. In fact, the acid esterification process successfully handles the raw palm oil despite its high FFA content of 16.9%, and biodiesel is produced from that feedstock with a yield of 90.4%. A two steps transesterification is carried out using potassium methylate 32% in methanol as a catalyst and anhydrous methanol too. Laboratory analyses have also been used to monitor the process and assess the final product quality. Furthermore, biodiesel cold filtering and top layer intake tank systems of a filling station, both proved to be efficient at helping to obtain a refined product by getting rid of suspensions appearing in biodiesel at room temperature due to sterol glucosides and waxes.

Keywords: Biodiesel, Palm Oil, FFA, Acid Esterification, Transesterification, Environmental Friendly, Biodiesel Sediments

1. Introduction

Biodiesel has attracted considerable interest in recent years as an alternative, non-toxic, biodegradable and environmentally friendly fuel. It is defined as a mixture of monoalkyl esters of long chain fatty acids derived from renewable lipid sources; including vegetable oils, animal fat and algae oils, among others. These feedstocks are mainly composed of 85-98%wt triglycerides; three long fatty acid chains joined to a glycerol molecule [1].

Today, the most common biodiesel feedstocks are edible vegetable oils such as palm oil, soya bean oil, corn oil, and rapeseed oil. The main advantage of edible oils as feedstocks is that plantations and infrastructure are well established in most of oil producing countries, making these edible oils production expansion easier to meet the increasing demand.

Palm oil is by far a major component of global vegetable oil market. For instance palm oil is the world's most supplied of all edible oils; Malaysia Indonesia and Colombia being the world largest producers totalizing almost 83% of global production [2]. According to the US Department of Agriculture, during the 2013/2014 period, palm oil dominated the world vegetable oil production (169.56MMT) with 59.06MMT, whereas soybean and rapeseed produced 44.66MT and 26.09MMT respectively [3]. In addition, palm oil per hectare production yield of 2,404l/acre/yr ranks among the highest comparatively to other biodiesel oil feedstocks [4]. As shown on Figure 1, in 2010 worldwide palm oil production was reported the most growing of the main biodiesel vegetable oils feedstocks for almost the previous six years.

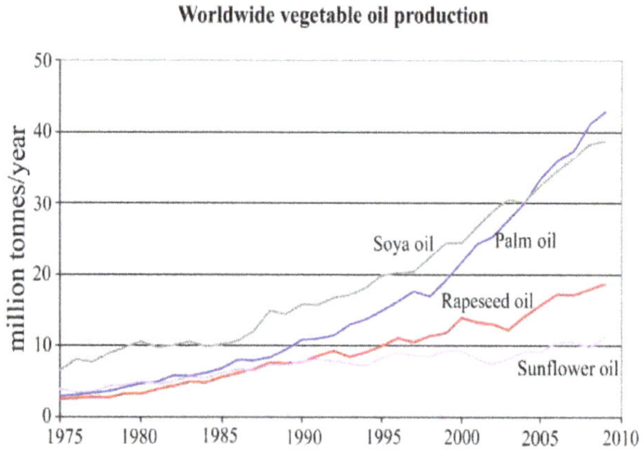

Figure 1. Annual growth of major edible oils production from 1975 to 2010 [4]

Depending on availability of feedstocks in different countries, biodiesel feedstocks preferences vary from country to country. For instance in the United States and Argentina soybean oil is the most used. European Union countries use rapeseed oil, whereas tropical countries such as Malaysia, Indonesia, Nigeria and Colombia prefer palm oil [2, 5].

What is more, the economic aspect of biodiesel production has shown to be the main hurdle. This is mainly due to high cost of feedstocks and a relatively low price of fossil fuels [6, 7].

As a result, evaluation of economic viability of biodiesel production is the most critical issue, due to high price of virgin and/or refined vegetable oils. The cost of feedstocks is revealed to account for 60–75% of the total cost of biodiesel fuel production. This is why feedstocks of low prices are essential to establish commercially viable biodiesel projects [8].

As it can be seen from the map on Figure 2, palm oil is actually the most abundant oil in the tropical regions where Rwanda is located.

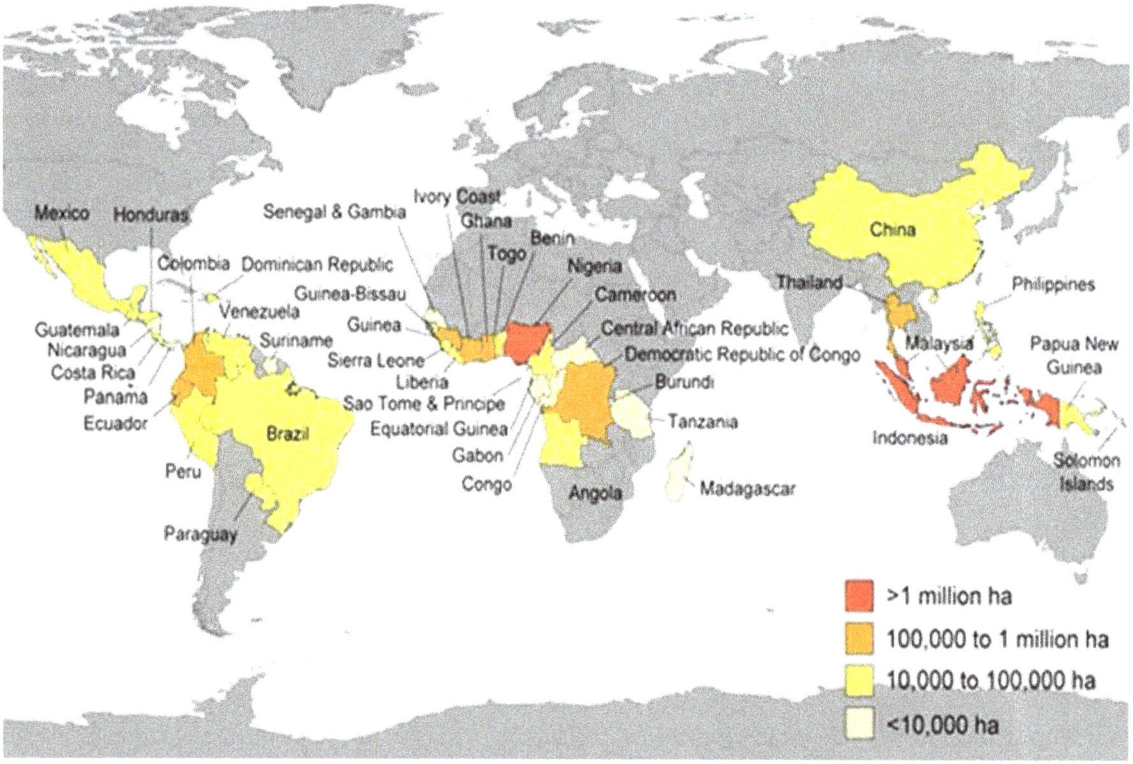

Figure 2. Palm oil cultivation in 43 countries worldwide [9]

In fact, as in many countries such as Malaysia for instance; the world second palm oil producer, the choice of biodiesel feedstock is driven by its affordability and availability. Therefore, being surrounded by palm oil producing countries, Rwanda is prone to use palm oil as biodiesel feedstock. Though the country targets 2nd generation biodiesel mainly based on jatropha oil as a feedstock, palm oil has been used ever since the outset of biodiesel production in 2007. In spite of palm oil inestimable role in biodiesel production in Rwanda, refined palm oil is generally recommended for industrial biodiesel production, but this leads to the most controversial biodiesel disadvantage - that of being not cost effective when produced from refined feedstocks [10, 11]. To ensure that biodiesel remains competitive over petroleum diesel, using crude palm oil may be the best option. However, conventional methods of using alkali catalyst to convert oils into biodiesel are not compatible with crude palm oil, given its inherent chemical composition. This is due to its high FFA content, high water content, insoluble impurities, etc [12, 13]. Crude palm oil is a low cost feedstock but associated with high content of FFA.

In fact, basic catalysts used to transesterify oils react with

FFA to generate soaps and water, thus consuming the catalyst (Figure 3.a). The consumption of the catalyst leads to incomplete transesterification, which results in a mixture of diacylglycerol, monoacylglycerol, methyl esters, glycerol, water and methanol. This mixture is difficult to separate and therefore prevents formation of usable biodiesel. Furthermore, the water formed by saponification exacerbates hydrolysis of formed alkyl esters into FFA as shown in Figure 3.b [14].

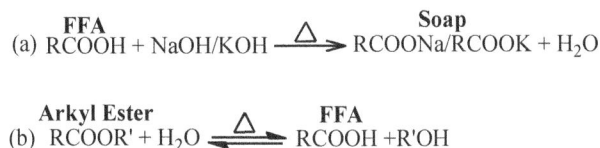

$$\text{(a)} \quad \underset{\text{FFA}}{RCOOH} + NaOH/KOH \xrightarrow{\triangle} \underset{\text{Soap}}{RCOONa/RCOOK} + H_2O$$

$$\text{(b)} \quad \underset{\text{Arkyl Ester}}{RCOOR'} + H_2O \underset{\triangle}{\rightleftharpoons} \underset{\text{FFA}}{RCOOH} + R'OH$$

Figure 3. *Parasitic reactions encountered in biodiesel production from oil high FFA and water contents [14, 15].*

Traditional technologies mainly used for crude palm oil extraction in developing countries are responsible for high water content, which consequently increases FFA content with oil aging [13, 16]. The common solution to overcome this issue is the use of refined oil. However, a system processing biodiesel from refined palm oil is technically successful, but not economically viable compared to using crude raw materials with pretreatment. In fact, refining will bring in FFA reduction, deodorization and bleaching, and require additional manpower and equipments, which will definitely increase the production cost [17]. Moreover, Dominik R. and Rainer J. argued that FFA which should have been converted into FAME is lost through refining, thus resulting in a decrease of the overall yield, especially if the feedstocks FFA content is high [12]. To this add the issue of cold flow properties of biodiesel which is a function of fatty acid composition. Since the type of feedstock plays an important role regarding the impurities and ways they must be removed, biodiesel from unrefined palm oil must be thoroughly purified and handled carefully. In fact, monoglycerides and sterol glucosides are at the origin of precipitations in biodiesel. Sterol glucosides are not soluble in biodiesel and crystalize slowly as the biodiesel cools down, and initial biodiesel seems to meet specification but after few days of storage filterability issue shows off. Sterol glucosides which are considered as dispersed fine solid particles as low as 35 ppm may accelerate crystallization and coprecipitation of other compounds and creation of deposits in the biodiesel. Those deposits are at the origin of filter clogging at temperatures even above cloud point [18]. Furthermore, unrefined palm oil produced by traditional presses contains too much solid materials like sludges, fiber, palm seeds, leaves, and cakes which need to be removed before any pretreatment.

There are different technologies and processes applied in biodiesel production, each having its pros and cons:

- *Base catalyzed transesterification* process is the most preferred technology and most economical process since it has a conversion rate up to 98%. It requires low temperatures and pressures, which leads to the use of usual equipments [19].

The drawback is that this technology needs raw oil of high purity, particularly with low water concentration and low FFA content. Otherwise saponification reaction sets in and competes with transesterification. The yield is proportional to the amount of catalyst but for economic purposes, the minimal amount of catalyst has to be used together with an optimal ratio of oil/alcohol [20].

- *Acid catalysed process* can produce biodiesel through both esterification and transesterification, but it is a very slow reaction requiring too much time up to one day for completion. It is also more corrosive than base-catalysed transesterification, requiring equipment of high cost[19, 21]. Moreover, water generated when FFA react with alcohol to form esters limits the completion of acid-catalysed esterification [20]. To overcome problems associated with above mentioned processes, saponification in case of high FFA content for base catalysed process and slow reaction for acid catalysed, a two steps reaction is recommended; an acid esterification in case of raw material with high FFA content, followed by a transesterification [2].

- *Enzymatic transesterification* has gained interest recently, particularly because it produces less wastes, consumes less energy, and has got high efficiency and high selectivity. However, its wide spread application is hindered by the fact that involved enzymes are costly to produce, and reactions require delicacy in parameters control [19].

- *Biodiesel production using solid catalysts* is also interesting due to the catalysts reusability, the fact that it produces purer biodiesel and glycerol and has very high yields. This process is also less selective with regards to the raw material water and FFA contents. Nevertheless, it requires hash reaction conditions and the solids catalysts are deactivated over a number of cycles [14].

- *Supercritical methanol method* is another technique of making biodiesel but this time without using any catalyst. It offers high efficiency up to 100% and it is environmental friendly. It eliminates the steps of neutralisation, washing and drying. The feedstocks water and FFA contents have no effect on the reaction and both triglycerides and FFA are converted simultaneously into biodiesel. However the reaction tank should be oversized due to a high quantity of alcohol used with respect to oil(up to 42:1), high pressure over 80 atmospheres, and high temperature up to 350-400°C. These conditions set the supercritical methanol process to harsh operating conditions and high costs [19].

There are *other emerging biodiesel production technologies* which do not use any catalyst. Among them there are:

Ultrasonic and microwaves enhanced transesterification show prominent advantages over base catalysed transesterification but these technologies need deep investigation before being applied to industrial production [19].

It's worth mentioning that biodiesel production from unrefined palm oil involves a set of chemical reactions which convert FFA and triglycerides into alkyl esters. These reactions are acid esterification and transesterification respectively.

1.1. Esterification

Esterification is a reversible reaction in which FFA are converted to alkyl esters via acid catalysis usually H_2SO_4. Its general equation is as follows:

$$\underset{\text{Acid Cat.}}{\overset{\triangle}{\underset{\longleftarrow}{\longrightarrow}}}$$

FFA   **Alkyl Ester**
$$RCOOH + R'OH \overset{\triangle}{\underset{\text{Acid Cat.}}{\rightleftharpoons}} RCOOR' + H_2O$$

Figure 4. Acid catalysed esterification general equation [14].

1.2. Transesterification

The triacylglycerols (TAG) react with a light molecular mass alcohol in presence of a catalyst under heat through the following general reaction equation:

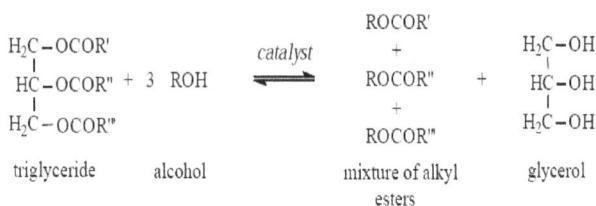

Figure 5. Transesterification general equation [14, 22]

Usually, methanol is the preferred alcohol because of its low cost. Ethanol is used in countries where its production cost is lower than that of methanol like in Brazil [14].

Transesterification happens in a three step reaction (Figure 6) in which three moles of FAME and 1 mole of glycerol are produced from every mole of TAG and 3 moles of alcohol [14, 17].

In this paper we report a successful biodiesel production from unrefined palm oil on a pilot plant scale by acid esterification of FFA, followed by a two steps alkali transesterification. For the sake of purification, this produced biodiesel is refined by a cleaning agent and then passed through a high speed separator followed by water evaporation under vacuum.

This study is a showcase of an economically viable process which optimizes production while ensuring quality standards, including biodiesel proper handling and storage up to delivery.

2. Materials and Methods

Rwanda is a landlocked country situated between 1°04' and 2°51' latitude south, and between 28°53' and 30°53' longitude east. It is a mountainous and lush country with a tropical weather and diverse ecosystems [23].

Research on Biodiesel production has been performed at IRST Biodiesel and Bioethanol Laboratory where a biodiesel processor is installed. It is designed to process 1,000 kg of raw oil per batch in 15h. The raw palm oil used is imported from neighboring countries, viz, Congo and Tanzania.

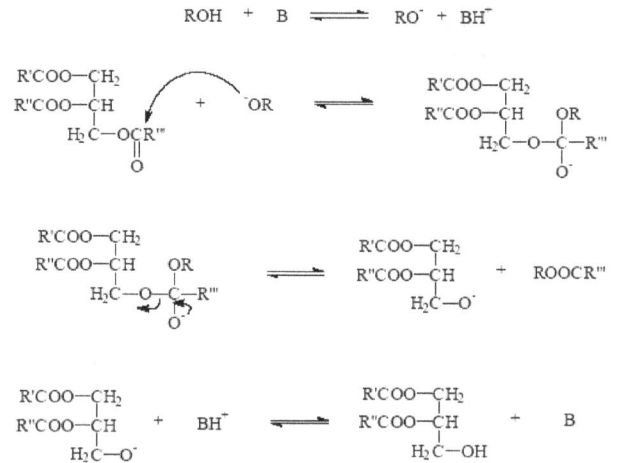

Figure 6. Transesterification reaction mechanism.

Chemicals used are mainly methanol 99.9%, concentrated sulfuric acid (96%), potassium methoxide 32% in methanol, manufactured by EVONIK, and ACA90 cleaning agent from AGERATEC. An installed SCADA program which monitors the process steps was designed by AGERATEC. Locally available materials and equipments have been used to solve issues raising from day to day production activities, namely a palm oil melting and impurities removal tank, a cold filtering system, and a biodiesel top layer intake storage tank. Regular analyses performed all along the process as well as on the finished biodiesel permit to comply with standards.

2.1. Biodiesel Production from Unrefined Palm Oil by Acid Esterification Followed by a Two Steps Alkaline Transesterification

2.1.1. Laboratory Scale Production of Biodiesel

Preliminary steps started in laboratory by determining appropriate triglycerides /catalyst/methanol proportions. The objective was to find optimum recipes which give higher triglicerides-methylesters conversion rates. Different types of oils such as refined and unrefined palm oil, soya oil, and jatropha oil were tested, and each type got its correct ratio determined for better efficiency. Incremental variations of NaOH amounts by +/- 0.05g were investigated for 100g oil samples.

During the laboratory oil tests for biodiesel, optimal proportions for the refined palm oil were found to be 0.35g of NaOH, 100ml of oil, 20ml of CH_3OH. This gave two layers in the test tube, with biodiesel on top and glycerol at the bottom. Unrefined palm oil behaved differently. In the same conditions, using base catalyzed process to produce biodiesel from unrefined palm oil without changing catalyst-methanol-oil ratio didn't give any positive results; the reaction formed a thick compound which didn't separate into distinctive layers.

2.1.2. Oil Purification

The crude palm oil comes packed in plastic bags from farmers. At the laboratory, the oil is stored in a heat-insulated PVC container (*F*), as detailed on Figure 7. At room temperature this palm oil is semi-solid. A heating system has been designed to melt the oil for the processor loading convenience. The heating system consists of a welded aluminium container *(A)* full of water, with heat insulation and located on top of *F*. Four electric heaters of 2,000W each, set to heat the water up to 80°C, are installed in the container. Inside the oil tank, coils of galvanized steel pipes are arranged at an angle of 120° from one another, and joined to two collectors, one at the bottom of *F* for water outlet, and another at its top for water inlet. The inlet pipe is connected to *A* and the outlet pipe is connected to a 50 liters tank *(L)* having a floater switch and a return pump connected to it.

Mode of Operation

Hot water from container *A* flows by gravity through the coiled pipes in the oil tank. That hot water heats the oil by conduction and radiation. A valve mounted on *A* is tuned to adjust the flow so that maximum heat is transferred to the oil. The water from three pipes collects in the collector at the bottom of F and flows into *L*. When *L* is full, the floater switch activates the pump which then returns back the water to *A*. The cycle repeats continuously until a needed amount of oil is liquefied. Hot Water leaves *A* at 80°C and reaches *F* bottom at 40°C. Liquefied palm oil in container *F* is at about 35°C.

Liquid oil is transferred to the processor via a pneumatic volumetric pump and a hose whose end floats in *F* on the oil top layer. This pipe is equipped with an inlet nylon solid impurities filter. While oil is being liquefied, heavy solid impurities settle at bottom of *F*.

The processor is equipped with a heating system combining a heat pump and an electric boiler to generate heat and cold wherever they are needed along the process. In the preheating tank made of stainless steel with a whole bottom, oil is heated up to 65°C. The entire floor of the tank is a heat exchanger that can give or get heat from the common hot water manifold. Then the hot oil passes through a 25μ polypropylene filter bag, sucked by a vacuum created in the pretreatment tank by a 500W water sealed vacuum pump before proceeding to a vacuum water evaporation at 80°C, which precedes the acid esterification. This vacuum water evaporation system reduces the interior pressure in the evaporation tank thus lowering the boiling point of the water and avoiding thermal decomposition of oil as well. In addition, the oil by passing through an atomizing nozzle on top of the tank, its evaporation surface increases, resulting in accelerating the evaporation. In addition, the water evaporation temperature is reached by means of a fusion heat exchanger in which hot water from the above mentioned heating system exchanges heat with the oil from the pretreatment tank.

2.1.3. Acid Esterification Process

Alkaline catalysts cannot directly catalyze the transesterification of oils containing high FFA. It has been reported that for alkaline transesterification to take place, FFA level in the oil should be below a desired level (ranging from less than 0.5% to less than 3%) [23]. Acid catalyzed esterification intends to transform FFA present in the raw oil

into methyl esters to prevent them from reacting with the base catalysts to form soaps.

In the acid esterification step, reactants proportions are of 11% of methanol and 0.8% of sulfuric acid (96%), by raw oil weight. The reaction takes place at 80°C in a pressure proof tank under 2 bars, within a mixing time of 4 hours maximum.

This mixing is carried out by a static mixer coupled with a centrifugal pump. The static mixer has helical mixing element which directs the flow of the mixture radially towards the pipe wall and back to the center. The additional velocity reversal and flow division results from combining alternating right- and left -hand elements, increasing mixing efficiency of methanol and oil phases which are normally immiscible [10].

At the end of the mixing, sedimentation is required to let the glycerol settle from the rest of the mixture. After glycerol is drained out, three laboratory tests are performed to determine the conversion rate (methanol solubility- Jan Warnqvist test), acid number test and water content.

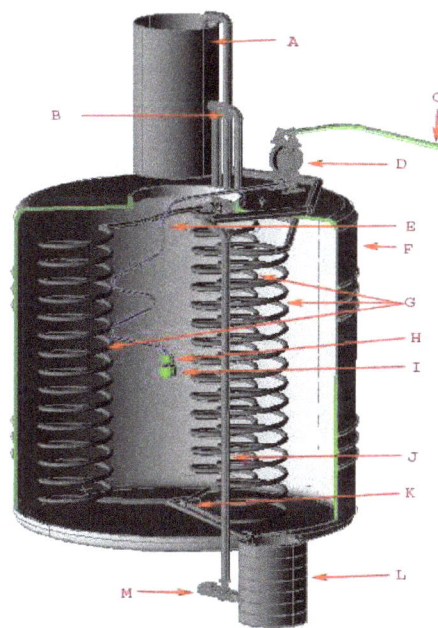

Figure 7. Raw oil hot filtering, sedimentation, storage and melting system designed in AUTOCAD (A: welded aluminium water tank, B:hot water outlet, C: oil outlet pipe, D: oil pump to the biodiesel processor, E: oil inlet pipe, F: PVC container, G: hot water coils, H:floating ball, I: solid matter filter, J: return water pipe, K: water collector box, L: water collection tank, M: water return pump.

These tests results are fed to the processor's computer and from them it computes the remaining amount of methanol and potassium methylate to be used for the transesterification step.

2.1.4. Alkaline Transesterification Process

The triacylglycerols (TAG) react with a light molecular mass alcohol (methanol) in presence of a catalyst under heat as shown by the general reaction equation (Figure 5).

As soon as glycerol is drained out and routine tests are performed, a second intake of methanol is pumped to the batch. The whole content is then heated up to the reaction temperature (55 °C) and the required amount of potassium methylate is pumped in.

In order to shift the reaction equilibrium to the right, methanol is used in excess and transesterification is split into two steps, each step using half of the catalyst with a glycerin drain scheduled at the end of each step.

The excess methanol is recovered by static and dynamic evaporation under -0.9 bar vacuum and condensation. The static evaporation is done at rest in sedimentation phases, while dynamic is performed at the time biodiesel is being mechanically sprayed under vacuum. It's worth mentioning that before dynamic evaporation a deactivation step is carried out. This deactivation helps to prevent reversion of methyl esters and glycerol back to mono- and diglycerides in presence of residual methylate during dynamic evaporation of methanol. To this end, 6wt% citric acid solution in methanol is added to deactivate traces of methylate present after transesterification. After its dynamic evaporation, methanol vapors are then sucked and condensed. Almost all the excess methanol is recovered, and it is reused after purification. This results in big savings on methanol costs.

2.1.5. Biodiesel end Users Survey

The produced biodiesel end users were required to submit a report detailing their experience using it in their cars. This was done by answering to a number of questions concerning

any problem they might have encountered from filter premature change, engine performance, to joint usage and any fuel line issues.

2.1.6. Biodiesel Purification

After pumping the last glycerol phase out, the formed biodiesel may still contain traces of ions such as potassium, sodium, magnesium, calcium, glycerol residue, sediments and other solids. All these impurities must be removed, along with any water that might still be present after the reaction. After settling and glycerol draining, biodiesel is passed through a 9,512 rpm separator for glycerol separation. Then a cleaning agent, named ACA90, containing an inorganic chelating agent combines with impurities in the biodiesel to produce large, dense gum-like complexes which are then removed by a separator. Actually, within the separator the force of gravity is replaced by centrifugal forces, which can be thousands of times greater, and consequently achieving in few seconds the separation which could take many hours in a tank under the influence of gravity [24]. Finally, the washed biodiesel undergoes vacuum evaporation to remove excess water, and an antioxidation agent is added to improve the biodiesel oxidative stability. Then, the moisture free biodiesel is pumped to the biodiesel storage tank through a 0.8μm pore size cellulose cartridge filter to remove any residual solid particles.

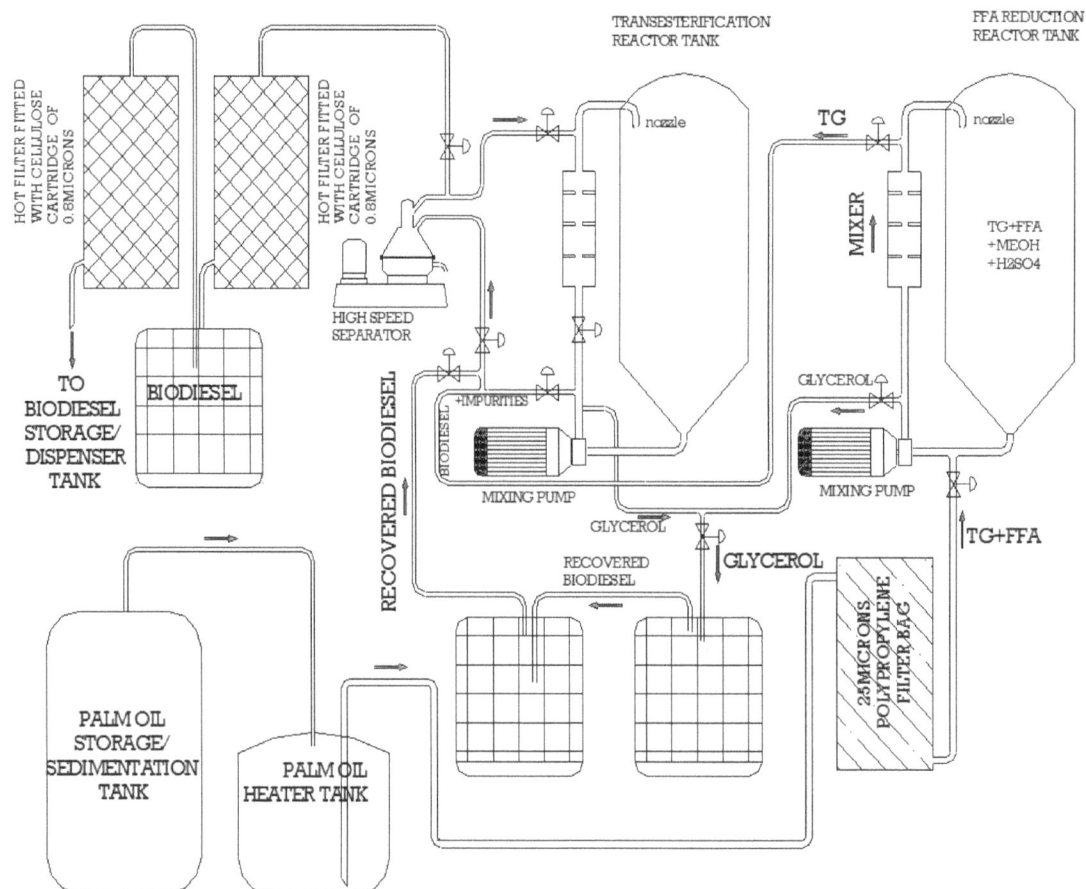

Figure 8. Flowsheet of the programmable biodiesel processor assembled by AGERATEC and used by IRST.

3. Results and Discussion

3.1. Efficacy of Reducing Solid Impurities in Raw Palm Oil

One of the main roles of the palm oil purification unit is mainly to reduce solid impurities so that palm oil can be easily filtered. Crude palm oil is known to have a high amount of solid impurities as shown in Table 1. The purification unit allows the separation of melted oil from solids and this facilitates the oil flow through the pipes, and the purification equipments as well. The total amount of heavy solids contained in the raw palm oil was found to range from 5 to 10 percent of gross palm oil, as it is shown in Table 1.

Table 1. Some Physical-chemical properties of crude palm oil used (Laboratory analysis results by Rwanda Bureau of Standards)

Analytes tested	Results	Method	Requirements
Volatile matter at 105° ,m/m, max	0.75	AOAC926.12	0.2%
Acid value, mgKOH/g Oil	31.70	AOAC940.28	10
Peroxide value milli-equivalent of active O_2/kg oil	1.90	AOAC965.33	Up to 15
Saponification value	227.00	AOAC920.160	NA
Insoluble impurities, % m/m, max.	6.6	ISO 663	0.05%

3.2. Process Analysis

3.2.1. Evaluation of Free Fatty Acid Reduction by Acid Number Test

Feedstocks high in free fatty acids are not easily converted by homogeneous base transesterification, because of the concurrent soap formation reaction of the free fatty acids with the catalyst. The excessive amount of soap formed significantly interferes with the washing process by forming emulsions, thus leading to substantial yield losses [25, 26].

FFA are released naturally in crude palm oil and can be increased by the action of enzymes in the palm fruit and by microbial lipases. During storage, FFA are produced by hydrolysis of triglycerides. Approximately 98% of crude palm oil is made up of fatty acids as shown in Table 2.

Table 2. Palm oil fatty acid composition [27-29].

Fatty acid	Formula	Structure (x, y)	Palm Oil (% by weight)
Palmitic acid	$C_{16}H_{32}O_2$	C16:0	32 to 47
Oleic acid	$C_{18}H_{34}O_2$	C18:1	39.1 to 52
Linoleic acid	$C_{18}H_{32}O_2$	C18:2	2 to 11
Stearic acid	$C_{18}H_{36}O_2$	C18: 0	1 to 6
Myristic acid	$C_{14}H_{28}O_2$	C14:0	0.8 to 6.0
Lauric acid	$C_{12}H_{24}O_2$	C12:0	0.3 to 1.0

For the palm oil feedstock used in this research, whose FFA content is particularly high as confirmed by its acid number (Table 1), typical homogenous base catalyst (methoxide) cannot be effective due to the unwanted side reaction as shown in Figure 3.a.

Acid pretreatment (acid esterification) is used to lower its FFA content prior to transesterification. In this research, since the esterification step is catalyzed by a strong acid; H_2SO4 96% whose main part is drained out together with glycerol, some residual acidity stays in the oil/biodiesel mixture. Thus, the total acid number includes the residual acidity from both of the catalyst and FFA contribution.

The acid number test is used to determine the acidity (FFA and residual acid catalyst concentration) of the mixture to be neutralized by the potassium methylate before transesterification. Therefore, the base catalyst required for transesterification will be increased by the amount of the base to neutralize the above mentioned acidity. The test is done by titration of a sample taken just after the first glycerol drain, with 0.1M potassium hydroxide aqueous solution and phenolphthalein as indicator.

The acid number is reported as mg KOH/g according to the following formula:

$$A = \frac{56.1 x V x c}{m}$$

V is the volume (ml) of the KOH solution used for the titration of the sample
C is the concentration in moles per litre (M) of the KOH solution
m is the mass of the sample in grams (g)
56.1 is the molar mass of KOH

Note that the acid number (A) can be converted into FFA value. This FFA value is reported as a percentage of oleic acid according to the following formula : $FFA = \frac{A}{1.875}$ [30].

Table 1 and Table 2 show that the acid esterification process reduces FFA content from 16.91% for crude palm oil to 2.44%; which is a reported average of 71 batches. The difference being the amount of FFA converted into FAME (biodiesel) through the esterification reaction. It's worth mentioning that 2.44% FFA is lower than 3 wt.%, the value above which base catalysed transesterification reaction will be significantly hindered due to the unwanted reaction in which the catalyst reacts with FFA to form soap and methanol, thus irreversibly quenching the catalyst and resulting in an undesirable mixture of FFA, unreacted TAG, soap, diacylglycerol (DAG), monoacylglycerol (MAG), biodiesel, glycerol, water, and/or methanol [31].

The average acid number from a series of tests for 71 batches is 4.57, and as shown in Table 1 and Figure 10, the water content of samples taken after the first glycerol drain, a subsequent step to the acid esterification, increases with the acid number also measured after this step. This suggests that the higher the FFA content, the higher the water content at this stage. From Figure 9, the water content before transesterification start is proportional to the water content of palm oil before acid esterification. Therefore, the higher the water content before acid esterification, the higher the acid value and the higher the amount of potassium methylate required to neutralize this acid. It is very important to evaporate water from raw oil in order to optimize biodiesel production.

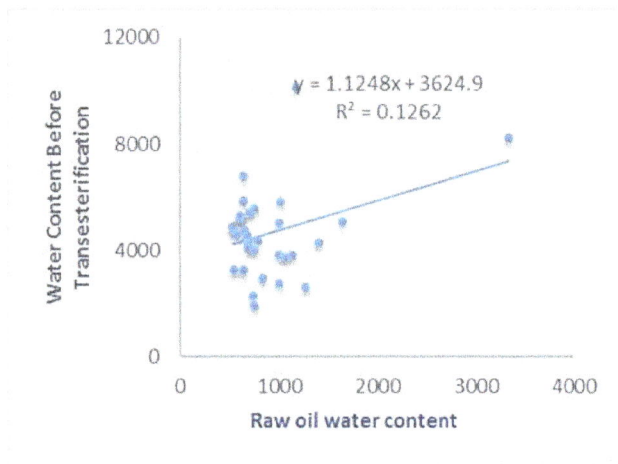

Figure 9. *Relationship between water content of raw oil before and after acid esterification.*

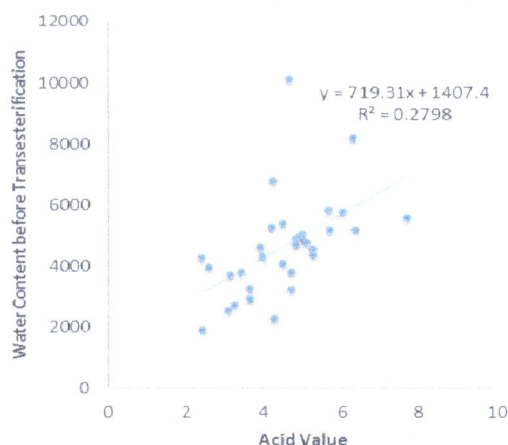

Figure 10. *Relationship between water content before transesterification and acid number.*

3.2.2. Determination of Oil Conversion Rate by Methanol Solubility Test (Jan Warnqvist test)

Methanol solubility test reveals the rate of oil conversion just after esterification, and the transesterification level of the produced biodiesel. The test assumes that the methyl esters are fully soluble in anhydrous methanol and that the oil is not. The test is performed by mixing a sample of oil & biodiesel mixture with anhydrous methanol in a separatory funnel. After mixing, the two layers are allowed to separate, the bottom layer (oil) is collected, dried and then weighed .The methanol solubility expressed as a percentile is then calculated as follows: $\left(1 - \dfrac{m_2}{m_1}\right) x100$, where m_1 represent the sample weight (g) and m_2 the weight (g) of the oil phase [32].

With the use of this test, data collected from August 2008 to December 2011 for methanol solubility tests showed that the acid esterification average conversion rate is of 57%. This means that the acid catalyzed transesterification also happened simultaneously with the acid esterification, since the average raw oil FFA content is of 16.9% only. This concurs with the fact that acid transesterification is also possible even though it occurs very slowly. For samples

taken after transesterification, the rate is 99.8%, meaning a nearly complete conversion of oil into methyl esters (biodiesel).

3.2.3. Water Content Measurements [33]

The water content either in oil or in biodiesel, is determined by Karl Fischer coulometric titration as described by the standard EN ISO 12937.

On one hand, from observing the raw oil remaining water content as shown on Figure 11, it's obvious that this quality of oil contains so much water that even after three to four hours of water evaporation, its average water content stays above 1,000ppm; the recommended threshold before chemicals input. On the other hand, water content increases at the end of the esterification step (Figure 4&11), with an average of 4,646ppm, and this is due to the water produced by the FFA esterification reaction.

Water content of the produced biodiesel is also measured for each batch. For 71 batches, the average water content was calculated to be of 395ppm, a value that falls far under the EN14214 standard (500ppm). This means that the process is very efficient on this parameter in spite of the high water content of the starting raw oil.

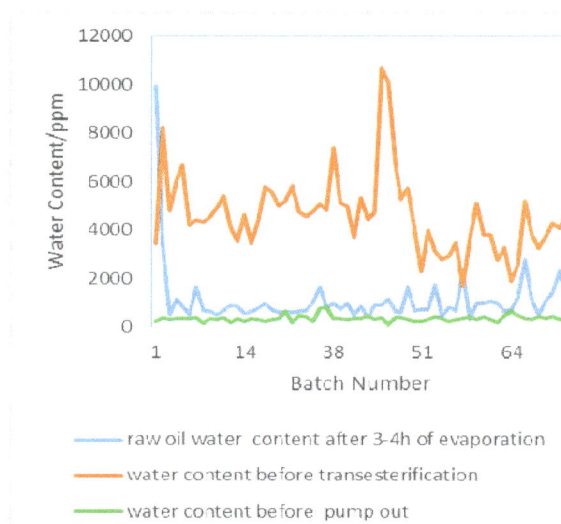

Figure 11. *Water content trends along the process*

3.2.4. Palm Oil Biodiesel Cold Flow Behavior

In some cases, biodiesel 100% (B100) used into diesel engines may clog filters. Table 3 summarizes results of the biodiesel end users survey.

Table 3. *End users survey results.*

Issue	Report
Any problem while using biodiesel	6% changed filters unexpectedly and 94% didn't face any problem
Joint usage	100% didn't experience any joint wear nor change
Fuel lines	100% didn't notice any wearing nor change
Fuel filters changing	9.6% changed filters, 90.4%didn't
Filter clogging	92.8% reported filter didn' t show any problem, 7.2% reported clogged filters

Note that when biodiesel cools down, solid deposits are formed. These precipitates may be sterol glucosides [18], waxes, mono and diglycerides, which are liquid at 80° C and not detectable by methanol solubility test, but unfortunately solidify and become visible at room temperature.

3.2.5. Solutions to Biodiesel Sediments

In order to get rid of those occurred impurities, a cold filtering and a top layer intake biodiesel tank systems were designed.

3.2.5.1. Cold Filtering System

This system comprises a tank, a 0.8μ cellulose filter, a tightening nut, a cover disk with bolts, an inlet, an outlet and a pump.

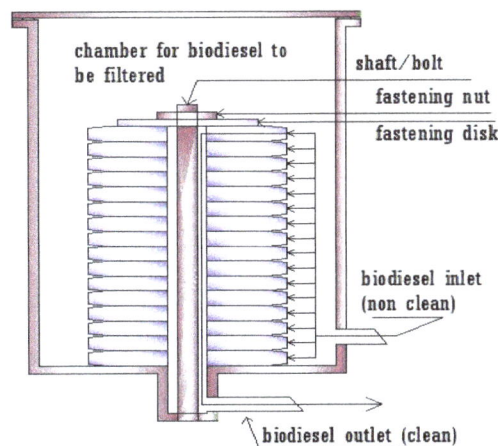

Figure 12. Biodiesel filter

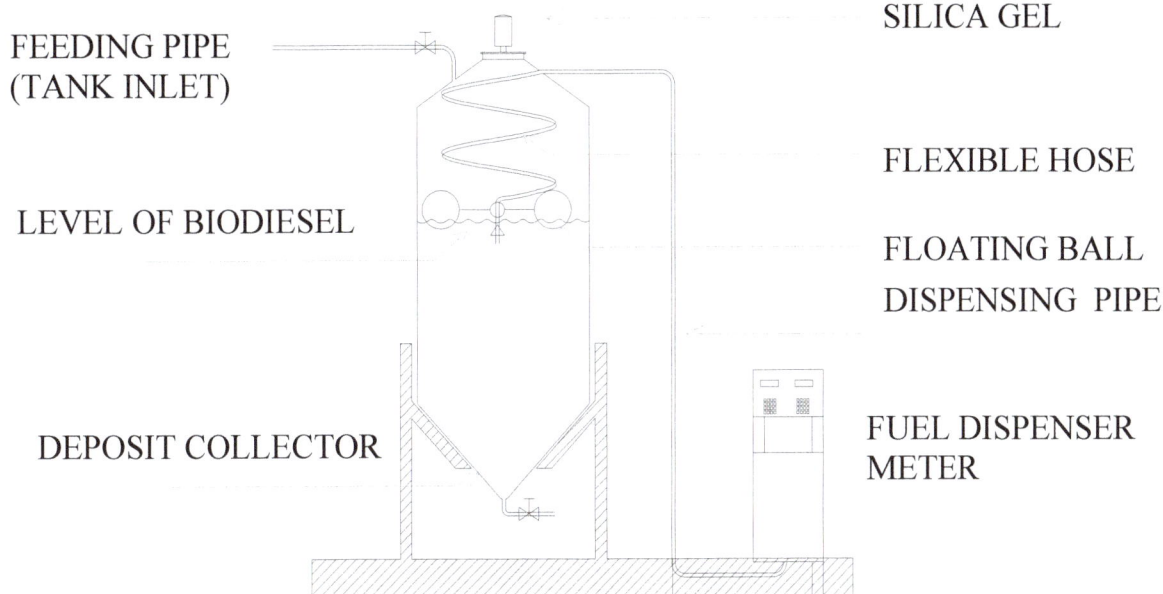

Figure 13. Biodiesel settling tank system

Mode of operation: The pump pushes biodiesel with suspensions through the room between the filter and the tank wall. The only way by which the biodiesel has to get out is through the cellulose filter and via the outlet, leaving behind suspended impurities. Figure 12 depicts the cold filtering system.

The pump stroke is kept low to minimize the pressure that would force precipitates through the filter. This system is so efficient that it removes almost all impurities (99%) and cars using biodiesel filtered this way never faced any problem of filter clogging at all.

3.2.5.2. Biodiesel Filling Station Tank System

A new biodiesel storage and delivery tank (Figure 13) was specifically designed to let any remaining sediments settle at the bottom, and biodiesel to be taken from the top layer. The tank has a conic bottom with a 45° angle slope. The biodiesel intake pipe is attached to a flexible hose, itself always kept at the biodiesel top layer by floating spheres which also permit the intake pipe to follow biodiesel level.

A minimum level of biodiesel is kept in the tank (5% of tank capacity) to prevent the filling pump from sucking presumed dirty biodiesel at the bottom. The arrangement is also equipped with a non-return valve which keeps the filling pump inlet pipe always full, to prevent the pump from sucking air and allow the pump to start sucking since it is of centrifuge type.

Furthermore, the filling pump itself has filters which may retain unexpected impurities.

Additionally, since biodiesel is hygroscopic, a silica gel cartridge is installed on the tank top cover, to retain air moisture when the filling tank is sucking air.

4. Conclusions

The production of biodiesel from low-cost raw materials which generally contain high amounts of free fatty acids (FFAs) is a fascinating alternative that would make their production costs more competitive than petroleum-derived fuel. A good quality biodiesel can be produced from crude palm oil by base catalysed transesterification, after a simple and cheap oil purification accompanied by acid esterification pretreatment to reduce FFA content. This technology was applied on a pilot plant scale with an automated biodiesel processor machine in Rwanda and showed to be promising. In this plant, biodiesel is produced with a yield of 90.4% and laboratory analyses have proved its quality to be acceptable on common biodiesel parameters. This was also confirmed by end users survey which showed their high satisfaction in various dimensions such as low carbon emissions, engine performance, and local cold weather operability. In all, the findings are industrially applicable, economically viable and repeatable. For a landlocked country like Rwanda, which doesn't possess any petroleum resources, biodiesel production from low cost raw materials (crude palm oil) even abundant in the region, may constitute a promising alternative to its energy security.

References

[1] Adriana G., Marious R., Monica T., Csaba P. and Florin D. I.(2012). *Biodiesel production using enzymatic transesterification e Current state and perspectives.* Renewable Energy . 39: 10-16.

[2] Rincón L.E., Jaramilo J.J., Cardona C.A.(2014). *Comparison of feedstocks and technologies for biodiesel production: An environmental and techno-economic evaluation.* Renewable Energy. 69: p. 479-487.

[3] USDA. *Oilseeds: World Market and Trade.* (2014).P: 7.

[4] Frank R.C., Luc. P., and Arnaldo W. *A global overview of vegetable oils, with reference to biodiesel.* (2009). IEA, London. P: 9,19.

[5] FOE.(2006).*The use of palm oil for biofuel and as biomass for energy.* Available from: www.foe.co.uk/sites/default/files/.../palm_oil_biofuel_positio n.pdf. Accessed on 12/8/2014.

[6] Atadashi I.M., Aroua M.K., Abdul Aziz A.R., Sulaiman N.M.N. (2012). *Production of biodiesel using high free fatty acid feedstocks.* Renewable and Sustainable Energy Reviews. 16: 3275-3285.

[7] Antolin G., Tinaut V.F., Briceno Y., Castano V., Perez C., and Ramırez A.I.(2002). *Optimisation of biodiesel production by sunflower oil transesterification.* Bioresour Technol. 83:111-114.

[8] Huang G., Chen F., Wei D., Zhang X.W., and Chen G.(2010). *Biodiesel production by microalgal biotechnology.* Appl Energy . 87: 38-46).

[9] ACET. (2013). *The Oil Palm Value Capture Opportunity in Africa.* Accra, Ghana. P: 11.

[10] Gerhard K., Jon V.G., and Jurgen K. *The Biodiesel Handbook, 2005.* AOCS Press. Illinois-USA. pp.9,15,55 and 236.

[11] Szulczyk K.R., *The economics of malaysian palm oil industry and its biodiesel potential.(2013).* Social Science Research network.

[12] Rutz D., Rainer J., *Biofuels Technology Handbook.(2007).* WIP renewable energies: Munich-Germany. p. 73,78,87.

[13] Poku K., *Small Scale Palm Oil Processing in Africa.(2002).* FAO.p: 9,11.

[14] Caye M. D., Nghiem P.N., Terry H.W., *Biofuel Engineering Process Technology 1st ed.(2008).* McGraw-Hill Professional, New York. p. 197-200,202-204.

[15] Jawad N., Syed K., Farrukh N.(2008). *Palm Biodiesel an Alternative Green Renewable Energy for the Energy Demands of the Future.* ICCBT. F(07): 79-94.

[16] Nyanjou R.N.(2008). *Modernisation and Innovation of Palm Oil Extraction Process: The Palm Nut, Its By-products and Its Properties,* in *IAALD -AFITA -WCCA World Conference On Agricultural Information And IT .* Tokyo.

[17] Moser B.R., *Biodiesel production, properties, and feedstocks.(2009).* In Vitro Cell .Dev.Bio.-Plant. 45:229-266.

[18] Inmok L, Lisa M.P., George B.P., Erica P., and Troy H.(2007). *The Role of Sterol Glucosides on Filter Plugging.* Biodiesel Magazine. Available on http://www.biodieselmagazine.com/articles/1566/the-role-of-sterol-glucosides-on-filter-plugging/. Accessed on 15/09/2014.

[19] Refaat A.A.(2009). *Different techniques for the production of biodiesel from waste vegetable oil.* International Journal of environmental science and technology.7 (1):183- 213 .

[20] Alemayehu G., Abile T. (2014). *Production of biodiesel from waste cooking oil and factors affecting its formation: A review.* International Journal of Renewable and Sustainable Energy. 3(5): 92-98.

[21] Khalid Khalizani, Khalisanni Khalid.(2011). *Transesterification of Palm Oil for the Production of Biodiesel.* American Journal of Applied Sciences. 8 (8): 804-809.

[22] Roger A., Isabel A., Ulf H., *Green Chemistry and catalysis,*2006.wiley. p373.

[23] Sharma Y.C., Sing.B.(2009). *Development of biodiesel: current scenario.* Renewable and Sustainable Energy Reviews. 13:1646:51.

[24] Falconer A. (2003).*Gravity separation: old technique/ new methods.* Physical separation in science and engineering. 12(1): 31-48.

[25] Godlisten G.K., Abraham.K.T., Hassan M.R., Godwill D.M., Jibrail K.,and Keat T.(2013). *Pre-Treatment of High Free Fatty Acids Oils by Chemical Re-Esterification for Biodiesel Production: A Review.* Advances in Chemical Engineering and Science.3: 242-247.

[26] Freedman B., Pryde E.H., and Mounts T.L. (1984).*Variables Affecting the Yields of Fatty Esters from Transesterified Vegetable Oils.* Journal of the American Oil Chemists Society. 61: 1638-1643.

[27] Dennis Y.C.L., Xuan W., Leung M.K.H.(2010). *A review on biodiesel production using catalyzed transesterification.* Applied Energy. 87 :1083–1095.

[28] Kansedo J, Lee K.T, Bhatia S.C.O. (2009). *Oil as a promising non-edible feedstock for biodiesel production.* Fuel. 88:1148–50.

[29] Srivastava A., Ram. P.(200). *Triglycerides-Based Diesel Fuels.* Renewable and Sustainable Energy Reviews.4:111-33.

[30] Europian Committee for Standardization.(2003). *Fat and oil dereivatives-Fatty Acid Methyl Esters (FAME)-Determination of acid Value,* in *EN14104: 2003E.* Swedish Standards Institute. P: 3-7.

[31] Lotero E, Liu Y., Lopez D.E.,Suwannakarn K., Bruce D.A, Goodwin J.G. Jr.(2005). *Synthesis of biodiesel via acid catalysis.* Ind.Eng. Chem. Res. 44:5353-5363.

[32] Young R.(2008). *Field Test Equipments Enhances Quality Assuarance. Biodiesel Magazine* .p :3.

[33] ISO. European Standards (2000). *Petrolium products – Determination of water – Coulometric Karl Fischer titration method(ISO 129:2000).*CEN.

3

Enhancement of Biogas Yield from Cow Dung and Rice Husk Using Guano as Nitrogen Source

P. A. Nwofe[*], P. E. Agbo

Division of Materials Science and Renewable Energy, Department of Industrial Physics, Ebonyi State University, Abakaliki, Nigeria

Email address:

patricknwofe@gmail.com (P. A. Nwofe), ekumaagbo@gmail.com (P. E. Agbo)

Abstract: The study reports on the influence of nitrogen source on the biogas yield from cow (N'Dama) dung and rice husk. The digester performance for both feedstocks were evaluated using standard parameters such as; initial PH, water dilution, nitrogen source (guano and poultry droppings) and heavy metals. The source of innoculum used was cow rumen fluid. The result show that for feedstock to water dilution ratio of 1:6 w/v and initial pH of 7.0, the maximum biogas yield for rice husk was 430 mL/day and 350 mL/day for cow dung. The heavy metals (Ni^{2+} and Zn^{2+}) increased the biogas yield while Fe^{2+} (100 ppm) shows no effect. Addition of guano results in maximum production rate of 85 mL/day and 60 mL/day in rice husk and cow dung respectively. The use of guano indicates more biogas production rate in both feedstocks compared to poultry droppings.

Keywords: Cow Dung, Biogas, Guano, Poultry Droppings, Rice Husk

1. Introduction

Renewable energy technologies are currently competing favourably in the energy sector, all geared toward achieving a more sustainable and efficient energy use. Amongst other renewable sources: solar, hydro, geothermal, wind and biomass, biomass is amongst the most vastly explored in terms of research and applications both in the developed and in under-developed countries. It is a common knowledge that fossil fuels-based conventional grid extension constitutes the major centralised power systems from urban areas to rural areas in most under-developed countries. A scenario that is not only capital intensive but also economically unrealistic in most cases. It has been established that more than a quarter of the human population experiences an energy crisis, especially those living in the rural areas of developing countries such as Nigeria [1, 2]. Biomass is generally considered as a suitable alternative to energy from fossil fuels since it can be easily converted to other forms of energy such as biogas and biofuels [3, 4]. It has been reported that Nigeria has an installed capacity of 8,425 MW of electricity but the available capacity is only about 50% of installed capacity [5]. In Nigeria, energy conservation and energy efficiency is strongly needed, thus renewable energy technologies will be essential to the solution and are likely to play an increasingly important role for providing enhanced energy access, reduced over-dependence on fossil fuels, and to help Nigeria meet her vision 20-20-20 clean energy program.

Privatisation of PHCN (Power Holding Company of Nigeria) has made many States in Nigeria to be pursuing other source of energy vigorously and Ebonyi State is playing an active role in that direction. Energy from biomass will play a significant role in that Nigeria environment is largely polluted with huge amount of wastes due to rapid urbanisation and poor waste management practices [6-8].

Ebonyi State is popularly known for rice production (Abakaliki rice) in Nigeria and thus has large reserves of rice husks from rice mills which are randomly and strategically located in various towns. These large reserves of rice husks are only utilised as cooking fuels (on a low scale) and source of income for some poor rural dwellers who scavenge those rice husk hills to earn a living. Waste from agricultural, municipal, industrial and household are also common in the study area but these wastes are not presently utilised in a sustainable manner [8]. It has been shown that about 227,500 tons of fresh animal wastes is produced on daily basis in Nigeria [9], implying that Nigeria can produce 6.8 million m^3 gas/day since 1kg of fresh animal wastes yields up to 0.03 m^3 gas. Ogwueleka [10] noted that the waste density of municipal solid waste in Nigeria ranged from 280 to 370 kg/m^3 with the waste generation rates in the range 0.44 to 0.66 kg/capita/day. The generation rate of solid waste in the capital cities of some South-east States of Nigeria (Ebonyi and Imo) is in the range 9.580 to 9.74 x 10^{-3} m^3 [6, 11]. Biogas production from wastes is commonly achieved through anaerobic digestion. Biogas can be utilised as fuels both in low (household, village, community) and large scale (industry) or as fertilisers [12-15]. The use of suitable nitrogen supplement or other parameters to enhance biogas yield has been a subject of research for years. Some authors [16, 17] have used different supplement/parameters in this regard.

This paper investigates the use of locally available and abundant wastes to produce biogas using anaerobic digestion. It also reports on the effect of different parameters (initial PH, water dilution, guano and poultry droppings, heavy metal effect) on the biogas yield and it was observed that the use of guano (bat droppings) enhanced biogas production substantially.

2. Materials and Methods

2.1. Feed Materials

The feed materials were a batch of cow dung and rice husk. The cow dung was sourced locally from cow farms, and the rice husk from Abakaliki Rice Mill Industry, Ebonyi State, Nigeria. The cow dung from N'dama species was used because of its availability compared to other species in the study area. Fig. 1 gives a picture of a typical rice husk hill from Abakaliki rice mill industry. The digester feed included a batch of cow dung and rice husk. Each of these was dried using open sun-baking and then carefully stored in a stoppered polyethylene container. In order to ensure efficient stabilisation of the wastes during anaerobic digestion, innoculum obtained from the rumen of cows slaughtered at Abakaliki main market abattoir was used. The micro-organisms were maintained in an anaerobic environment by straining the innoculum in cheesecloth and then stored in an airtight container. The tip of the digesters was sealed off with clip/cello-tape in order to maintain complete anaerobic condition during the anaerobic digestion for the cow dung and rice husk respectively.

Figure 1. Picture of a typical rice husk hill in Abakaliki rice mill.

2.2. Experimental Set-up for the Anaerobic Digestions

Fig. 2 gives the experimental set-up for anaerobic digestion of cow dung (CD) and rice husk (RH). As shown in Fig. 2,

each set-up included 1.5 L bottle that served as the anaerobic digester, an inverted 50 mL graduated burette containing acidified water as the biogas collector, and a rubber container which was used to collect the water discharged from the biogas collector. A rubber pipe was used to convey the gas produced in the digester to the collector. The anaerobic digestion was maintained in the mesophilic range in that the room temperature was $31 \pm 1^{\circ}C$ [18-21]. It has been reported that stratification can lower the production rate of biogas [22, 23], thus this was avoided by mixing each digester once daily for 30 days. The water displacement method used in [24] was utilised in recording the biogas production rate for each anaerobic digester.

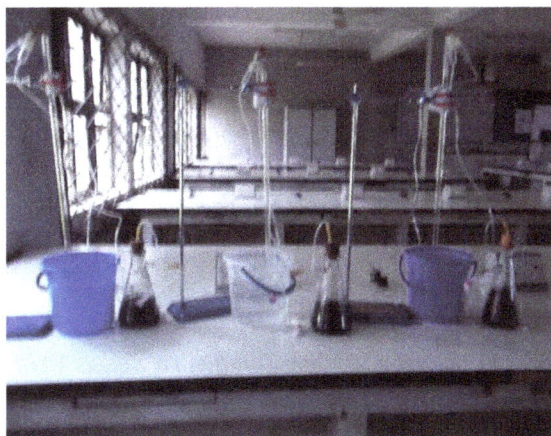

Figure 2. Experimental set-up.

2.3. Initial pH and Water Dilution

The buffer solutions: citrate buffer of pH 4, phosphate buffer of pH 7, and borax buffer of pH 10 were used to buffer the biogas digesters containing the feedstock. This was done to investigate the impact of the initial pH on the biogas production rate. A 55 mL cow rumen innoculum was added in each case to induce the digestion. A 40 g of feeds each of cow dung and rice husks were mounted in 1.5 L conical flasks, a total of ten stands (five stand for CD and RH respectively). The feedstock to water ratio of 1:2, 1:4, 1:6, 1:8, 1:10 (w/v), were created by moistening each case with water whose volume is in the range 250-450 mL. The fermentation temperature was maintained as before $(31 \pm 1^{\circ}C)$ with a hydraulic retention time (HRT) of 30 days. The same volume of cow rumen innoculum was introduced in each digester and anaerobic condition was ensured by sealing the tip of the flask appropriately. The control experiment had no water included in the digester for the cow dung and rice husk respectively.

2.4. Nitrogen Sources and Trace Metals

A 100 g of each feedstock (CD and RH) was included in conical asks and then mixed with varying quantities of nitrogen sources (poultry droppings and guano). The poultry droppings were sourced from poultry farms while the bat droppings were obtained from aged-buildings roofed with corrugated iron sheets in the rural areas of Ebonyi State. The controls had the same quantity of feedstock without nitrogen source. A feed to poultry droppings/guano ratio of 2.5:1 (w/w) was formed by mixing a 100 g of CD and RH respectively with 40g of poultry droppings/guano in different digesters. The fermentation process in each case was induced by adding 350 mL of water to the same volume of cow rumen as before. The hydraulic retention time was 30 days while the incubation temperature was maintained at $31 \pm 1^{\circ}C$.

Concentrations of Fe_3SO_4, $ZnSO_4$ and $NiCl_2$ in the range 40-250 ppm were added to each digester containing 55 g of CD and RH respectively, to investigate the effect of the heavy metal on the biogas production rate. In particular, 250 mL of each solution was added to the digesters. The control in each case had 250 mL of water included only. Fermentation was induced by innoculating each digester with a 55 mL of freshly strained cow rumen liquor.

2.5. Data Acquisition and Analysis

In the experiment, each set of data was taken in quadruplicates and the average was used for the analysis. The analysis was done using the Origin Pro 8 software (trial version).

3. Results and Discussion

3.1. Effect of Initial pH and Water Dilution on Biogas Yield

It is generally known that pH plays a substantial role in the biogas yield under different conditions. This is because the activities of various microbes depend strongly on the pH of the medium amongst other factors. The best pH values in biogas production is in the range 6.5-7.5 [25, 26], though some authors have reported pH values in the range 7.6-8 [27-29]. In the literature, there are varying reports on the effect of pH on the biogas yield involving same or different feedstocks. Result obtained in this study indicate that a pH of 7 gave the best yield of biogas. This value is in agreement with the reports of other authors [16, 27, 28, 30-37] as shown on Table 1.

Table 1. Literature of the effect of pH on biogas production

Feedstock	Initial pH	Biogas yield	Ref
Bioethanol waste	6 - 8	increase	[27]
Chicken droppings	7.0	increase	[28]
Chicken droppings	7.2	decrease	[28]
Cow dung	7.0	increase	[16]
Rice straw	6.9 - 7.28	increase	[30]
Rice straw	5.0-5.5	decrease	[30]
Rice husks and cow dung	7.0	increase	This study
Coal	8.0	increase	[31]
Pig manure and maize silage	8.0	increase	[32]
Apple waste with swine manure	7.81 – 7.85	increase	[33]
Cow manure	7.3 – 7.6	decrease	[34]
Saccharina japonica ethanol fermentation	8.0	increase	[35]
Diary manure with three crop residue	5.25 – 6.80	decrease	[36]
Raw and detoxified mahua seed cake	5.9 – 7.2	decrease	[37]

Figs. 3 and 4 give the effect of water dilution on the biogas production rates for rice husk and cow dung respectively. The feed to water dilution ratios of 1:2, 1:4, 1:6, 1:8 and 1:10 w/v all yielded more biogas than the control for the cow dung and rice husk. The maximum biogas production rates for rice husk were observed to be 42, 148, 193, 148 and 55 mL/day respectively at day two. However for cow dung, a maximum biogas production rate of 35 mL/day and 60 mL/day were observed at day two for dilution ratios of 1:2 and 1:10 w/v. On day four, 95 mL/day and 130 mL/day for dilution ratios of 1:4 and 1:6 w/v were observed. The decrease of biogas yield observed in the case of rice husk at lower (1:2 w/v) and higher (1:10 w/v) water dilution ratios could be due to a reduction in the cluster formation of the necessary bacteria needed for biogas production. The variation in days at which maximum biogas yield was observed for the cow dung and rice husk was attributed to the difference in the C/N (carbon/nitrogen) ratio of the feedstocks. Similar behaviour has been observed by other authors [17]. The maximum cumulative biogas yield occurred at a feed to water ratio of 1:6 w/v as shown on Fig. 5.

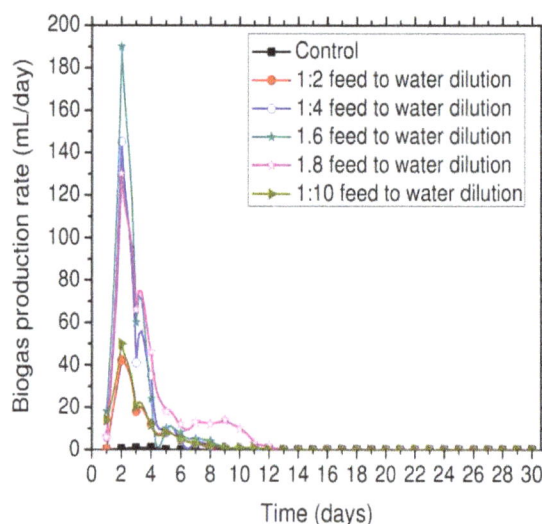

Figure 3. Biogas production rate for different feed to water dilution ratios (rice husk).

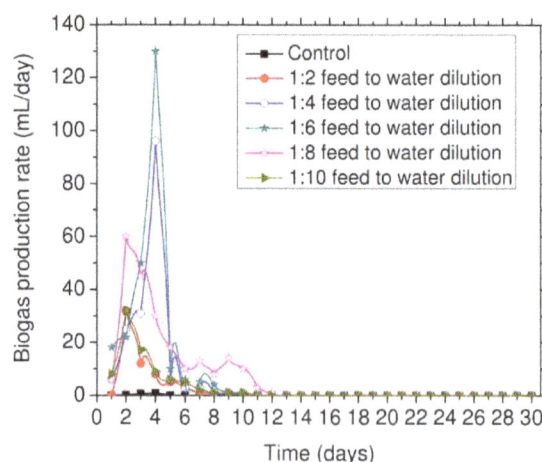

Figure 4. Biogas production rate for different feed to water dilution ratios (cow dung).

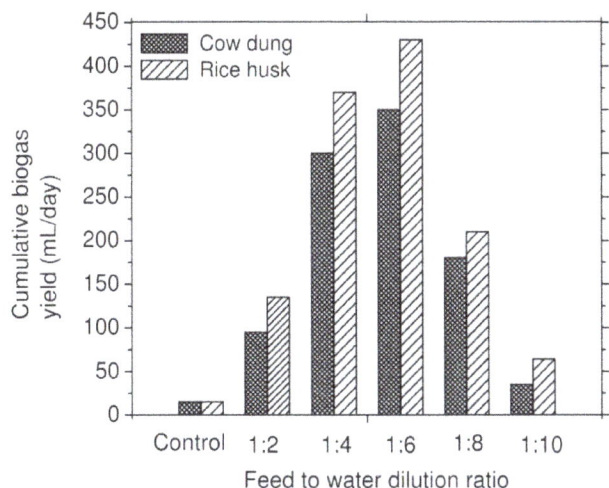

Figure 5. Cumulative biogas yield for different feed to water dilution ratios.

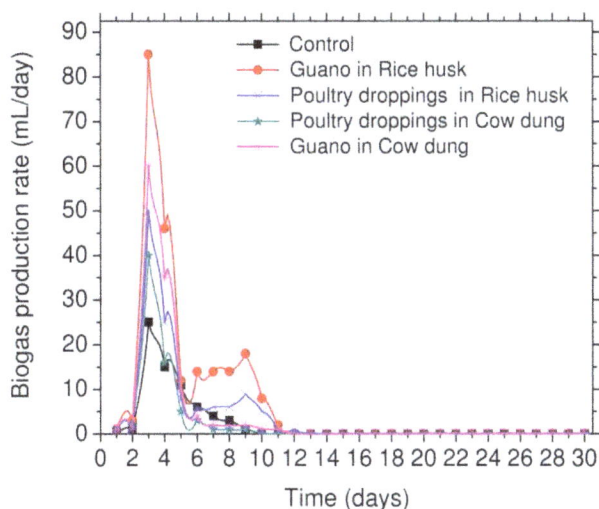

Figure 7. Influence of nitrogen source on digester performance.

3.2. Effect of Nitrogen Source and Trace Metal on Biogas Yield

Fig. 6 gives the effect of nitrogen source on biogas production rate while Fig. 7 gives the cumulative biogas yield. As shown on Fig. 6, the biogas production rate peaked up in both cases compared to the control. A maximum biogas production rate of 85 mL/day and 60 mL/day were observed for the guano supplement in RH and CD respectively. The biogas production rate dropped to 50 mL/day and 40 mL/day for RH and CD respectively when poultry droppings was used as the nitrogen supplement. This could be attributed to a better balance of C:N ratio from the bat droppings. Ahmadu et al. [38] and Ojolo et al. [39] noted that the organic matter content of poultry wastes is a factor that affects the digestion environment and the microbial habitat, hence this could also be responsible for the lower yield observed for the poultry droppings supplement in this study. Table 2 gives the literature of the effect of nitrogen supplement on the biogas yield for same or different feedstocks.

Trace metals are known to affect the yield of biogas independent of the feedstock. This is because of their effect on the microbial contents of the digester during the anaerobic digestion [47-49]. Demirel and Scherer [50] argued that there is no direct formula for optimum composition of trace metals needed for maximum biogas production. Our result shows that for rice husk, Ni^{2+} (100 ppm) gave 300 mL/day while Zn^{2+} (100 ppm) gave 110 mL/day. For cow dung, Ni^{2+} gave 30 mL/day while Zn^{2+} gave 74 mL/day. No significant effect was observed with Fe^{2+} in both RH and CD respectively. Table 3 gives the effect of trace metals on the digester performance/biogas yield for different/same or different feedstock according to the literature.

Table 2. Literature of the effect of Nitrogen source on biogas digester

Feedstock	Nitrogen source	Biogas yield	Ref
Chemical source	not given	not given	[40]
Rice husks	poultry droppings	increase	[41]
Dairy cattle manure	urea	decrease/increase	[42]
Rice straw	urea	increase	[43]
Boiled rice	human urine	increase	[44]
Rice	poultry droppings	increase	[45]
Rice husks and cow dung	guano and poultry droppings	increase	This study
Cattle dung	brassica compestries	increase	[46]
Rice wastes	poultry droppings	increase	[41]

Table 3. Literature of the effect of trace metals on biogas production

Feedstock	Trace metal	Biogas yield	Ref
Molasses silage	trace metal	decrease	[48]
Palm oil mill effluent	Ni and Co	increase	[49]
Food waste	K, Ni, Co, Mo, Se and W	decrease/increase	[51]
Palm oil mill effluent	Fe	no effect	[52]
Food waste	Co, Mo, Ni, Se and W		[53]
Azolla piñata R.Br and Lemna minor L	Cd and Ni	increase	[54]

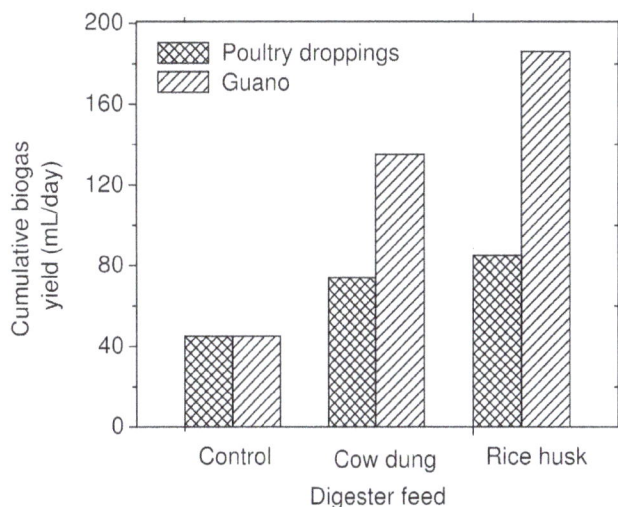

Figure 6. Influence of nitrogen source on biogas yield.

Municipal waste	Fe, Co and Ni	increase	[55]
Rice husks	Ni^{2+}	increase	This study
Cow dung	Zn^{2+}	increase	This study
Rice husk and cow dung	Fe^{2+}	no effect	This study
Holy Basil (ocmium sanctum)	Se	increase	[56]
Chicken manure	Fe^{2+}	decrease	[57]
Municipal waste	Cd, Cr, Cu, Pb, Ni, and Zn	no effect	[58]
Municipal waste	ash	increase	[58]
Municipal waste	Ni, Co and Fe	increase	[59]
Chicken manure	Trace metals	increase	[60]

4. Conclusion

Locally abundant wastes have been investigated to establish the possibility of utilising them as alternative source of energy, thus the effect of different nitrogen sources (poultry droppings and guano) on the biogas yield from cow dungs from N'dama species and rice husk has been reported. The results show that the use of guano as nitrogen supplement is more promising than poultry droppings. A maximum biogas production rate of 85 mL/day and 60 mL/day were obtained for guano supplement in RH and CD respectively while poultry droppings in both feedstocks gave lower biogas yield. The other parameters: initial PH, water dilution, nitrogen source (guano and poultry droppings) and trace metals also affected the digester performance. Our findings show that a feed to water dilution ratio of 1:6 w/v yielded maximum cumulative biogas of 450 mL/day for RH and 350 mL/day for CD. At initial pH of 7.0 for both feedstocks, Ni^{2+} (100 ppm) for rice husk, and Zn^{2+} (100 ppm) for cow dung gave the best biogas yield while Fe^{2+} show no effect in RH and CD respectively. All the parameters tested increased the biogas yield without affecting the methane content. The findings reported herein could serve as a useful guide for further research to optimise the conditions needed to improve biogas yield from locally available and abundant wastes.

Acknowledgements

Dr P.A. Nwofe and Dr P.E. Agbo would wish to thank the staff of Ishieke Annex, Ebonyi State University, Abakaliki, for permission to use some equipments/chemicals from their laboratory. The authors also thank the staff of Ebonyi State Hatchery, Nkaliki and Rice Mill Industry, Abakaliki for permission to procure the feedstocks (poultry droppings and rice husks).

References

[1] C.E. Nnaji, C.C. Uzoma, and J.O. Chukwu, "The role of renewable energy resources in poverty alleviation and sustainable development in Nigeria", Continental J. Social Sciences, vol. 3, pp. 31 – 37, 2010.

[2] Y.Y Babanyara and U.F. Saleh, "Urbanisation and the Choice of Fuel Wood as a Source of Energy in Nigeria," J. Hum. Ecology, vol. 31(1), pp. 19-26, 2010.

[3] A. Converti , R.P.S Oliveira, B.R. Torres, A. Lodi, and M. Zilli "Biogas production and valorisation by means of a two-step biological process," Bioresource Technology vol. 100(23) pp. 5771-5776, 2009.

[4] C.I. Marrison and E.D. Larson, "A preliminary analysis of the biomass energy production in Africa in 2025 considering projected land needs for food production" Biomass and Bioenergy, vol. 10(5/6), pp. 337-351, 1996.

[5] B. Nnaji, Investment Opportunities in the Nigerian Power Sector. Nigeria Business and Investment Summit, London. July 30, 2012. http://www.newworldnigeria.com/pdf/HMPPowerSectorRefor msPresentationBoINOCLondon30JULY2012.ppt, Accessed January 31, 2015.

[6] P.A. Nwofe, "Determination of the Generation rate of solid waste in Abakaliki Metropolis, Ebonyi State, Nigeria", Continental J. Environmental Sciences, vol. 7(2), pp. 1-3, 2013.

[7] N.I. Elom, "Healthcare solid wastes management protocols in Nigeria and implications for human health risk A review", Continental J. Environmental Sciences, vol. 7(1), pp. 11-19, 2013.

[8] P.A. Nwofe, "Waste management and environmental sustainability: A case study of selected cities in Ebonyi State," Continental J. Environmental Sciences, vol. 7(1), pp. 20-28, 2013.

[9] O.C. Eneh, "Managing Nigeria's environment: the unresolved issues," J Environ Sci Technol, vol. 4(3), pp. 250-263, 2011.

[10] Tch. Ogwueleka, "Municipial solid waste characteristics and management in Nigeria", Iran. J. Environ. Health. Sci. Eng. Vol. 6(3), pp. 173-180, 2009.

[11] H.U. Nwoke, "Generation rate of solid wastes in Owerri metropolis, Imo State Nigeria," Continental J. Environmental Sciences, vol. 7(1), pp. 8-10, 2013.

[12] J. Abubaker, H. Cederlund, V. Arthurson, and M. Pell, "Bacterial community structure and microbial activity in different soils amended with biogas residue and cattle slurry," Applied Soil Ecology, vol. 72, pp. 171-180, 2013.

[13] P. Weiland, "Biogas production: Current state and perspectives", Appl Microbiol Biotechnol, vol. 85(4), pp. 849-860, 2010.

[14] T. Bond and M.R. Templeton, "History and future of domestic biogas plants in the developing world", Energy Sustain Dev, vol. 15(4), pp. 347-354, 2011.

[15] W. Zhong, Z. Zhang, W. Qiao, P. Fu, and M. Liu, "Comparison and chemical pretreatment of corn straw for biogas production by anaerobic digestion," Renewable Energy vol. 36, pp. 1875-1879, 2011.

[16] A.S. Sambo, B. Garba, and B.G. Danshehu, "Effect of some operating parameters on biogas production rate," Renewable Energy, vol. 6(3), pp. 343-344, 1995.

[17] C.E.C. Fernando, "Factors which affect biogas production", Nigerian Journal of Solar Energy, vol. 4, pp. 150-154, 1985.

[18] K.J. Chae, Jang Am, S.K. Yim, and S. In Kim, "The effects of digestion temperature and temperature shock on the biogas yields from the mesophilic anaerobic digestion of swine manure", Bioresource Technology, vol. 99(1), pp. 1-6, 2008.

[19] P. Bolzonella, P. Battistonic, and F. Cecchi, "Mesophilic anaerobic digestion of waste activated sludge: Influence of the solid retention time in the wastewater treatment process" Process Biochemistry, vol. 40, pp. 1453-1460, 2005.

[20] A. Bonmat'i and X. Flotats, "Air stripping of ammonia from pig slurry: characterisation and feasibility as a pre- or post-treatment to mesophilic anaerobic digestion," Waste Management, vol. 23(3), pp. 261-272, 2003.

[21] C. Gallert and J. Winter, "Mesophilic and thermophilic anaeorobic digestion of source-sorted organic wastes: effect of ammonia on glucose degradation and methane production" Appl Microbiol Bitechnol, vol. 48, pp. 405-410, 1997.

[22] M. Saidu, A. Yuzir, M.R. Salim, S. Salmiati, Azman, and N. Abdullah, "Influence of palm oil mill effluent as inoculum on anaerobic digestion of cattle manure for biogas production," Bioresource Technology, vol. 141, pp. 174-176, 2013.

[23] Zhu J, Zheng Y, Xu F, Li, Y. Solid-state anaerobic co-digestion of hay and soybean processing waste for biogas production. Bioresource Technol. 2014; 15:240-247.

[24] Chen G, Zheng Z, Yang S, Fang C, Zuo X, Luo, Y. Experimental co-digestion of corn stalk and vermicompost to improve biogas production. Waste Management 2010; 30:1834-1840.

[25] Tong Z, Linlin L, Zilin S, Guangxin R, Yongzhong F, Xinhui H, Gaihe Y. Biogas Production by Co-Digestion of Goat Manure with Three Crop Residues. PLoS One 2013;8(6):e66845.

[26] Chandra R. Studies on production of enriched biogas using jatropha and pongamia de-oiled seed cakes and its utilisation in I.C. Engines. PhD thesis 2009; Centre for Rural Development and Technology, IIT Delhi.

[27] Budiyono, Igbal S, Siswo S. Biogas production from bioethanol waste: the effect of pH and urea addition to biogas production rate. Waste Tech 2013;1(1):1-5.

[28] Oyewole OA. Biogas production from chicken droppings. Science World Journal 2010;5(4):11-14.

[29] Parkin G, Owen F. Fundamentals of anaerobic digestion of waste water sludges. Journal of Environmental Engineering 1986;112(5):867-920.

[30] Ye J, Li D, Sun Y, Wang G, Yuan Z, Zhen F, Wang Y. Improved biogas production from rice straw by co-digestion with kitchen waste and pig manure. Waste Management 2013;33:2653-2658.

[31] Gupta P, Gupta A. Biogas production from coal via anaerobic fermentation. Fuel 2014;118:238-242.

[32] Strik DPBTB, Domnanovich P, Holubar P. A pH-based control of ammonia in biogas during anaerobic digestion of artificial pig manure and maize silage. Process Biochemistry 2006;41:1235-1238.

[33] Kafle GK, Kim SH. Anaerobic treatment of apple waste with swine manure for biogas production: Batch and continuous operation. Applied Energy 2013;103:61-72.

[34] Sánchez-Hernández EP, Weiland P, Borja R. The effect of biogas sparging on cow manure characteristics and its subsequent anaerobic biodegradation. International Biodeterioriation and Degradation 2013;83:10-16.

[35] S.M. Lee and J.H. Lee, "Effect of sludge treatment on biogas production from Saccharina japonica ethanol fermentation by-products", Journal of Industrial and Engineering Chemistry, vol. 21, pp. 711-716, 2015.

[36] Li J, Wei L, Duan Q, Hu G, Zhang G. Semi-continuous anaerobic co-digestion of dairy manure with three crop residues for biogas production. Bioresource Technol. 2014;156:307-313.

[37] Gupta A, Kumar A, Sharma S, Vijay VK. Comparative evaluation of raw and detoxified mahua seed cake for biogas production. Applied Energy 2013;102:1514-1521.

[38] Ahmadu TO, Folayan CO, Yawas DS. Comparative performance of cow dung and chicken droppings for biogas production. Nig. J. Eng. 2009;16(1):154-164.

[39] Ojolo SJ, Dinrifo RR, Yadesuyi KB. Comparative study of biogas from five substrates. Advance Materials Research Journal 2007;18(10):519-525.

[40] Wagner AO, Hohlbrugger P, Lins P, Illmer P. Effects of different nitrogen sources on the biogas production a lab-scale investigation. Microbiological Research 2012;167(10):630-636.

[41] Okeh CO, Onwosi CO, Odibo FJC. Biogas production from rice husks generated from various rice mills in Ebonyi State, Nigeria. Renewable Energy 2014;64:204-208.

[42] Sterling Jr MC, Lacey RE, Engler CR, Ricke SC. Effects of ammonia nitrogen on H_2 and CH_4 production during anaerobic digestion of dairy cattle manure. Bioresource Tchnol. 2001;77(1):9-18

[43] Zhang R, Zhang Z. Biogasification of rice straw with an anaerobic phased solids digester system. Bioresource Technology 1999;68(3):235-245.

[44] Sau SK, Manna TK, Giri A, Nnandi PK. Effect of Human Urine during Production of Methane from Boiled Rice. International Journal of Science and Research 2013;2(10):60-64.

[45] Ganiyu OT, Oloke JK. Effects of organic Nitrogen and Carbon supplementation on biomethanation of rice bran. Fountain Journal of Natural and Applied Sciences 2012;1(1):25-30.

[46] Satyanarayana S, Murkute P, Ramakant. Biogas production enhancement by Brassica compestries amendment in cattle dung digesters. Biomass and Bioenergy 2008;32:210 215.

[47] Mudhoo A, Kumar S. Effects of heavy metals as stress factors on anaerobic digestion processes and biogas production from biomass. Int. J. Environ. Sci. Technol. 2013;10:1383-1398.

[48] Espinosa A, Rosas R, Ilangovan K, Noyola A. Effect of trace metals on the anaerobic degradation of volatile fatty acids in molasses stillage. Water Science and Technology 1995;32(12):121-129.

[49] Matseh I. Effect of Ni and Co as Trace Metals on Digestion Performance and Biogas Produced from The Fermentation of Palm Oil Mill Effluent. International Journal of Waste Resources 2012;2(2):16-19.

[50] Demirel B, Scherer P. Trace elements requirements of agricultural biogas digesters during biogas conversion of renewable biomass to methane. Biomass and Bioenergy 2011;35:992-998.

[51] Faachin V, Cavinato C, Pavan P, Bolzonella D. Batch and continuous Mesophilic Anaerobic Digestion of Food Waste:Effect of Trace Elements Supplementation. Chemical Engineering Transact-ions 2013;32:457-462.

[52] Irvan. The Effect of Fe Concentration on the Quality and Quantity of Biogas Produced From Fermentation of Palm Oil Mill Effluent. International Journal of Science and Engineering 2012;3(2):35-38.

[53] Facchin V, Cavinato C, Fatone F, Pavan P, Cecchi F, Bolzonella D. Effect of trace element supplementation on the mesophilic anaerobic digestion of foodwaste in batch trials: The influence of inoculum origin. Biochemical Engineering Journal 2013;70:71-77.

[54] Jain SK, Gujral GS, Jha NK, Vasudevan P. Production of biogas from Azolla pinnata R.Br and Lemna minor L.: effect of heavy metal contamination. Bioresoure Technol 1992;41:273-277.

[55] Wanqin Z, Shubiao W, Qianqian L, Renjie D. Trace elements on influence of anaerobic fermentation in biogas projects. Transactions of the Chinese Society of Agricultural Engineering 2013;29(10):1-11.

[56] Swapnavahini K, Sumanth Kumar M, Appala Naidu G. Effect of Selenium on acceleration of biogas and trends of Nitrogen and Phosphorous. International Journal of Innovative Research and Practices 2013;1(7):24-28.

[57] Wanqin Z, Jianbin G, Shubiao W, Renjie D, Jie Z, Qian-qian L, Xin L, Tao L, Changle P, Li C, Baozhi W. Effects of Fe^{2+} on the Anaerobic Digestion of Chicken Manure: A Batch Study. 2012 Third International Conference on Digital Manufacturing & Automation, Guilin, China July 31, 2012 to Aug. 2, 2012, ISBN: 978-1-4673-2217-12012;364-368.

[58] Lo HM. Metals behaviors of MSWI bottom ash co-digested Anaerobically with MSW. Resources, Conservation and Recycling 2005;43(3): 263-280.

[59] Zitomer DH, Johnson CC, Speece RR. Metal Stimulation and Municipal Digester Thermophilic/Mesophilic Activity. Journal of Environmental Engineering 2008:134(1):42-47.

[60] M. Brule, R. Bolduan, S. Seidelt, P. Schlagermann, A. Bott, "Modified batch anaerobic digestion assay for testing efficiencies of trace metal additives to enhance methane production of energy crops," Environmental Technology vol. 34(13/14), pp. 2047-2058, 2013.

4

Simulating human occupancy in an experimental laboratory setting

Joseph Martin Petersen[1,2]

[1]Washington State University, School of Electrical Engineering and Computer Science, Richland, Washington, USA
[2]Pacific Northwest National Laboratory Energy Policy & Economics, Richland, Washington. USA

Email address:

Joseph.m.petersen@gmail.com

Abstract: Energy conservation within a residential home is a primary focus for both home owners and power utilities throughout the country. Developing a technology to model, detect, and measure human occupancy would allow for laboratory settings to more accurately model residential energy use without the need for actual human activity within the space. An accurate way to measure occupancy is through detecting the latent and sensible heat that is generated by activities within the home. As industry facilities move forward with research, innovative ways to model every aspect of a residential home comes into play. These research settings require the development of technology that appropriately models and detects human activity within a residential home.

Keywords: Alternative, Energy, Modeling, Robotics

1. Introduction

Laboratory settings that model whole house energy use have struggled to find an accurate and efficient way to simulate human occupancy within these closed loop facilities. Both the latent and sensible heat, in reference to human generation, can account for large changes in the whole house energy use over time. Generating these representative loads can be effected by many external parameters. In order to acquire an accurate representation of the whole house energy use, a system is needed to model the day to day humidity generation of a typical family. This paper proposes the development of a standalone occupancy simulating system.

2. Background

Many laboratory studies have been conducted on the whole house energy use of a residential home with the hopes of understanding how differing variables will alter the total energy use. Energy efficient and smart technologies have been implemented in ways that dynamically shift the power use of the homes. Collected by precise data logging technology, the total load of the homes can be determined. Though most of the loads are implemented through load and occupancy profiles, the elements of human interaction

on humidity and latent heat in the homes are neglected for the most part. The absence of human presence and load alters the amount of stress that is exerted on the home. Daily activities such as cooking, cleaning, and exercising generate an environment that is currently not being modeled within the Pacific Northwest Nation Laboratory (PNNL) lab homes. The ability to use this technology will, "allow us to observe the occupancy of the home by operating it in unexpected ways" [2]. With the main goal of energy efficiency, this system will allow for a more realistic view of the typical home.

2.1. Problems with Previous Technology

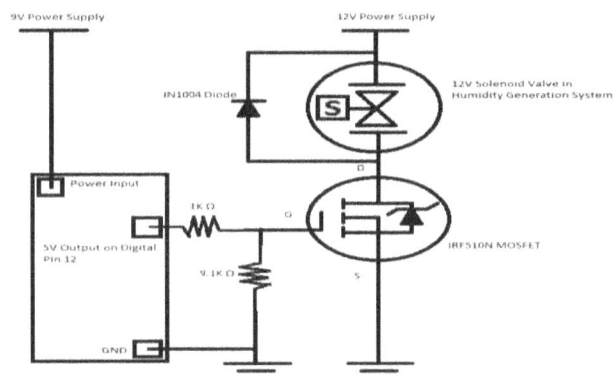

Figure 1. PNNL humidity generator [4].

PNNL has developed an occupancy simulation system to address this problem. At PNNL, Austin Winkelman developed standalone humidity generator. Below is the schematic detailing the specifics of the generator.

Controlled by an Arduino microcontroller, a solenoid is turned off and on based on the ASHRAE generation schedule. The following problems were encountered:

1) The hot plate needed to remain hot and on the entire time. This wasted energy and was a potential fire hazard.
2) Another issue was that the generator had no feedback system or I/O interface. The amount of humidity generated and the frequency of generation was assumed to be correct.
3) The system had no way of differentiating from the latent and sensible heat loads of the environment.
4) Not expandable or portable. The location and the reservoir size remain static throughout the experiment.

3. Ultra Sonic Controllable Humidifier

With the idea of portability and flexibility in mind, a new humidity generation system has developed. It is able to differentiate between the sensible and latent heat loads, able to run for extended periods of time, and interface with the Campbell system to generate a feedback loop.

3.1. Latent and Sensible Generation Profiles

To facilitate research and promote consistent standards with analyzing residential homes, the Department of Energy has developed a "Building America Analysis Spreadsheets" [6]. These documents are specifically designed to provide a set of standard operating conditions. Specific building parameters are entered into the workbook. These include square footage, number of bedrooms, and predicated number of occupants. With this information, expected load profiles can be generated to reflect the input parameters. Specifically, this workbook is used to generate the latent and sensible electrical loads for new or existing homes.

From this workbook, an occupancy schedule was generated. This did not give numerical values for the total latent and sensible heat generated by occupants within the homes. This only gave the probability that the occupants will be in the home at specific hours of the day. The sensible and latent heats were calculated by PNNL engineers from the following equations.

$$\frac{Gal.H20}{hour} = \frac{TC}{\rho L} \qquad (1)$$

T = total latent heat
C = conversion from BTU to Watts
ρ = density of H20
L = latent heat of evaporation
Latent Occupancy Load- reflected in heating water [6]

$$Latent\ Occupancy\ Load = \frac{PNO}{C} \qquad (2)$$

P = probability of occupant within home
N = number of occupants
O = Occupant load per person per BTU per hour
C = conversion from BTU to Watts

3.1.1. Humidity Generation System- Latent Load

From the DOE building America house simulation protocol and latent heat calculations, it was determined that a total of 1.56 Gal H20/day needed to be generated with a max generation hour of .085 Gal. This max hourly generation rate falls outside the max generation of the ultrasonic humidifier. To facilitate the total generation required, two human occupancy generators are installed within the PNNL lab homes, one in the kitchen and the other in the master bathroom. This promotes better penetration of the humidity throughout the house and will eliminate strain on a single unit.

The ultrasonic humidifier has been modified to generate both the latent and sensible heat. The transducer produces humidity that is independent of a heat source. This allows for complete controllability of both the latent and sensible heat to be measured and controlled. Through experiential measurements, the average generation rate of 4.55 ml of H2O vapor/min was found. This was generated through the experimental measurement of the evaporated water over a given 10 minute interval of time. The data collected can be seen below.

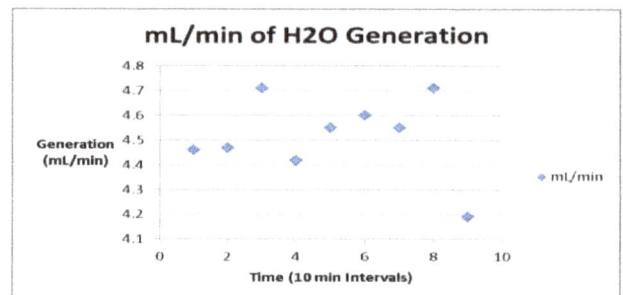

Figure 2. *Experimentally measured latent heat load generation by the occupancy simulator.*

3.1.2. Humidity Generation System- Sensible Load

Similar to the occupancy latent heat load generation, the sensible load was calculated by PNNL engineers through the use of Eq. 3. The sensible load will be generated through heating the water that is stored within the ultrasonic humidifier reservoir. From the given dynamics equations the final temperature of the water was found.

$$Q = mc\Delta T \qquad (3)$$

m = mass in grams
c = specific heat of water
ΔT = change in H2O and ambient temperature
Q = Sensible heat in Joules

In this equation it is assumed that the transducer reservoir is always full. This has and expected weight of 64g of H2O. Ambient air and water temperature are recorded and used to generate the change in temperature. The max sensible load seen within a 24 hour period is 193 W. This needs to be generated over the course of 1 hour. From the above equation the water temp will need to be heated to 147.2 F.

The heating of the water is done through a metal heating element that has been installed into the transducer reservoir. A feedback loop is generated between the temperature sensor within the reservoir and the heating element. It is attached with water tight gaskets to prevent water penetration into the control housing. The water will be constantly heated until it matches the required heat to compensate for the sensible heat load. Due to the limited size of the reservoir, an equilibrium temperature will penetrate through the full liquid without the required assistance of a mixing mechanism. Using a 3D printer, a housing for the heater was made to expose the maximum surface area, of the element, to the surrounding water. Below is a schematic of the housing.

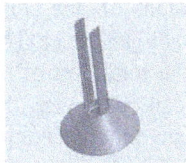

Figure 3. Heating element housing

4. Control System

The control systems were chosen based on previous engineering knowledge of the technology, a limited budget, and the ability to interface with the existing technology within the lab homes.

4.1. Control System- Data Acquisition and Wireless Control

A control system has been integrated into the existing enclosure of the humidifier. A 3D printer has been used to expand the enclosure of the humidifier to include all controlling peripherals. The main control system is the Arduino Uno rev3 micro controller. The controller uses the Atmel Atmega328 microprocessor and has a limited memory of 32Kb. The main peripheral mods that are installed are the relay and xbee shields. This allows for 4 relays to control via the Arduino. Above is a picture of the control system.

The 4 relays can be controlled by the 5v analog pins on the Arduino. The ultrasonic transducer and heating element are attached to individual relays. 120V is supplied to the Arduino and allows for the switching to control the transducer and heating element. This is supplied by the "master" relay. The transducer and heating element remain on only as long as necessary to generate the required sensible and latent heat for the given hour.

The Arduino is equipped with an Xbee shield that will allow for wireless communication between the Arduino and existing Campbell data acquisition system. The Xbee has a large band width for wireless communication. The frequency of communication can be altered to not interfere with other wireless technology within the lab homes. Below is a schematic of the Xbee and Arduino interconnections.

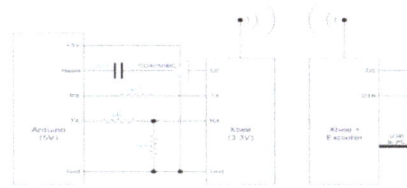

Figure 4. Xbee Arduino architecture schematic

The integrated Campbell scientific control system is the industry standard for data logging. Currently, this data logger is used within the lab homes to aggregate all of the sensor data. It is equipped with 12v and 5v channels that can power the Xbee. A transmitter Xbee is installed to interface between the Campbell and humidifier. The Campbell reads the internal relative humidity of the home or responds to a humidity schedule. If humidity is needed, the Campbell sends bits of information to the Xbee which will then transmit this data to the Arduino. Below is a picture of the described feedback loop.

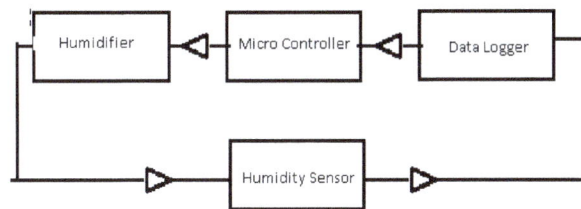

Figure 5. Control system feedback loop for the occupancy generator

4.2. Water Control

Based on the location of installation, a simple bypass or float system has been installed into the occupancy generator. Below is a simple drawing of the two methods used to control the water flow and reservoir water level.

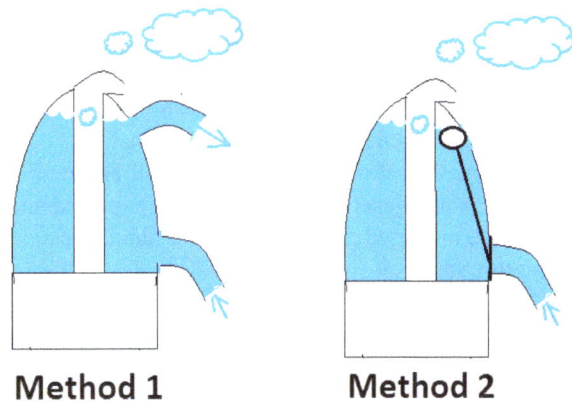

Figure 6. Reservoir control and fill methods

Method 1: Within the kitchen, predefined water draws have been programmed to happen on the hour via the Campbell control system. Method 1 utilizes this almost constant stream of water to ensure that the water reservoir is constantly full. Due to the fact that the occupancy generator only produces

1.56 gallons of H2O per day, there is no possible way for the reservoir to run out of water within this scheme. Because the water is being bypassed, there is no possible way for the Reservoir to be over filled causing water damage in the lab.
Method 2: This method has been implemented into the master bedroom. This area is not subject to water draws based on the Campbell's occupancy schedule. A simple float mechanism is installed into the second human occupancy generator. Installed within the master bathroom faucet, the reservoir will remain constantly full while preventing over flow.

4.3. Control System- Safety mechanisms

Both PNNL and Washington State University put a large emphasis on the safety culture with the laboratory settings. This has been continued within this project with the belief that a system product is useless if it is unsafe for the user and environment. Within the human occupancy generator, there are a few redundancies that promote safety.

1) Relay control- from experimental trial and error, the relay for the heating element should be in the on position no longer than an hour. If the relay remain on for greater than a predefined time this could cause damage to the transducer reservoir or become a fire hazard. As a safety control, the relay will periodically cycle off and on with a frequency of 1 time per 5 minutes. This ensures that the relay is working properly. If during this cycle, the relay remains closed for a period of greater than 10 minutes, the relay is assumed to have failed and the master relay will trip.

2) Temperature control – equipped with a temperature sensor, the control system monitors the temperature of the water that is within the reservoir. Knowing that the temperature should never surpass 147.2 F, the arduino can trip the master relay if the water is heated above this set point for any sustained period of time for any reason.

3) Humidity Control – taking advantage of the installed humidity sensors within the lab homes aids in the protection feedback loop. The max humidity set point is known. If too much humidity is detected by the Campbell data logging technology, a signal will be sent wirelessly to the arduino. This will cause the arduino to trip the master relay.

4) Testing – It is recommended that the master relay be tested bi-monthly to ensure that it is fully functional.

The implementation of these 4 safeguards will ensure that the human occupancy generator is operating as expected and will not generate increased risk of fire or over humidification of the laboratory space.

5. Benefits

The benefits of the new occupancy generator are as follows:
1) The distribution of the sensible and latent heat loads. This allows for a more accurate representation and manipulation of an already dynamic system.
2) External control by data logging technologies.

3) Transducer eliminates the need for a hot plate.
4) Accurate and measureable load generation.
5) Portability due to expanded water reservoir.

Using this system, PNNL are able to achieve fairly accurate results for both the sensible and latent heat loads.

5.1. Cost Estimation

Table 1. Cost estimation for the project

Item	Cost ($)
Arduino Uno	24.51
Relay Shield	14.9
Xbee	19
Humidifier	29.99
Heating Element	10.99
PSU	6.99
Total	106.38

Above is an aggregation of the total price of the components of the occupancy generator. This is only for one unit and does not reflect the total cost of the Campbell data logging technology. It can be seen that the price is substantially lower than many other technologies currently on the market that allow for expanded control of the laboratory environment. It would be difficulty to find a comparable ultrasonic humidifier that allows for any advanced user input for less than $150. This product undercuts competing products on programmability, flexibility, and price point.

6. Conclusion

Adaptation of a ultrasonic humidifier with a built in control system has allowed for a more accurate means of controlling and measuring the dynamic environment generated in a residential or laboratory setting. Collaboration between PNNL and WSU has allowed for the further expansion of this topic into uncharted areas.

References

[1] ASHRAE (2008). "ASHRAE Standard 160P" Retrieved 12/15/20213

[2] Fang, X. et al. (2011). "Field Test Protocol: Standard Internal Load Generation for Unoccupied Test Homes". NREL retrieved 12/17/2013

[3] Lee, Yoo. (2011). "Simulating Human Behavior and Its Impact on Energy Uses". University of Pennsylvania. Retrieved 12/15/2013.

[4] TenWolde, A. et al. (2007). "The Effect of Indoor Humidity on Water Vapor Release in Homes" ASHRAE. Retrieved 12/17/2013

[5] Winkelman, A (2012) "Humidity Generation System" PNNL. Retrieved 12/15/2013.

[6] DOE, (2012). "Building America Workbook." Department of Energy. Retrieved 3/12/2014

Characterization of Oil and Biodiesel Produced from *Thevetia peruviana* (Yellow Oleander) Seeds

Ana Godson R. E. E.[*], **Udofia Bassey G.**

Department of Environmental Health Sciences, Faculty of Public Health, College of Medicine, University of Ibadan, Ibadan

Email address:
anagrow@yahoo.com (Ana G. R. E. E.), basseyudofia@yahoo.com (Udofia B. G.)

Abstract: Background: There is increasing emphasis on renewable energy following recurrent economic crises and environmental concerns associated with the use of fossil fuels such as petrodiesel. Research into biodiesel production from oil-bearing renewable biomass sources can provide a more sustainable alternative to petrodiesel. This study evaluated the biodiesel yielding potential of *Thevetia peruviana* seeds. Methods: Oil was extracted from the seeds using Soxhlet and Cold-solvent extraction methods. Hexane-only (H-only) was used in the Soxhlet while Hexane/Ether (H/E) mixture and H-only were respectively used in the Cold extraction. The oil was processed using Methanol/Ethanol (M/E) mixture and Methanol-only (M-only) respectively to biodiesel via transesterification with sodium hydroxide as catalyst. The oil and biodiesel physicochemical parameters such as density, viscosity at $40^{\circ}C$, Saponification value, Flash Point (FP) and Acid Value (AV) were determined using the American Standard for Testing and Material (ASTM D6751) methods. Results: The oil yields from Soxhlet, H/E and H-only extractions were: 62.3%, 51.9% and 45.8% respectively. The biodiesel yield in the M/E and M-only transesterifications were: 78.4% and 85.20% respectively. The density at $40^{\circ}C$, viscosity, and saponification value of the oil were: $0.868g/cm^3$, $21.50mm^2/s$ and 120mgKOH/g respectively. The density at $40^{\circ}C$, viscosity, FP and AV of the biodiesel were: $0.760g/cm^3$, $4.70mm^2/s$, $130^{\circ}C$ and 0.441mgKOH/g respectively. Conclusion: The seeds of *Thevetia peruviana* are viable sources for biodiesel production, and quality parameters of the biodiesel met the American Standard for Testing and Materials limits. However, further work to explore the optimization of the process and sustainability of the model is recommended.

Keywords: Biodiesel Production, *Thevetia peruviana*, Transesterification

1. Introduction

The World's energy demand is increasing geometrically as evidenced in the increasing demand for fuels for transportation, industrial as well as domestic activities. This sky-rocketing energy demand continues despite the attendant environmental pollution and global warming effect resulting from the use of petroleum-based fuels [1].

Within the last 20 years about 75% of human made CO_2 emissions were from burning of fossil fuels. Nigeria's oil, for example, has not guaranteed ecologically and socially acceptable development in the country. At present, there are over 11 oil companies operating 1,481 wells from 159 oil fields in the Niger Delta producing 2.7 million barrels of crude oil each day and flaring about 17 billion cubic metres of associated gas, spewing 2,700 tons of particulates, 160 tons of sulphur oxides, 5,400 tons of carbon monoxide, 12 and 3.5 million tons of methane and carbon dioxide, respectively in the process [2].

Modern biofuels have even been reported as a promising long term renewable energy source, which has the potential to address both environmental impacts and security concerns posed by current dependence on fossil fuels. In comparison to petroleum-based fuels, biodiesel offers reduced exhaust emissions, improved biodegradability, reduced toxicity and higher cetane rating which can improve performance and clean up emissions [3].

There is currently no commercial biodiesel plant that exists in Nigeria, except for a few production facilities that are notably not well documented. Production and consumption are still at their infancy stage. With the increasing emphasis on renewable energy following the global trend in the automobile industry, and biodiesel gaining increasing popularity, this is paving the way for increased consumer confidence in automobile engines' ability to utilize biodiesel of which Nigeria cannot be isolated. With an estimated population of about 150 million people and a population growth rate of 2.38% (2007 estimate) and an average of 12 vehicles to 1000 people (1997 estimate), the potentials of biodiesel cannot be underestimated [4].

A variety of biolipids can be used to produce biodiesel. These include (a) virgin vegetable oil feedstock (rapeseed and soybean oils are most commonly used, though other crops such as mustard, palm oil, sunflower, hemp, and even algae show promise) (b) waste vegetable oil; (c) animal fats including tallow, lard, and yellow grease; and (d) non-edible oils such as algal oil, jatropha oil, neem oil, castor oil, and tall oil [5].

The use of edible vegetable oils from biomasses like those of soybean, sunflower, cotton seed, safflower, canola, palm fruits, fish oil and also animal fats for biodiesel production has recently been of great concern. This is because of the major criticism against large-scale fuel production from agricultural crops that it will consume vast expanse of farmlands and native habitat; compete with food materials and drive up food prices [6]. Hence, in view of these concerns and no reliable prospects for a massive compensatory scale-up in food production capacity (especially in developing countries like Nigeria), there is the need to intensify effort in exploring other potentially viable inedible oil-rich biomasses (such as algae, thevetia, jatropha, rice bran, neem, castor, etc.) for biodiesel production.

Thevetia peruviana is a plant that is grown as hedges and kept for its bright and attractive flowers, and has been grown in Nigeria for over fifty years basically as an ornamental plant in homes, schools and churches by missionaries and explorers. To date, despite the fact that there is high level of oil content in its kernel, about 60-65% [7] and valuable protein content in the seed, about 40-45% [8], it remains non-edible because of the presence of cardiac glycoside (toxins), hence the plant remains a plant of no significant economic value whereas it has a lot of potentials such biofuel production.

Few works have been carried out using *Thevetia peruviana* seeds as feedstock for biodiesel production in Nigeria. Ibiyemi et al [8] studied the variation in oil composition between the different seed varieties of Yellow oleander while Olisakwe et al [9] carried out a comparative study of *Thevetia peruviana* and *Jatropha curcas* seed oils as feedstock for grease production amongst others. The present study was carried out to extract oil from *Thevetia peruviana* seeds using two methods (Soxhlet and Cold extraction), and process the oils to biodiesel via transesterification with methanol using sodium hydroxide as catalyst. The biodiesel produced from the seed oil was further characterized through ASTM standard tests for basic fuel properties such as density, viscosity, flash point and acid value.

2. Materials and Methods

Thevetia fruits were harvested from thevetia plantations grown as hedges in and around the University of Ibadan campus.

2.1. Feedstock Preparation

The harvested thevetia fruits were manually decorticated with stainless knife to reveal the kernels. Both thevetia and palm kernels were sundried under 5hours continuous sunlight to ease the removal of the seeds from them. After the seeds were removed from the kernels, they were milled using a hand-powered bench grinder. The milled seeds were subjected to 24hour continuous oven-drying period at 105°C until a constant weight was recorded.

Phosphorous, calcium, and magnesium are minor components typically associated with phospholipids and gums that may act as emulsifiers [10] or cause sediment, lowering yields during the transesterification process [11]. Hence, their percentage composition in the biomass was determined to establish if there would be a significant reduction in their composition upon taking the substrate through the solvent extraction process and the extracted oil through transesterification process.

The proportion of Total Phosphorus (%P), Sodium (%Na), Calcium (%Ca), and Sulphur (%S) in the milled biomasses were analyzed chemically according to the official methods of analysis described by the Association of Official Analytical Chemists [12]. The moisture content was determined using a Moisture analyzer equipment (AИD MX-50 model) and the density of the milled biomasses was determined gravimetrically.

2.2. Oil Extraction

The milled oven-dried samples were used for the extraction process and two extraction methods were experimented *viz*: Soxhlet extraction and Cold solvent extraction (H/E mixture and H-only alcohol). For the Soxhlet extraction, 250g each of milled thevetia seeds were respectively placed in the thimble of a Soxhlet extractor with the use of about 800ml hexane for one extraction and a mixture of hexane and ether in ratio 1:1 v/v for another extraction. A round bottom flask containing the extraction solvent was fixed to the end of the extractor and a condenser was tightly fixed to the other ends of the extractor.

Once the sample for a particular extraction period was placed in the thimble of the extractor, the flask was heated at 60°C with the use of an electric mantle. As the solvent was heated in the boiler, the pure vapor rose through a by-pass and into the top part of the Soxhlet container (thimble) where the sample to extract was contained. In the condenser, the vapors condensed and drip into the sample-containing thimble. When the level of liquid reaches the same level as the top of the siphon, the liquid containing the extracted material was siphoned back into the boiler.

Each of the extraction processes carried out underwent a minimum of 40 cycles within the 8 hours period, which is considered necessary to complete an extraction [13]. After the extraction period the solvent was recovered at 65°C under vacuum using a rotary evaporator (Buchi Rotavapor: R-210 model), and the residual oils obtained thereafter were measured and recorded.

For the Cold Solvent extraction, the method used by Hossain et al [14] was adopted. 250g of the milled dried thevetia seeds were mixed with 600ml hexane and another 250g was mixed with 600ml H/E, 1:1, v/v mixture. The two separate sample/solvent mixtures were kept to settle in their respective labeled and well-sealed containers for 48 hours, with intermittent shaking (every 3-5hours) of the containers to enhance optimum percolation/breakage of the solvent into the cell wall of the ground seeds.

The 48hour period was followed by the separation of the sample/solvent mixtures by "squeeze-filtration" using two muslin cloths inserted into each other as a precaution to better reduce the amount of sediment that may probably be small enough to pass through the sieve pores into the solvent/oil mixture. The extracted oil/solvent mixtures were left to settle and air-dry for 24hours.

The clear oils were then decanted through a No 1 Whatman filter paper into labeled bottles, leaving behind the residual paste/sediments. Each of the decanted oils were individually evaporated under vacuum for about 5minutes at 60°C using the Buchi type Rotavapor (R-210 model). This was to ensure that all the extraction solvents in the oils are evaporated off. The residual oils obtained after evaporation was left to air-dry for about 2hours; the volume and weight of the oils were subsequently measured and recorded. The oils were kept for characterization and further processing via transesterification process.

The percentage of oils extracted from the milled samples by both the Soxhlet extraction and Cold extraction systems respectively was determined using equation below:

$$\% \text{ Oil content} = (W_o/W_u)\ 100\%$$

Where: W_o = weight of oil extracted (g)

W_u = weight of oven-dried biomass used for the extraction process (g)

2.3. Degumming of the Extracted Oil

This was accomplished by mixing 2% distilled water with the combined 482.2cm^3 thevetia oil obtained from the three different extraction processes; the mixture was agitated slightly. The water combined with the gums and precipitated out after allowing the mixture to settle for approximately one hour, and the water was drained off at the bottom of the oil. About three washings were carried out to remove a good quantity of any gum that might be present. The oil was heated on a heating mantle to a temperature above 100°C, and allowed to cool to a temperature of 60°C and then separated.

2.4. Characterization of the Extracted Oil

The physicochemical properties of the oil were determined following standard methods. Kinematic viscosity (KV) was determined using a calibrated Cannon-ubbelohde viscometer at a temperature of 40°C according to ASTM D445 (2006) [15]; density at 25°C was determined gravimetrically; Saponification value was determined by estimating the number of milligrams of KOH that neutralized the fatty acids resulting from the complete hydrolysis of 1g of oil; Fatty Acid Profile (FAP) was done according to the method described by Christie [16]; and Free Fatty Acid (FFA) was determined by selective formation of diethyl amide derivatives according to the method described by Kangani et al [17].

FFA (or fatty acids that have been unbound from the original triglyceride) occur in vegetable oils either because of contact with water or poor storage or because of the presence of enzymes that rapidly cleave the fatty acids from the glycerol backbone. When a homogeneous alkali catalyst is used, Gerpen [3] recommends that the maximum FFA content of the feed oil should be 5%. Otherwise, soaps will form, making separation of the glycerine difficult. Hence, an additional step to remove the FFA or to convert them via an esterification step would be necessary before using the feed oil in transesterification reaction.

2.5. Biodiesel Production by Transesterification Process

The process flow chart for a typical biodiesel production is presented in Figure 1. The transesterification reaction was carried out with Methanol-only (M-only) and Methanol/Ethanol (M/E) mixture (1:1, v/v) respectively in the presence of NaOH as catalyst. The molar ratio of alcohol to oil was 6:1 and 1% weight of the oil of NaOH catalyst was used at 65°C reaction temperature. Transesterification is a reversible reaction, thus the methanol quantity is required to shift the equilibrium favorably. The methanol to oil molar ratio, the weight percent of catalyst and the reaction temperature were chosen since they have been found to give optimal yields of methyl ester from seed oils [18].

An Erlenmeyer flask (500ml capacity) was charged with 250g of the thevetia oil and warmed to a desired temperature of about 55°C, which is less than the boiling point of methanol (65°C) in a water bath. While the oil was being warmed, a methanol quantity of 6:1 molar ratio methanol to oil and an optimal weight of NaOH pellets (1% weight of the oil) were mixed and heated in a separate flask to a desired temperature of 50°C on the magnetic stirrer until the NaOH pellets were completely dissolved.

After the complete dissolution of the NaOH pellets, the beaker was taken-off the magnetic stirrer, and the Erlenmeyer flask containing the warm oil was removed from the water bath and placed on the stirrer. The alcohol-NaOH mixture (i.e. sodium methoxide and/or sodium ethoxide) in the beaker was then added to the oil in the flask (including a corrode-resistant stir bar), the temperature of the hot plate was immediately increased to 65°C and the revolution of the stirrer was set at level four (i.e. 400rpm). The mouth of the flask was sealed with an aluminum foil to minimize alcohol evaporation during the conversion process. The reaction was allowed to continue for 1 hour, after which the stirrer was turned off, the stir bar was removed, and the content of the flask was immediately poured into a separatory funnel.

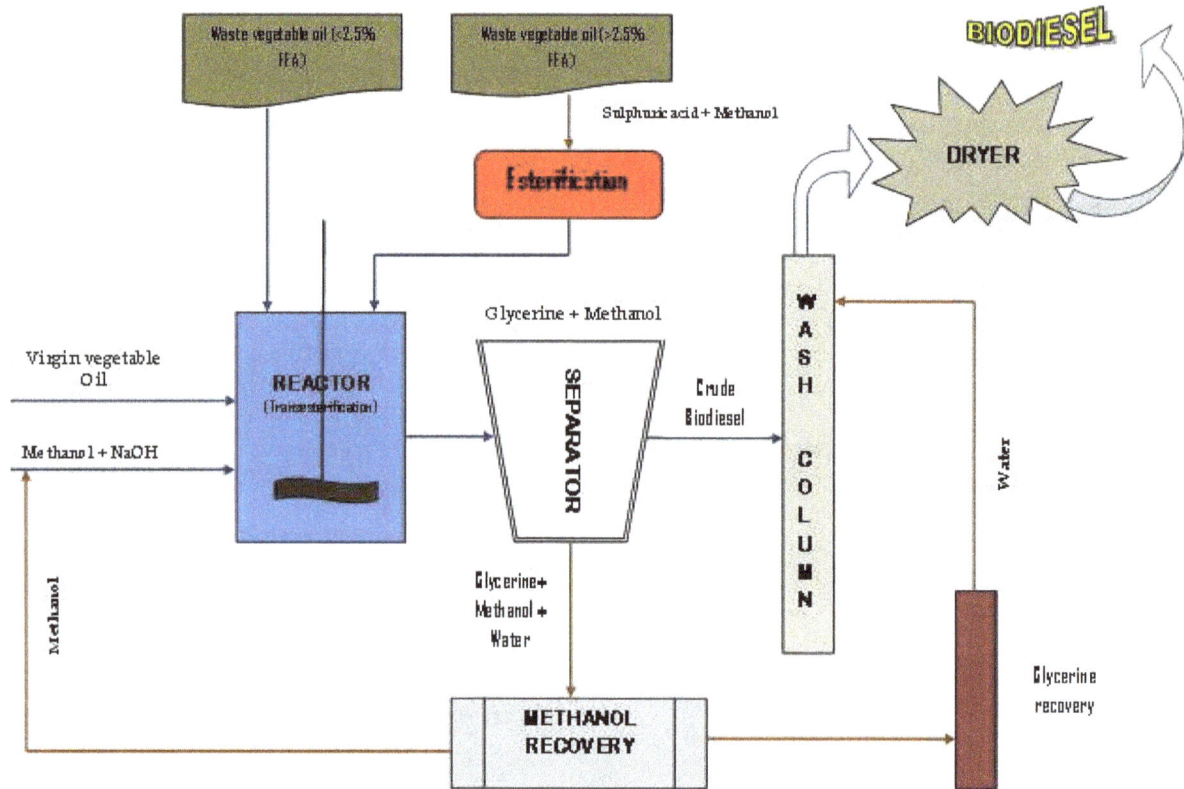

Figure 1. *Process flowchart for typical biodiesel Production.*

2.6. Phase Separation and Washing Procedure

The transesterification reactions produced a mixture of glycerol and methyl esters when they were completed. These, being completely insoluble with one another, separated into two distinct phases when poured into a separatory funnel. The impure glycerol settled at the bottom part of the funnel and was thus drained out by the stopper at the bottom of the separator. The quantity of the glycerol byproduct was measured using a measuring cylinder.

A sample of the biodiesel (methyl ester layer) in the flask was thereafter taken and the pH determined. If found to be caustic i.e. alkaline (say like pH 8 and above), the biodiesel in the flask was washed with hot water (about 55°C) and 0.1% acid solution. However, if found to be in normal pH range (of say like 7.0-7.5), then only warm distilled water was used in washing. The thevetia biodiesel was within normal pH range of 7.0-7.5 and hence was washed with hot water only. One third ($^1/_3$) as much hot distilled water as there is biodiesel was added in a stepwise fashion to the biodiesel in the flask. The water settles quickly at the bottom of the flask and was subsequently drained out as it settles. The washing continued in the stepwise fashion until the water settling at the bottom of the flask was visibly clear.

The biodiesel yield (% wt) after the post-treatment stage, relative to the amount of the different substrate oils poured into the flask was calculated from the equation below:

$$Biodiesel\ yield = \frac{Volume\ of\ biodoesel\ produced}{Volume\ of\ oil\ used} \times 100\%$$

2.7. Characterization of biodiesel

The parameters analysed for in the biodiesel were: Kinematic viscosity, which was determined using a calibrated Cannon-ubbelohde viscometer at a temperature of 40°C according to ASTM D445 [15]; Flash Point, determined using a Manual Pensky-Martens closed cup apparatus according to ASTM D93 [19]; Density at 25°C, determined gravimetrically; Acid value and Percentage elemental composition (P, Ca, Na and S levels) according to the official methods of analysis described by the Association of Official Analytical Chemists [12].

3. Results

A highlight of the major steps involved in the production of biodiesel from thevetia seeds is presented in Figure 2 below. The moisture content of the milled thevetia seeds determined gravimetrically and by the moisture analyzer equipment was 6.64 + 0.01% and 6.63 + 0.05% respectively. The density of the milled biomass was 0.750g/cm^3 and 0.881g/cm^3 respectively. The proportion of elements in the milled biomass was: TP (0.046%), Ca (0.04%), Na (0.017%) and S (0.08%).

The percentage oil yield in the Soxhlet, Cold extraction (H-only) and Cold extraction (H/E) methods were: 62.32%, 51.87% and 45.81% respectively. Thevetia oil was observed to have a deep yellow colour and a very sweet butter smell. The result of viscosity measurement and estimation of saponification value of the oil was $21.5 mm^2/s$ and 120.1mgKOH/g respectively. The density at 25^oC and 40^oC were both $0.9g/cm^3$ respectively. The result of the Fatty Acid Profile (FAP) analysis revealed that thevetia oil had a rather high level of unsaturation (53.7%) (Table1). The FAP also showed that the oil's principal fatty acid constituent was oleic acid (42.3 + 0.0%).

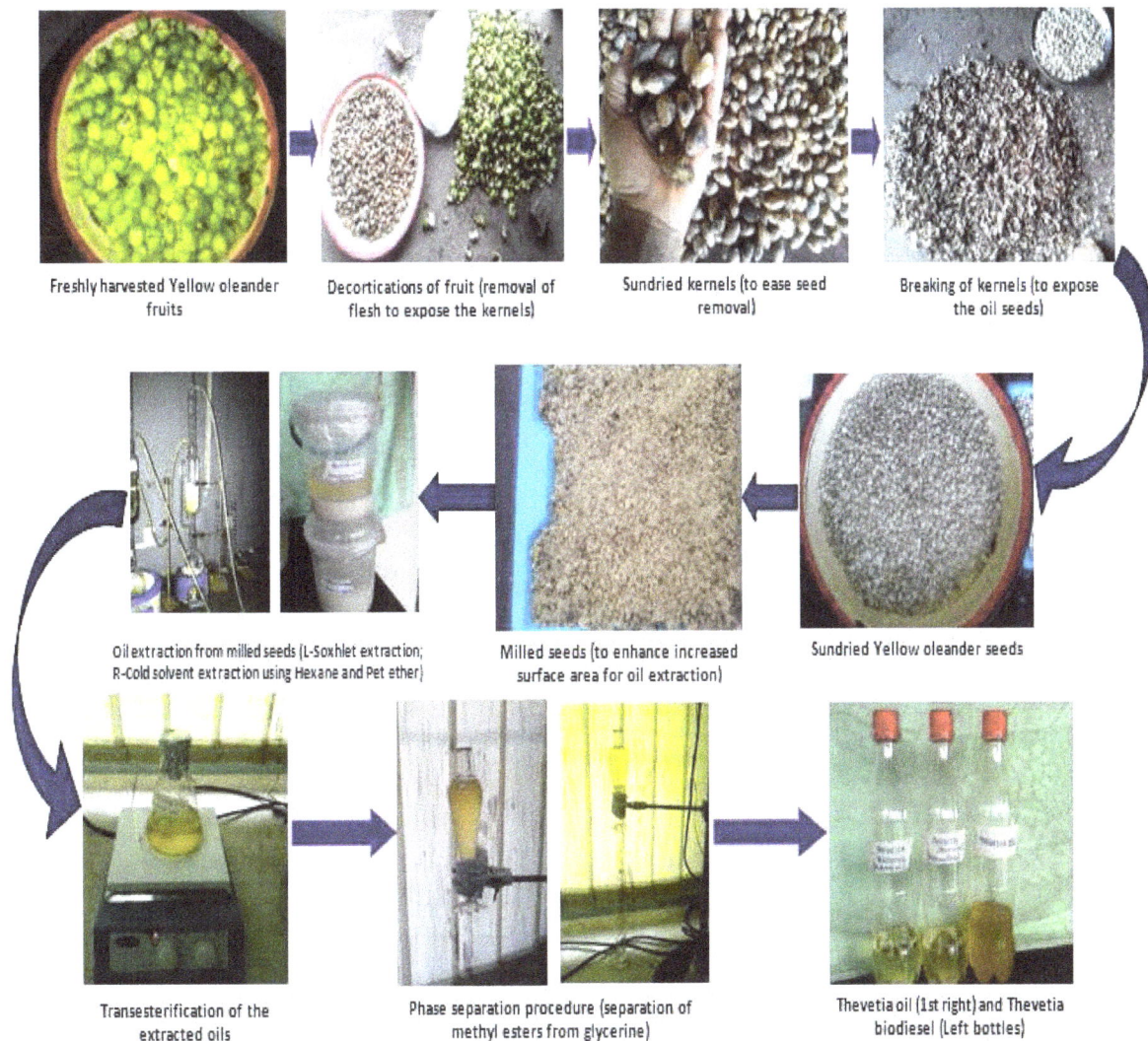

Figure 2. Highlights of the major steps involved in the production of biodiesel from Yellow oleander fruits.

Table 1. Fatty Acid Profile (FAP) or Percentage Fatty Acid Composition (FAC) of Thevetia Oil.

Test parameter[a]	Name	Thevetia(%) x+S.D
C8:0	Caprylic	-
C10:0	Capric	-
C12:0	Lauric	-
C14:0	Myristic	0.19+0.01
C15:0	Pentadecanoic	-
C16:0	Palmitic	19.50+0.01
C16:1	Palmitoleic	0.25+0.01
C17:0	Margaric	0.10+0.01
C18:0	Stearic	6.39+0.01
C18:1	Oleic	42.25+0.01
C18:1-9c, 12 (OH)	Ricinoleic	0.05+0.01
C18:2	Linoleic	10.50+0.00
C18:3	Linolenic	0.50+0.01
C18:3-9c,11t, 13t	α-Eleostearic	0.01+0.01

Test parameter[a]	Name	Thevetia(%) x+S.D
C20:0	Arachidic	1.25+0.00
C20:1	Gadoleic	0.13+0.01
C20:1-11c,14(OH)	Lesquerolic	-
C20:2	Eicosadienoic	-
C20:5	Timnodonic	-
C22:0	Behenic	0.82+0.01
C22:1	Erucic	-
C24:0	Lignoceric	1.15+0.00
C24:1	Nervonic	-
Unknown	=	17.30
Total known	=	82.70
Total saturated	=	29.02
Total unsaturated	=	53.68

[a] Numbers denote the number of carbon atoms and double bonds in one molecule. For example, in Linoleic acid, 18:2 indicates that each molecule contains eighteen carbon atoms and two double bonds.

The biodiesel yield from the oils of thevetia after the transesterification reaction process was: M-only (85.2%) and M/E (78.4%) respectively. The biodiesel still maintained the colour of the parent oil but was much lighter. The results of the fuel quality parameters tested in the biodiesel is as follows: Density at $25^{\circ}C$ ($0.9g/cm^3$), Density at $40^{\circ}C$ ($0.8g/cm^3$), Kinematic viscosity ($4.7mm^2/s$), Flash point ($130^{\circ}C$) and Acid value ($0.4mgKOH/g$).

The proportion of elements in the biodiesel was: TP (0.001%), Ca (0.003%), Na (0.002%) and S (0.008%). Correlations between the percentage TP in the biomass and biodiesel yield from both the M-only and M/E transesterification processes are presented in Figures 3 and 4 respectively. Also, a correlation table showing the relationship between mean elemental composition and biodiesel yield is presented in Table 2 below.

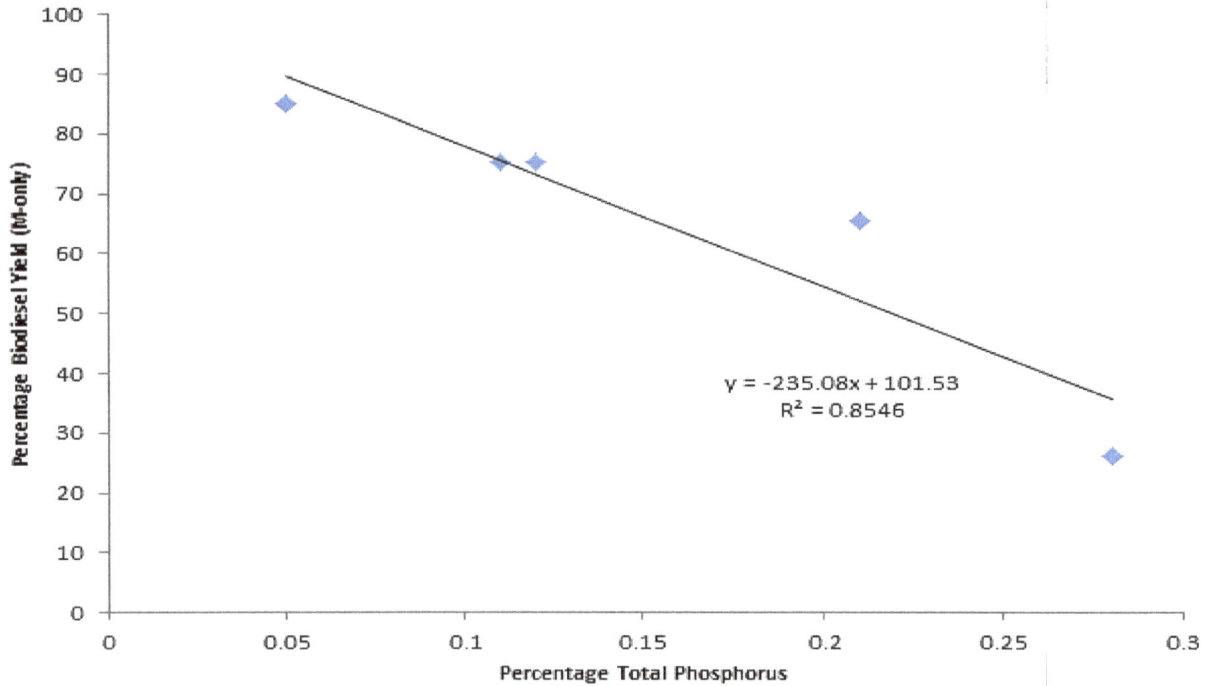

Figure 3. Relationship between percentage Total Phosphorus in biomasses and Biodiesel yield (M-only transesterification).

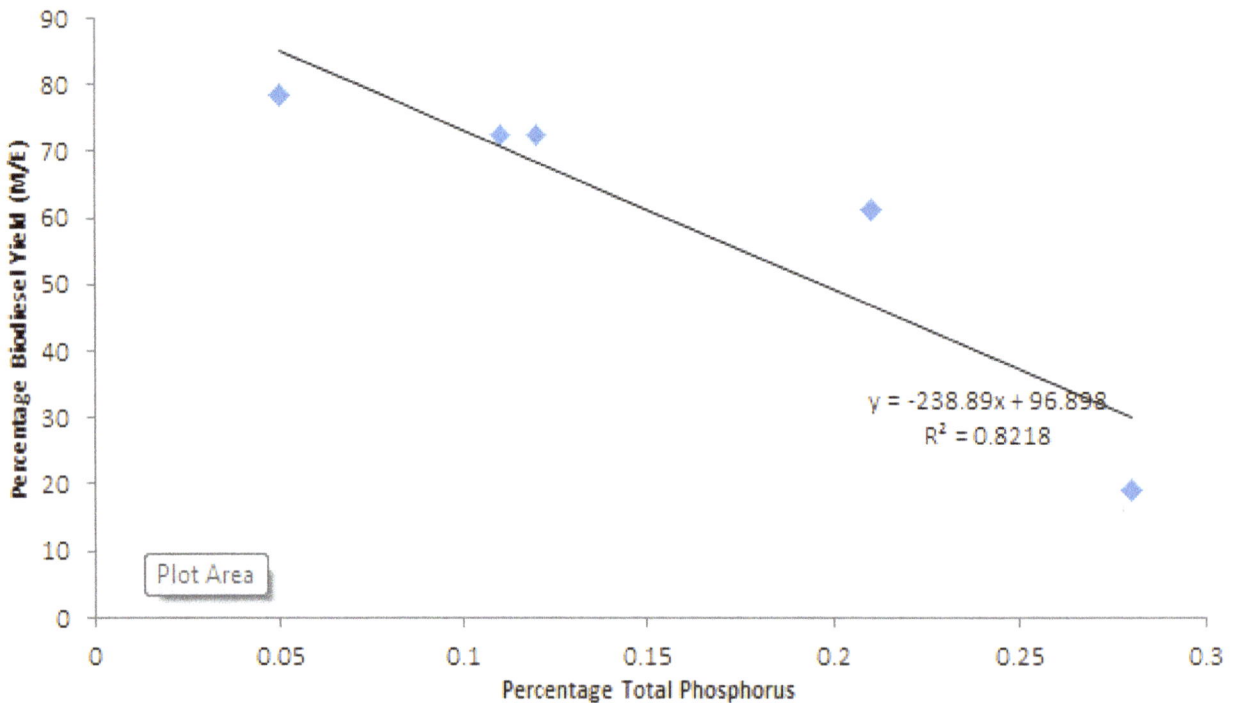

Figure 4. Relationship between percentage Total Phosphorus in biomasses and Biodiesel yield (M/E transesterification).

Table 2. Spearman correlation between mean elemental composition and biodiesel yield.

Parameter		TP (%)	Ca (%)	Na (%)	S (%)	M-only iodiesel yield (%)	M/E biodiesel yield (%)
TP (%)	Correlation coefficient	1.00					
	Sig. (2-tailed)	.					
Ca (%)	Correlation coefficient	0.79*	1.00				
	Sig. (2-tailed)	0.02	.				
Na (%)	Correlation coefficient	0.27	-0.20	1.00			
	Sig. (2-tailed)	0.52	0.64	.			
S (%)	Correlation coefficient	0.34	0.02	0.39	1.00		
	Sig. (2-tailed)	0.40	0.60	0.34	.		
M-only biodiesel yield (%)	Correlation coefficient	-0.99**	-0.80*	-0.30	-0.29	1.00	
	Sig. (2-tailed)	0.00	0.02	0.47	0.48	.	
M/E biodiesel yield (%)	Correlation coefficient	-0.99**	-0.80*	-0.30	-0.29	1.00**	1.00
	Sig. (2-tailed)	0.00	0.02	0.47	0.48	.	.

* Correlation is significant at $p < 0.05$
**Correlation is significant at $p < 0.01$

4. Discussion

The fresh milled thevetia seeds had a moisture content of $6.6 + 0.1\%$. Chindo et al [20] reported 2.2% moisture content for thevetia seed cakes, which was at slight variance with the results obtained in this work. In conventional transesterification of fats and vegetable oils for biodiesel production, water and free fatty acids always produce negative effects since the presence of these components cause soap formation, increased catalyst consumption, and reduces catalyst effectiveness.

Kusdiana and Saka [21] are of the opinion that water can pose a greater negative effect than presence of free fatty acids and hence the feedstock should be water free. Canakci and Gerpen [22] insist that even a small amount of water (0.1%) in the transesterification reaction will decrease the ester conversion from vegetable oil. That is why effort was made to properly dry the biomasses and heat the oils extracted from them for any residual water to evaporate completely. The density of the fresh milled thevetia seeds was $0.750 + 0.001 g/cm^3$.

Soxhlet extraction gave a higher oil yield in the sample when compared to that of the two Cold extraction procedures employed. The resistance to flow and shear under the forces of gravity (that is, Kinematic viscosity) of thevetia oil clearly exceed the $1.9\text{-}6.0 mm^2/s$ limit set by ASTM D6751 [10]. A high Kinematic viscosity indicates that the oil cannot be used directly in diesel engines, and there would be need for refining it. This is because oil and/or biodiesel viscosity are one of the most important properties of these liquids because it brings out a fuel's capacity to lubricate moving parts. Incorrect viscosity leads to poor lubrication, and poorly lubricated machinery can quickly break down.

The higher the unsaturation levels of oils, the greater the tendency for FFAs to cleave from their parent triglyceride molecules, thereby making the oil high FFA-containing oil. The rather high unsaturation of thevetia oil could easily predispose it to chemical degradation, and thereby cause rancidity. However, this could also imply a more reactive substrate for chemical manipulation such as in the transesterification reaction.

Saponification value, which is the number of milligrams of KOH required to neutralize the fatty acids resulting from the complete hydrolysis of 1g of fat or oil, gives an indication of the nature of the fatty acids constituent of oil and thus, depends on the average molecular weight of the fatty acids constituent of the oil. If a feedstock has a high free fatty acid (FFA) content, excess of alkali causes loss of the free fatty acids as their insoluble soaps. The resulting soaps can induce an increase in viscosity, formation of gels and foams, and make the separation of glycerol difficult. This decreases the final yield of methyl ester (biodiesel) and consumes alkali. The saponification value of the thevetia oil was considerably low as compared to an oil such as Palm Kernel Oil (PKO), which has a value of 230-254 mgKOH/g as the general saponification range for PKOs [23].

The viscosity of the biodiesel, which was $4.70 mm^2/s$, fell within the $1.9\text{-}6.0 mm^2/s$ limit of ASTM D6751[10]. The higher the viscosity of a fuel, the poorer the atomization of the fuel would be. Thus, operation of the injection would be less accurate. Moreover, at decreased temperature viscosity of the biodiesel increases [24] thereby aggravating the situation. Hence, it was observed that the biodiesel produced possesses the right viscosity to perform effectively in a diesel engine. The significant reduction ($p < 0.05$) in the viscosity from $21.50 mm^2/s$ (in the oil) to $4.70 mm^2/s$ (in the biodiesel) is an indication of the effectiveness of transesterification process in reducing the viscosity of oils.

Flash point is used in shipping and safety regulations to define flammable and combustible materials. The flash point of biodiesel is generally higher than those of petrol diesel (which is usually between 60-80°C). A minimum flash point for diesel fuel is normally required for fire safety. If methanol, with its flash point of 12°C is present in a biodiesel the flash point can be lowered considerably. The flash point of the biodiesel produced (130°C) was above ASTM D6751 [10] minimum limit. This implies that the biodiesel is safer than conventional diesel in terms of storage and transportation from the standpoint of fire hazard.

ASTM D6751 [10] stipulates 0.8mgKOH/g as the maximum limit of acid value for a biodiesel fuel. A high acid value will cause an increase in the FFA which will in turn lead to corrosion of engine parts as water is increased in the fuel [25]. The methyl esters obtained from thevetia possessed an acid value of 0.441mgKOH/g indicating that the biodiesel underwent proper production process and has not undergone any oxidative degradation as at the time of testing.

The percentage composition of the elements-P, Ca and Na in the thevetia biomass was seen to be significantly ($p<0.05$) reduced when estimated for in the biodiesel produced thus indicating that the processing that the biomass went through for biodiesel production helped reduced the levels of these undesirable elements. Although the reduction in percentage sulphur was not significant, it was still considerably reduced below the 0.05% maximum limit set by ASTM Standard D6751 [10] in the final thevetia biodiesel produced.

5. Conclusion

The oil obtained from *Thevetia peruviana* seeds was considerably high in relation to the feedstock quantity used in the extraction process. Soxhlet extraction provided a better yield than the Cold solvent extractions. The alkaline transesterification reaction that was employed in processing the oils to biodiesel was observed to be an effective refining method as certain key parameters of the oil were significantly improved after the process.

The results of the experimental analyses carried out in this work indicate that thevetia biomass is a high-yielding and good source of oil for biodiesel production. The quality parameters of the biodiesel were found to be within international acceptable standard. However, further research into the extraction and processing of thevetia oil using other solvent mixtures is recommended to establish if, peradventure, the solvents could prove to be more effective than the ones explored in this work.

Acknowledgements

The authors wish to appreciate the technical support provided by members of staff of the Nigerian Institute of Science Laboratory Technology (NISLT), Samonda, Ibadan and the Multidisciplinary Central Research Laboratory (MCRL), University of Ibadan.

References

[1] Rodrigues S, Mazzone ICA., Santos FFP. Optimization of the Production of Ethyl Esters by Ultra-sound Assisted Reaction of Soybean Oil and Ethanol. Brazilian Journal of Chemical Engineering 2009; 26 (2): 361-363.

[2] Olaniyi A. Biofuels Opportunities and Development of Renewable Energies Markets in Africa: A Case of Nigeria. A paper presented during the Biofuels Market Africa 2007 Conference, in Cape Town, South Africa, November 5-7; 2007.

[3] Gerpen JV. Biodiesel processing and production. Fuel Processing Technology, 2005; 86(10), 1097-1107.

[4] Idusuyi N, Ajide OO, Abu R. Biodiesel as an Alternative Energy Resource in Southwest Nigeria. International Journal

of Science and Technology 2012. Volume 2 No.5: ISSN 2224-3577. http://www.ejournalofsciences.org, 2012.

[5] Demirbas A. New liquid biofuels from vegetable oils via catalytic pyrolysis. Energy Educ. Sci. Technol 2008; 21:1–59.

[6] Patil V, Tran KQ, Giselrød HR. Towards sustainable production of biofuels from microalgae. International Journal of Molecular Sciences. 2008;9(7):1188–1195.

[7] Azam MM, Waris A, Nahar NM. 2005. Prospects and potential of fatty acid methyl esters of some non-traditional seed oils for use as biodiesel in India. Journal of Biomass and Bioenergy 2005;29, 293-302.

[8] Ibiyemi SA., Fadipe VO, Akinremi OO, Bako SS. Variation in oil composition of Thevetia peruviana Juss (Yellow Oleander) fruits seeds, J. Appl. Sci. Environ. Mgt. (JASEM), 2002;6 (2): 61-65.

[9] Olisakwe HC, Tuleun LT, Eloka-Eboka AC. Comparative Study of Thevetia peruviana and Jatropha curcas seed oils as feedstock for grease production. International Journal of Engineering Research and Applications (IJERA) 2009;Vol. 1, Issue 3, pp.793-806; ISSN: 2248-9622.

[10] ASTM Standard D6751:2009. Standard Specification for Biodiesel Fuel Blend Stock (B100) for Middle Distillate Fuels. ASTM International, West Conshohocken, PA, 2009.

[11] Gerpen VP, Clements D, Knothe G, Shanks B, Pruszko R. Building a successful biodiesel business; Biodiesel Technology Workshop, Chapter 28, Iowa State University, 2004.

[12] AOAC standard. Association of Official Analytical Chemists Book; English, 16th edition; 1998.

[13] Barthet VJ, Daun JK. Oil Content Analysis: Myths and Reality. Paper C76, Chapter 6, Canadian Grain Commission, Grain Research Laboratory; Copyright © 2004, AOCS Press; 2004.

[14] Hossain SABM, Salleh A, Amru NB, Partha C, Mohd N. Biodiesel fuel production from algae as renewable energy. Am. J. Biochem. and Biotechn 2008;4(3): 250-254; ISSN 1553-3468.

[15] ASTM D445-04:2006. Standard Test Method for Kinematic viscosity of transparent and opaque liquids (and the calculation of Dynamic viscosity); ASTM International, West Conshohocken, PA, 2006.

[16] Christie WW. Lipid Analysis: Isolation, Separation, Identification and Structural Analysis of Lipids. 3rd ed. Oily Press, Bridgwater, UK, 2003.

[17] Kangani CO, Kelley DE, DeLany JP. New method for GC/FID and GC-C-IRMS Analysis of plasma free fatty acid concentration and isotopic enrichment. J Chromatogr B Analyt Technol Biomed Life Sci 2008;15; 873(1): 95–101.

[18] Berchmans BJ, Hirata S. Biodiesel Production from Crude Jatropha Curcas L. Seed Oil with a High Content of Free Fatty Acids. Bioresource Technology 2008;99: 1717.

[19] ASTM Standard D93:2003. Standard Test Methods for Flash Point by Pensky-Martens Closed Cup Tester. ASTM International; West Conshohocken, PA, 2003.

[20] Chindo IY, Danbature W, Emmanuel M. Production of biodiesel from Yellow oleander (Thevetia peruviana) oil and its biodegradability. Journal of the Korean Chemical Society

2013; Vol. 57, No. 3;
http://dx.doi.org/10.5012/jkcs.2013.57.3.377.

[21] Kusdiana D, Saka S. Effects of water on biodiesel fuel production by supercritical methanol treatment. Bioresour Technol 2004; 91:289–95.

[22] Canakci M, Gerpan JV. 1999. Biodiesel production via acid catalysis. Trans Am Soc Agric Eng 1999;42:1203–10.

[23] CODEX-STAN 210:1999. Codex Standards for Fats and Oils from Vegetable Sources. Produced by Agriculture and Consumer Protection; FAO Corporate Document Repository. Available at www.fao.org/docrep/004/y2774e/y2774e04.htm, 2003.

[24] Antony SR, Robinson DSM, Lindon CRL. Biodiesel production from jatropha oil and its characterization. Res. J. Chem. Sci. 2011;1(1):81-87.

[25] Knothe G, Steidley KR. Kinematic viscosity of biodiesel fuel component and related compounds: Influence of compound structure and comparison to petrodiesel fuel components, Energ. Fuel 2005;1059–1065.

6

Annealing Effect on Efficiency of Aspilia Africana Flowers Dye Sensitized Solar Cells

Adenike Boyo[1], Henry Boyo[2], Olasunkanmi Kesinro[1]

[1]Department of physics, Lagos State University, Ojo, Lagos state, Nigeria
[2]Department of Physics, University of Lagos, Lagos State, Nigeria

Email address:

nikeboyo@yahoo.com (A. Boyo), princeboyo@yahoo.com (H. Boyo), olakesinro02@gmail.com (O. Kesinro)

Abstract: Energy was generated by using methanol as a solvent to extract dye from Aspilia africana Flowers. The maximum absorption of the extracted dye was observed at different wavelengths (350-1000nm). TiO_2 was annealed at different temperatures and phytochemical screening was done. We observed insignificant presence of anthocyanin compared to flavonoids in the flowers. The solar energy conversion efficiency changes from 0.21% to 0.52%, due to the sintering of the TiO_2 at different temperatures. The increase in solar energy conversion efficiency can be attributed to the changes in the morphology, crystalline quality, and the optical properties caused by the sintering effect.

Keywords: Aspilia Africana, Methanol, Anthocyanin, Flavonoids and Dye Sensitized Solar Cell

1. Introduction

The technology of harvesting solar energy in producing photovoltaic systems has attracted worldwide attention. Several researches have been done on photovoltaic technologies which includes dye sensitized solar cells [1], [5],[8], organic solar cells [4], and multijunction solar cells [9]. Dye sensitized solar cells (DSSCs) have played an important role in the development of photovoltaic technology, although encouraging efficiency of the DSSCs have been presented by many authors, but the use of the DSSCs is still not feasible for commercial applications. These low efficiencies of DSSCs might be due to challenges faced in obtaining suitable materials for its production. Thus, careful importance is placed in harnessing dyes that have a long lifespan, cheaper, and environmental friendly solar cells.

The nano-crystalline dye sensitized solar cell is a photo electrochemical cell. The basic working principle of DSSCs is that photons strike the monolayer of dye which gives up energy. This excites electrons from the dye, instead of generating electron – hole pairs. As shown in 'Fig.1' the solar cell consist of two conductive glass slides (ITO or FTO) with nanoparticle TiO_2 sintered on one of the conducting glass slides, stained with the dye and the other conducting glass slide is carbonized with soot. DSSCs are placed in the category of third generation photovoltaics where new trends in the photovoltaic technology are applied. In the first generation PV cells, the electric interface is made between doped n-type and p-type bulk silicon. First generation PV cells provide the highest so far conversion efficiency. The second generation PV cells are based on thin film technology. These cells utilize less material and they thus drop the production cost, however, they are less efficient than the bulk cells. Both first and second generation cells are based on opaque materials and necessitate front-face illumination and moving supports to follow sun's position. Thus they may be either set up in PV parks or on building roofs. Third generation solar cells, are based on nanostructured (mesoscopic) materials and they are made of purely organic or a mixture of organic and inorganic components, thus allowing for a vast and inexhaustible choice of materials.

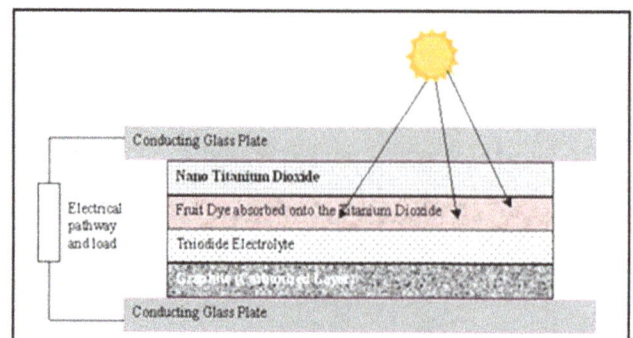

Figure 1a. Dye sensitized solar cell.

A liquid triiodide electrolyte is added in between the glass slide.. The absorption of photons causes dye molecules to be attached to the surface of the nanocrystalline titanium dioxide which produces photoelectrons. The titanium dioxide is the photoanode, photoexcitation of the dye results in the transfer of electrons into the conduction band of the nanocrystalline titanium dioxide.

Annealing nanoparticle titanium dioxide at different temperature plays an important role in the solar energy conversion performance of Dye Sensitized Solar Cells. These affect the physical, chemical and optical characteristic of TiO_2. Titanium dioxide (TiO_2) can crystallize in three different phases namely: rutile (tetragonal), anatase (tetragonal) and brookite (orthorhombic) [10]. The rutile-TiO_2 is thermodynamically stable at all temperatures and can be obtained in most crystal growth processes.

The anatase-TiO_2 and brookite are less dense and less stable in comparism to the rutile-TiO_2.phase [10].

We fabricated DSSCs using Aspilia africana flowers dye extracted from methanol and study the effect of annealing of TiO_2 films at different temperature intervals from 100^0C, to $500\,^0C$ for 30 minutes on the performance of DSSCs.

This study was borne out of the availability of these plants in Nigeria. Another important factor is that Nigeria uses about 3 trillion kWh of energy each year, and with energy requirement growing every year, it is the responsibility of the government and other institutions to develop energy harvesting technologies from renewable energy sources for a sustainable future.

Hence, in order to improve the generation of electricity in our country and beyond, it would be necessary that devices like dye-sensitized solar cells are produced at commercial scale to meet energy demands in Nigeria.

2. Methodology

2.1. Sample Source and Preparation

Samples of the flower for Aspilia africana were obtained from Magbon community along Badagry expressway, Lagos State, Nigeria. The sample was pulverized with the aid of mechanical blender (liquidizer). 25g of the pulverized sample was weighed using OHAUS Electronic weighing balance model brain weight B1500 made in USA and soaked in 250ml methanol.

The extracting solvent and the mixture were placed in an orbital shaker (SLAUART SSL1 ORBITAL SHAKER at 25 rpm) for 4 hours .The extract of the sample was decanted to remove the residual part of the samples. Simple distillation was carried out at 650^oC in order to concentrate the dye of the samples.

2.2. Preparation and Construction of Cells

We used conductive ITO glass with dimension of 2.5 X $2.5cm^2$. The photo anode was prepared using a slide of the conductive ITO glass. A digital multimeter was used in detecting the conductive side of the ITO conducting glass and

the value was 22 Ω. Adhesive –tape were used to the face of the ITO conducting glass plate in order to create an opening of dimension 2.0 X $2.0cm^2$ at the centre of the glass where the TiO_2 paste will be applied. The cells were assembled and tested using the method reported by [2], [6].

2.3. Measurement of Photocurrent Voltage of the DSSC

The absorption spectra of natural dye of Aspilia africana flower and natural dye mixed with TiO_2 were recorded using a VIS Spectrophotometer (Spectrumlab 23A GHM Great Medical England). The photocurrent voltage was measured by using digital multimeters under indoor illumination. The electrical characteristics were taken for different sintering temperature.

After the TiO_2 paste was applied on each conducting glass. Each conducting glass was sintered at different temperatures $(100 – 500^oC)$ before the cells were assembled.

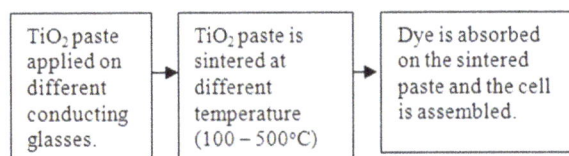

TiO₂ paste applied on different conducting glasses.	→	TiO₂ paste is sintered at different temperature (100 – 500°C)	→	Dye is absorbed on the sintered paste and the cell is assembled.

Figure 1b. Flowchart of methodology.

3. Results and Discussion

The absorption spectra for the extracts of Aspilia africana is presented in 'Fig.2'. The extract has optimal absorbance at 400nm. This indicates that extracts of Aspilia africana can be applied as an organic sensitizer in DSSCs

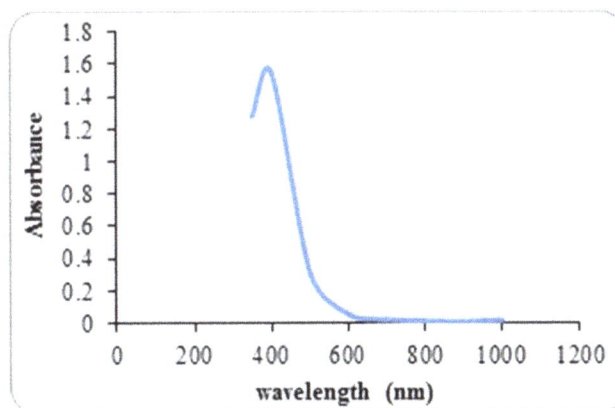

Figure 2. Absorbance graph of Aspilia africana extract (series removed).

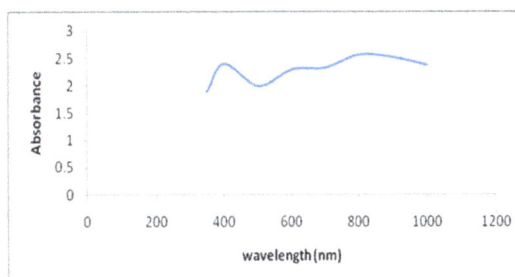

Figure 3. Absorbance of TiO₂ with dye.

'Fig. 3' shows the absorbance of dye with TiO$_2$ having the maximum peak at 800nm .The peak wavelength falls within the visible spectrum. This indicate that `the dye is complexed or chelated (attached) to the titanium dioxide and be able to absorb the photons' energy, exciting and freeing some of its electrons.

This research observed the effect sintering temperature has on cell efficiency by annealing the TiO$_2$ at different interval of temperatures from 100 $^\circ$ C to 500 $^\circ$ C. Each cell was fabricated using the same procedure except that the TiO$_2$ temperature was varied. Experimental results after electrical characterization are shown in Table 1

Table 1. Photovoltaic performances of cells at different temperatures.

Temp.$^\circ$ C	Isc (short circuit current) (mA)	Voc (Open circuit voltage) (mV)	FF (Fill factor)	Efficiency
100	0.62	140	0.59	0.21
200	0.7	138	0.63	0.24
300	1.06	232	0.45	0.44
400	0.9	198	0.68	0.49
500	0.98	220	0.6	0.52

This indicates that cell efficiency increased with increase in sintering temperature and maximum efficiency was observed at 500° C. It was observed that as the sintering temperature increased there was also an increase in the photocurrent of the cell except for the cells fabricated at 300 $^\circ$ C which had a higher photocurrent compared to other fabricated cells. Higher photocurrent at 300 $^\circ$ C could be as a result of reduction in the particle size and better absorption of dye. Finally, the cell fabricated at 400 $^\circ$ C showed higher fill factor than that of others. The change in fill factor could be as a result of the size of the TiO$_2$ layer.

An unstable behavior is observed in the I-V curve at different temperatures in 'Fig.4 to Fig.8'. This unstable behavior indicates that the liquid electrolytes do not diffuse fast to deliver adequately charges for the regeneration of oxidized dye molecules, these results in a significant reduction of photocurrent.

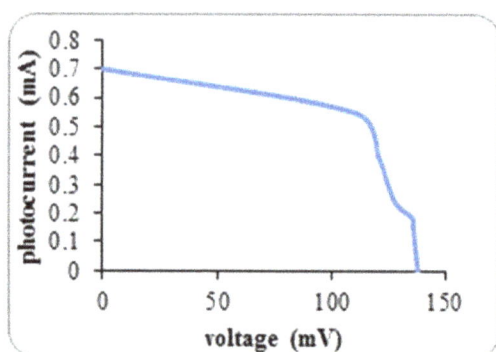

Figure 6. I-V curve at 300 $^\circ$ C (series removed).

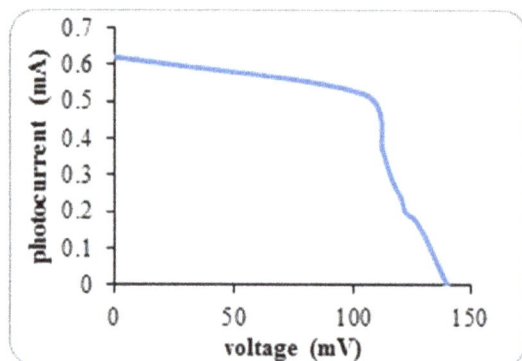

Figure 7. I-V curve at 400 $^\circ$ C (series removed).

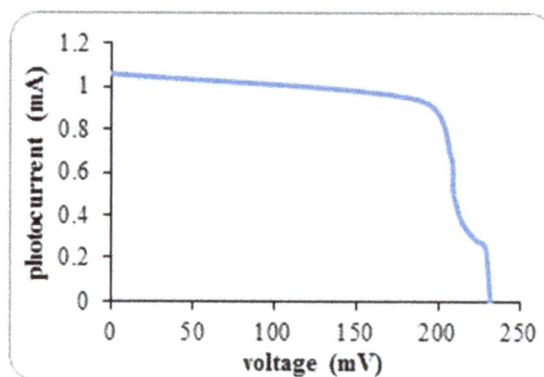

Figure. 4. I-V curve at 100 $^\circ$ C. (series removed).

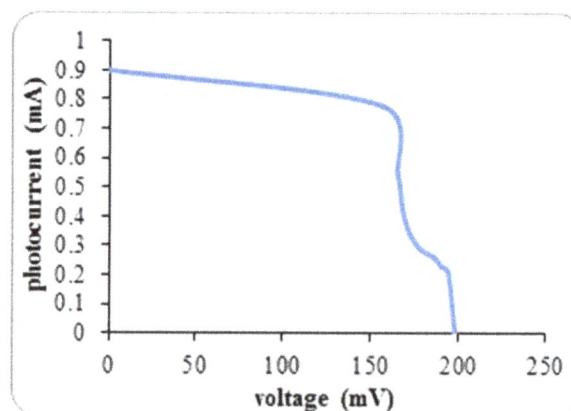

Figure 5. I-V curve at 200 $^\circ$ C (series removed).

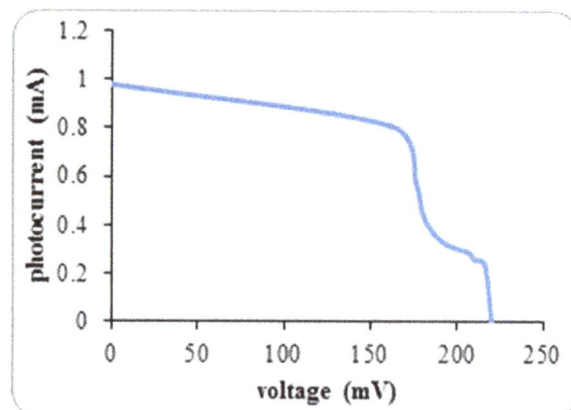

Figure 8. I-V curve at 500 $^\circ$ C (series removed).

The phytochemical screening was done for anthocyanins and flavonoids. Anthocyanins are used in DSSCs due to their ability to covert solar energy to electrical energy [3] due to its high presence in red plants. Most important plant pigments for flower colorations producing yellow or red/blue pigmentation are flavonoids. It was observed that the level of anthocyanin in the flower is low compared to flavonoids and does not improve the conversion of light energy to electrical energy.

4. Conclusion

Using screen printing method, TiO_2 nanocrystals were successfully deposited on Indium Tin oxide (ITO) glass. Sintering was carried out at different temperatures to improve the efficiency of the cell. The significant improvement of the photovoltaic performance obtained can be attributed to the improvement in the crystal quality of TiO_2 nanocrystal and enhanced absorption of the cell. This is because as sintering temperature increases there is a reduction in the number of open pores in the TiO_2 resulting in dye absorption.

References

[1] B. O'Regan and M. Gr"atzel, "A low-cost, high-efficiency solar cell based on dye-sensitized colloidal TiO2 films," *Nature*, vol. 353, no. 6346, pp. 737–740, 1991.

[2] Boyo A. O, Boyo H.O, Abudusalam I.T. and Adeola S. (2012). Dye Sensitized Nanocrystalline Titanium Solar cell using Laali Stem Bark (lawsonia inermis), Transnational Journal of Science and Technology Macedonia.

[3] Cherepy, Nerine J., Smestad, Greg P.; Gratzel, Michael; Zhang, Jin Z. (1997). "ultrafast Electron Injection: Implications for a photoelectrochemical Cell Utilizing an Anthocyanin Dye – Sensitized TiO2 Nanocrystalline Electrode". The Journal of Physical Chemistry B 101 (45): 9342 – 51.

[4] F. A. de Castro, F.N"uesch, C.Walder, and R.Hany, "Challenges found when patterning semiconducting polymers with electric fields for organic solar cell applications," *Journal of Nanomaterials*, vol. 2012, Article ID 478296, 6 pages, 2012.

[5] H. S. Chen, C. Su, J. L. Chen, T. Y. Yang, N. M. Hsu, and W. R. Li, "Preparation and characterization of pure rutile TiO2 nanoparticles for photocatalytic study and thin films for dye-sensitized solar cells," *Journal of Nanomaterials*, vol. 2011, Article ID 869618, 8 pages, 2011.

[6] How to build your own solar cell: A nanocrystalline dye-sensitized solar cell (http://www.solideas.com/solarcell/english.html)

[7] Lapornik B., Prosek M., Wondra A.G., Food J., Eng. 2005, 71, 214

[8] M. Gr"atzel, "Photoelectrochemical cells," *Nature*, vol. 414, no. 6861, pp. 338–344, 2001.

[9] M. Yamaguchi, "Multi-junction solar cells and novel structures for solar cell applications," *Physica E*, vol. 14, no. 1-2, pp. 84–90, 2002.

[10] R.C. Weast (Ed.), Hand Book of Chemistry and Physics, 67th Edition, (CRC Press, Boca Raton, FL, 1986–1987, p. B-140).

7

Study on Energy Efficiency and Measurement of CO_2 Emissions on Buildings

Lei Wen[1,2], Ye Cao[1,*]

[1]Department of Economics & Management, North China Electric Power University, Baoding, Hebei, China
[2]The Academy of Baoding Low-Carbon Development, Baoding, Hebei, China

Email address:

441477582@qq.com (Lei Wen), Wenlei0312@126.com (Ye Cao)

Abstract: Research on energy efficiency and measurement of CO_2 emissions on buildings, is crucial for taking countermeasures against climate change and identifying low carbon pattern in buildings of Hebei. Energy efficiency directly influence CO_2 emissions. This paper presents two measurement methods of CO_2 emissions, including the measurement from top-down using energy balance sheet and the measurement from bottom-up regarding structure decomposition of energy consumption of various service demand. Meanwhile, this study decomposes the energy consumption of buildings into detailed categories of service demands to explain energy efficiency, such as cooling, household appliance, domestic hot water and cooking in urban and rural residence except for central heating. Results reveal that energy consumption and CO_2 emissions in urban, rural and public buildings maintain continuous growth reaching a highest year-increasing rate 18.07% in 2010. Specifically, public buildings show an extreme increasing rate with a total CO_2 emissions of 208.45 million tons. Besides, CO_2 emissions in cooling and cooking reach higher than other service demand. Eventually, policy implications are provided to mitigate the growth of CO_2 emissions and identify energy efficiency strategies in Hebei.

Keywords: CO_2 Emissions, Energy Efficiency, Buildings, Measurement

1. Introduction

Since 2004, global CO_2 emissions burned by fossil fuels have accounted for 56.6% of greenhouse gas emissions. In addition, China's CO_2 emissions from fossil energy accounted for as high as 26.38% of the world's total and ranked the first with absolute predominance [1]. Copenhagen agreement in 2009 has urges China solemnly to pledge a significant cut in carbon intensity by at least 40% by 2020 from 2005 level. Since Copenhagen agreement, energy efficiency issue and greenhouse gases growth, has made China a focal point for criticism.

As a critical part, buildings released a lower CO_2 in the process of construction and a higher CO_2 due to the pull effect on majorities of other industries, accounting for 50% of the total CO_2 emissions. Therefore, the application of more accurate method to measure CO_2 emissions on buildings sector and then decompose impact components are crucial in promoting low-carbon buildings. Development and Reform Commission of China established a Low-carbon City policy and announced the selection of five provinces and eight cities to pilot low-carbon development work [2-6]. As an economy-developed province in China, Hebei ranks the sixth with GDP reaching 2019.71 billion yuan in 2010. Moreover, as economic development is highly dependent on energy consumption, the latter has also increased to 25418 million tons on standard coal in 2009. In this context, research on the measurement of CO_2 emissions on buildings of Hebei in China has experienced a remarkable impetus.

This paper presents two measurement methods of CO_2 emissions, including the measurement from top-down by the energy balance sheet and the measurement from bottom-up regarding structure decomposition of energy consumption of various service demand on building sector. The paper is structured as follows. Section 2 presents an overview of the main methods related to the measurement from top-down and bottom-up. In section 3, the data source is reported. Section 4 draws the result and discussions followed by conclusions and implications in section 5.

2. Methodologies

This paper conforms to international general classification and decomposes buildings sector into residential buildings and public buildings. Meanwhile, residential buildings can be divided into urban and rural residence buildings, seen as Figure 1.

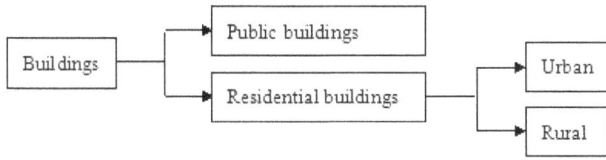

Figure 1. *Buildings sector after redivision.*

2.1. The Measurement From Top-Down

Energy balance sheet intuitively reveals balance between energy resources, conversion and final consumption judged by data. Accordingly, energy consumption in buildings involve in four factors, namely consumption in central heating, urban residence except for central heating, rural residence except for central heating, and public buildings except for central heating. In addition, energy consumption in central heating are overall derived from heating production sector thus been calculated in heating production sector. Table 1 shows the adjustment method of energy consumption in buildings of Hebei [7].

Table 1. *The measurement from top-down in buildings of Hebei.*

Sector	Sectors in energy balance sheet	Adjustment method
Urban Buildings	Final consumption in urban living	Various energy consumption except for all gas and 95% diesel oil
Rural Buildings	Final consumption in rural living	Various energy consumption except for all gas and 95% diesel oil
Public buildings	Final consumption of wholesale, retail, hotels and catering industry in tertiary industry	Various energy consumption except for 95%gas and 35% diesel oil
	Final consumption of other industry in tertiary industry	Various energy consumption except for 95%gas and 35% diesel oil
	Final consumption of transportation, storage, mail industry in tertiary industry	15% electricity consumption

2.2. The Measurement from Bottom-Up

The measurement from bottom-up, typically taking energy service demand and technology level into account, is to break down the structure of all sectors energy consumption [8]. Calculating steps and formulas are shown in Table 2.

(1) The combination of technologies for achieving specific energy service demand is calculated as follows:

$$P_i = \sum P_{t,i} \tag{1}$$

where P_i is the demand quantity for energy service demand I; $P_{t,i}$ presents the demand quantity for energy service demand i by technology t; i is the type of energy-service demand, such as building lighting in urban, steel demand quantity etc; t denotes the type of technology, such as building lighting in cities including incandescent, energy-saving lights, LEDs.

(2) Total energy consumption for all sectors energy-service demand is calculated as follows:

$$E = \sum_i \sum_t \sum_n F_{n,t,i} \times P_{t,i} \tag{2}$$

where E denotes the total energy consumption for all sectors energy-service demand (ton standard coal); $F_{n,t,l}$ presents the energy consumption of N during the process of meeting energy service demand. N is energy types, such as electricity, coal, natural gas etc.

(3) Emissions that satisfy the sector's energy-service demand are calculated as follows:

$$CE = \sum_i \sum_t \sum_n EF_{n,t,i} \times F_{n,t,i} \times P_{t,i} \tag{3}$$

where CE is the CO_2 emissions of this sector; $EF_{n,t,l}$ denotes emission factor of n in the process of meeting energy-service demand with technology t.

Table 2. *The measurement from bottom-up in buildings of Hebei.*

Sector	Various categiories	Adjustment method	Emission factor
Urban buildings	Cooling	$E_{cool,u} = \sum E_i = \sum H \times EQ_i \times T_i \times W_i$	0.9914 TCO2/MWh
	Household appliance	$E_{happ,u} = \sum E_i = \sum H \times EQ_i \times T_i \times W_i$	
	Domestic hot water	$E_{dome,u} = \sum P_u \times 1.3$	2.62 TCO2/MWh
	Cooking	$E_{cook,u} = \sum P_u \times 4.5$	
Rural buildings	Cooling	$E_{cool,u} = \sum E_i = \sum H \times EQ_i \times T_i \times W_i$	0.9914 TCO2/MWh
	Household appliance	$E_{happ,u} = \sum E_i = \sum H \times EQ_i \times T_i \times W_i$	
	Domestic hot water	$E_{dome,u} = \sum P_r \times 1.3$	2.62 TCO2/MWh
	Cooking	$E_{cook,u} = \sum P_r \times 4.5$	
Public buildings	Operational Area of Catering Services	$E_{cate,u} = \sum M_c \times 195$	0.9914 TCO2/MWh
	Operational Area of hotels	$E_{area,u} = \sum M_o \times 195$	
	Floor Space of Public Buildings	$E_{floo,u} = \sum M_f \times 95$	

Note: E is total energy consumption of various categories (ton standard coal); E_i is energy consumption of equipment i(ton standard coal); H is the total households; EQ_i is ownership of household appliance i; T_i is the time using household appliance i(hours); W_i is the power of household appliance i(watts); I is equipment type, such as air-conditioning etc. P is the total person; M is the total area. Per capita domestic hot water demand coefficient is 1.3; Per capita cooking heating demand coefficient is 4.5 kgce/a. Big hotels and restaurants electricity consumption per unit area is 195(kWh/(m2·a)); large schools、public library electricity consumption per unit area is 95(kWh/(m2·a)).

3. Data Sources

The data for buildings is obtained from the Hebei Statistic Yearbook (2003–2012). Moreover, various coefficients of energy conversion and CO_2 emissions are collected from the general principles of the comprehensive energy consumption calculation GB/T 2589-2008. In accordance with related departments, this study is accessed to Hebei statistical review and energy balance sheet, and then picks the methodology from IPCC guidelines for inventories. Besides, energy consumption multiplies emission factor is taken as 2.62 reference to China city greenhouse-gases inventory guide and per capita heating demand coefficient is reference to China city greenhouse-gases inventory Guide [9]. All unit is in 10^4 tons.

4. Results and Discussions

4.1. The measurement of CO_2 Emissions from Top-Down in Buildings

Investigating Table 3 and Table 4, energy consumption of buildings reach the highest year-increasing rate at 18.07% in 2010; CO_2 emissions of buildings approaches the highest annual growth rate at 18.08% in 2010. Both annual growth rate keep 6.75%.

Table 3. The energy consumption of buildings in 2003-2012.

	Urban	Rural	Public buildings
2003	525.666	585.988	195.905
2004	537.196	581.411	263.901
2005	544.338	598.850	459.806
2006	477.309	555.074	458.636
2007	462.097	556.941	509.988
2008	581.977	599.352	587.565
2009	613.533	586.167	602.870
2010	764.177	654.697	709.431
2011	787.182	664.551	773.103
2012	794.552	707.477	795.612

Table 4. The CO_2 emissions of buildings in 2003-2012.

	Urban	Rural	Public buildings
2003	1377.244	1535.289	513.271
2004	1407.454	1523.297	691.421
2005	1426.165	1568.988	1204.693
2006	1250.550	1454.294	1201.625
2007	1210.694	1459.186	1336.169
2008	1524.781	1570.303	1539.421
2009	1607.456	1535.758	1579.519
2010	2002.143	1715.307	1858.709
2011	2062.416	1741.124	2025.529
2012	2081.727	1853.591	2084.505

Figure 2 shows a fluctuated trend of CO_2 emissions in urban residence except for central heating, namely an upward trend in 2003-2005, followed by a little decreasing in 2006-2007 and finally a remarkable increasing. However, CO_2 emissions in rural residence except for central heating, fluctuates around 15,800,000 tons; More specifically, owing to a improving

living standard and then more utilization of heating radiator or small boiler etc, CO_2 emissions shows a sharply increasing trend since 2010. CO_2 emissions in public buildings except for central heating keeps increasing mainly in recent years, except for a little decline in 2006-2007.

Figure 3 presents the percentages of CO_2 emissions from urban, rural and public buildings in 2003-2012. The percentage of in urban residence except for central heating, shows a cyclical changing trend, namely a little decrease in 2003-2007, followed by a little increase in 2008-2012. The percentage of in rural residence except for central heating mainly keeps decreasing, while the percentage of public buildings except for central heating shows an upward trend continuously. Therefore, the orientation of emission reduction has been urged toward public buildings except for central heating.

Figure 4 demonstrates CO_2 emissions variations from various industries of public buildings in 2003-2012. Figure 5 shows the detailed percentage variations. Figure 4-5 illustrates that CO_2 emissions of transportation, storage, mail industry wholesale, retail, hotels and catering industry and other industry in tertiary industry keeps stead increasing. Besides, CO_2 emissions of transportation, storage, mail industry wholesale, retail, hotels takes up the majority of total CO_2 emissions, accounting for 70%.

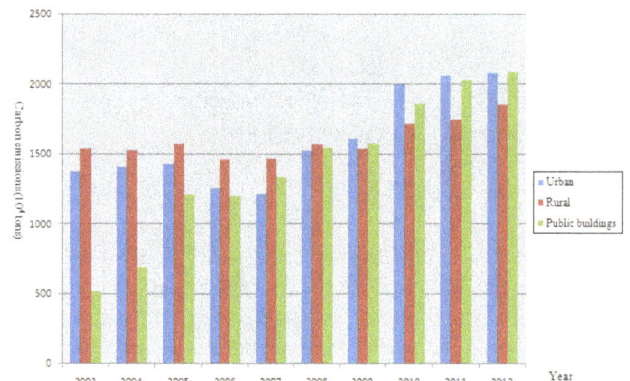

Figure 2. The bar chart of CO_2 emissions from urban, rural and public buildings in 2003-2012.

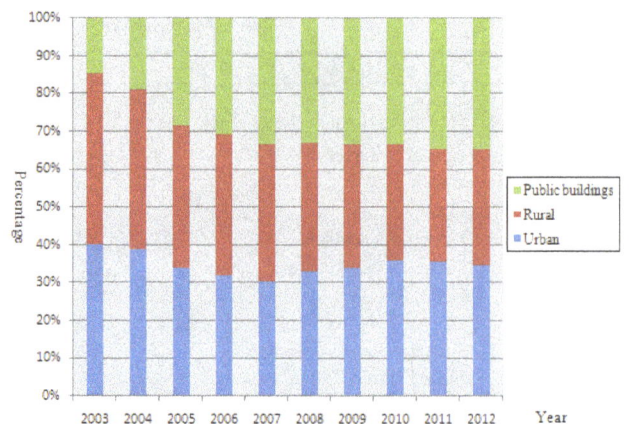

Figure 3. Percentages of CO2 emissions from urban, rural and public buildings in 2003-2012.

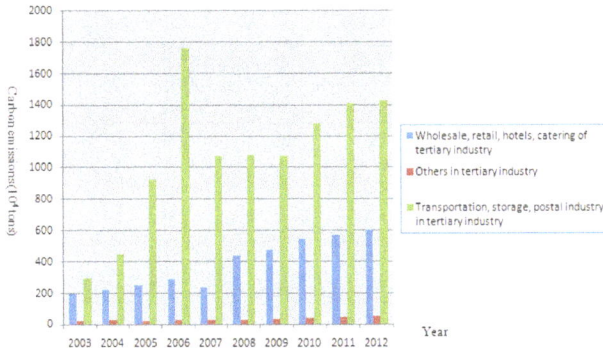

Figure 4. *CO_2 emissions from various industries of public buildings in 2003-2012.*

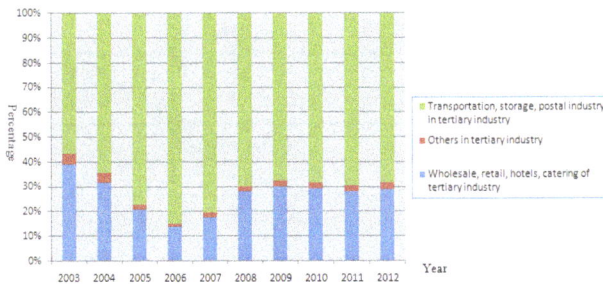

Figure 5. *CO_2 emissions percentages from various industry of public*

4.2. The Measurement of CO_2 Emissions from Bottom-Up in Buildings

4.2.1. CO_2 Emissions of Urban Residence Except for Central Heating

Table 5-8 demonstrate CO_2 emissions of urban residence except for central heating respectively, including cooling, household appliance, domestic hot water and cooking. Results show that CO_2 emissions in urban cooling and household appliance electricity, keep decreasing over a decade. Both decline rate reach the highest at 22.14% and 23.33%. However, domestic hot water and cooking keep increasing at an annual growth rate at 6.42% and 6.12% respectively.

Table 5. *Urban cooling energy consumption and CO_2 emissions in 2003-2012.*

Year	Air-conditioning energy consumption (MWh)	CO_2 emissions (10^4 tons)
2003	494394.900	49.014
2004	442064.700	43.826
2005	413951.966	41.039
2006	481367.374	47.723
2007	515421.709	51.099
2008	545687.863	54.099
2009	431984.383	42.827
2010	336300.914	33.341
2011	380700.000	37.743
2012	334568.000	33.169

Table 6. *Urban household appliance electricity consumption and CO_2 emissions in 2003-2012.*

Year	TV energy consumption (MWh)	Washing machine energy consumption(MWh)	Refrigerator energy consumption (MWh)	Total energy consumption of domestic appliance (MWh)	CO_2 emissions (10^4 tons)
2003	136384.800	51144.300	315.896	187844.996	18.623
2004	126304.200	45928.800	300.701	172533.701	17.105
2005	123680.770	38889.796	297.885	162868.452	16.147
2006	119215.838	36032.178	286.667	155534.683	15.420
2007	116447.127	35634.093	277.714	152358.935	15.105
2008	105175.219	35457.301	294.105	140926.625	13.971
2009	105462.854	34487.642	289.070	140239.566	13.903
2010	81029.024	26265.047	220.149	107514.221	10.659
2011	95175.000	25380.000	258.833	120813.833	11.977
2012	83642.000	19302.000	238.664	103182.664	10.230

Table 7. Urban domestic hot water energy consumption and CO_2 emissions in 2003-2012.

Year	People (persons)	Per capital heating demand (kgce/a)	Domestic hot water energy consumption (tons of standard coal)	CO_2 emissions (tons)
2003	19950000	1.3	25935.000	67949.700
2004	23870000	1.3	31031.000	81301.220
2005	25820000	1.3	33566.000	87942.920
2006	26990000	1.3	35087.000	91927.940
2007	27950000	1.3	36335.000	95197.700
2008	29280000	1.3	38064.000	99727.680
2009	30770000	1.3	40001.000	104802.620
2010	32010000	1.3	41613.000	109026.060
2011	33020000	1.3	42926.000	112466.120
2012	33870000	1.3	44031.000	115361.220

Table 8. Urban cooking energy consumption and CO_2 emissions in 2003-2012.

Year	People (persons)	Per capital heating demand (kgce/a)	Cooking energy consumption (tons of standard coal)	CO_2 emissions (tons)
2003	19950000	4.5	89775	235210.5
2004	23870000	4.5	107415	281427.3
2005	25820000	4.5	116190	304417.8
2006	26990000	4.5	121455	318212.1
2007	27950000	4.5	125775	329530.5
2008	29280000	4.5	131760	345211.2
2009	30770000	4.5	138465	362778.3
2010	32010000	4.5	144045	377397.9
2011	33020000	4.5	148590	389305.8
2012	33870000	4.5	152415	399327.3

4.2.2. CO_2 Emissions of Rural Residence Except for Central Heating

Table 9-12 demonstrate CO_2 emissions of rural residence except for central heating respectively, including cooling, household appliance, domestic hot water and cooking. Findings demonstrate that, these four parts show a decreasing trend in recent years, out of which, rural cooling has the largest annual decreasing rate at 1.04% and cooking decreases slowly at 1.59%.

Table 9. Rural cooling energy consumption and CO_2 emissions in 2003-2012.

Year	Air-conditioning energy consumption (MWh)	CO_2 emissions (10^4 tons)
2003	1248624.000	123.789
2004	1114267.000	110.468
2005	1002977.537	99.435
2006	1086830.049	107.748
2007	1172286.292	116.220
2008	1282680.891	127.165
2009	1029939.359	102.108
2010	998347.325	98.976
2011	922140.000	91.421
2012	804544.000	79.762

Table 10. Rural household appliance electricity consumption and CO_2 emissions in 2003-2012.

Year	TV energy consumption (MWh)	Washing machine energy consumption(MWh)	Refrigerator energy consumption (MWh)	Total energy consumption of domestic appliance (MWh)	CO_2 emissions (10^4 tons)
2003	344448.00	129168.00	797.81	474413.81	47.033
2004	318362.00	115768.00	757.95	434887.95	43.115
2005	299670.12	94227.34	721.76	394619.21	39.123
2006	269165.22	81353.36	647.24	351165.82	34.815
2007	264849.87	81046.95	631.64	346528.46	34.355
2008	247222.37	83345.09	691.32	331258.77	32.84
2009	251445.07	82225.61	689.20	334359.89	33.15
2010	240543.83	77970.77	653.54	319168.13	31.64
2011	230535.00	61476.00	626.95	292637.95	29.01
2012	201136.00	46416.00	573.92	248125.92	24.60

Table 11. Rural domestic hot water energy consumption and CO_2 emissions in 2003-2012.

Year	People (persons)	Per capital heating demand (kgce/a)	Domestic hot water energy consumption (tons of standard coal)	CO_2 emissions (tons)
2003	46010000	1.3	59813.000	156710.060
2004	43670000	1.3	56771.000	148740.020
2005	42690000	1.3	55497.000	145402.140
2006	41990000	1.3	54587.000	143017.940
2007	41480000	1.3	53924.000	141280.880
2008	40610000	1.3	52793.000	138317.660
2009	39570000	1.3	51441.000	134775.420
2010	39930000	1.3	51909.000	136001.580
2011	39390000	1.3	51207.000	134162.340
2012	39760000	1.3	51688.000	135422.560

Table 12. Rural cooking energy consumption and CO_2 emissions in 2003-2012.

Year	People (persons)	Per capital heating demand (kgce/a)	Cooking energy consumption (tons of standard coal)	CO_2 emissions (tons)
2003	46010000	4.5	207045.000	542457.900
2004	43670000	4.5	196515.000	514869.300
2005	42690000	4.5	192105.000	503315.100
2006	41990000	4.5	188955.000	495062.100
2007	41480000	4.5	186660.000	489049.200
2008	40610000	4.5	182745.000	478791.900
2009	39570000	4.5	178065.000	466530.300
2010	39930000	4.5	179685.000	470774.700
2011	39390000	4.5	177255.000	464408.100
2012	39760000	4.5	178920.000	468770.400

4.2.3. CO_2 Emissions of Public Buildings Except for Central Heating

Table 13 shows the construction areas of various public buildings. In addition, Table 14 illustrates energy consumption of various public buildings. As for this sector, CO_2 emissions keep an upward trend.

Table 13. Construction areas of various public buildings.

	Operational Area of Catering Services (sq. m)	Operational Area of hotels (sq.m)	Floor Space of Public Buildings (sq.m)
2003	278000	371000	249000
2004	299000	418000	251100
2005	310000	489000	259600
2006	311000	627000	260300
2007	411000	718000	265600
2008	543000	974000	267000
2009	654000	1222000	256500
2010	774000	974000	249000
2011	1062000	1374000	304000
2012	1209000	1614000	323000

Note: reference to Hebei Statistic Yearbook.

Table 14. Energy Consumption of various public buildings.

	Operational Area of Catering Services (MWh)	Operational Area of hotels (MWh)	Floor Space of Public Buildings (MWh)
2003	54210	72345	48555.0
2004	58305	81510	48964.5
2005	60450	95355	50622.0
2006	60645	122265	50758.5
2007	80145	140010	51792.0
2008	105885	189930	52065.0
2009	127530	238290	50017.5
2010	150930	189930	48555.0
2011	207090	267930	59280.0
2012	235755	314730	62985.0

5. Conclusion and Implications

Derived from energy efficiency and measurement from top-down, the increasing rate of CO_2 emissions in public buildings ranks the first, while total CO_2 emissions in urban and rural residence keeps a huge base value in the decade. Based on the measurement from bottom-up, CO_2 emissions in the service demand of cooling and cooking involved in technique combination, shows a huge growth potential.

Firstly, utilization of low-carbon material and low-carbon buildings technology should be promoted, including exterior energy-saving technology, energy-saving windows and doors, energy-saving roof, development of new energy as well as heating, refrigeration and lighting technology. Secondly, national policy support and supervision should be payed huge attention. Development of low-carbon buildings needs various energy-saving materials and technology as well as relevant policy support. China's energy conservation in buildings is backward, which is not merely technical problems, but also governmental supervision, industrial operation and public mind. Thirdly, low-carbon buildings concept popularizing education and professional education is significant. Meanwhile, courses are also necessary to teach low-carbon buildings technology in related majors such as architecture, civil engineering, energy, management.

Acknowledgment

This work is supported by the Fundamental Research Funds for the Central Universities (No. 12ZX12).

References

[1] Bo Peng. Green construction lifecycle energy consumption and carbon dioxide emission case study [D]. Tsinghua University, 2012.

[2] BP. Statistical Review of World Energy. http://www.bp.com/statisticalreview

[3] He, K. et al., Oil consumption and CO_2 emissions in China's road transport: current status, future trends and policy implications. Energy Policy, 2005, 33, 1499–1507.

[4] International Transport Forum, 2011. Top10 CO_2-emitting non-ITFeconomies China. Available online at: <http://www.in ternationaltransportforum.org/ jtrc/environment/CO_2/China.pdf> [accessed July 10th 2011].

[5] Lu, I.J., Lin, S.J.and Lewis, C., Decomposition and decoupling effects of carbon dioxide emission from highway transportation in Taiwan, Germany, Japan and South Korea. Energy Policy, 2007, 35, 3226–35.

[6] National Development and Reform Commission (NDRC). (2010). The notice of piloting low-carbon provinces and low-carbon cities. http://www.sdpc.gov.cn/

[7] Taoxin Zhang, Yueyun Zhou, Peng Lu. The concept of low-carbon city construction and the measurement model of carbon dioxide emission in China [J]. Hunan industial journal, 2011, 25(001): 77-80.

[8] Xiaolin Cao, Ting Qu, Study on green construction lifecycle cost measurement model [J]. Construction economics, 2011, 6: 92-95.

[9] Zhang, M. et al., Energy and energy efficiencies in the Chinese transportation sector, 1980–2009. Energy, 2011, 36, 770–76.

8

Evaluation of Indoor Environment System Performance for Airport Buildings

Abdulhameed Danjuma Mambo[1, *], Mahroo Eftekhari[2], Thomas Steffen[3]

[1]Department of Building, Federal University of Technology, PMB 65, Minna, Nigeria
[2]School of Civil and Building Engineering, Loughborough University, LE11 3TU, UK
[3]Department of Aeronautic and Automotive Engineering, Loughborough University, LE11 3TU, UK

Email address:
ad.mambo@futminna.edu.ng (A. D. Mambo)

Abstract: Airport terminals are energy intensive buildings. They are mostly thought to operate on a 24/7 scale and so indoor environment systems run on full schedules and do not have fine control based on detailed passenger flow information. While this assumption of round-the-clock operation may be true for the public areas of the airport building and so opportunity for complete shut-down of HVAC and lighting systems are limited especially in a busy airport terminals, there are many passenger exclusive area within the airport in which occupancy varies strictly with flight schedules. This paper presents the results of indoor environment measurement and flight schedules to identify such opportunities and to implement energy conservation measure in the passenger exclusive areas of the airport building. It also uses building simulation to assess the benefits of such energy saving interventions in terms of comfort, energy and carbon emission savings.

Keywords: Airport Terminal Building, Energy Conservation in Airport Terminal, Flight Schedule, Thermal Comfort

1. Introduction

Airports are major magnets of economic growth and development and because only about 5% of the population of the world has ever flown (2), it is an area with huge capacity for further growth. However, like all human activities, airports have great impact on the environment. These impacts includes water and air pollution, waste generation, noise pollution, extensive use of land resources and in direct relation to this paper, the use of fossil energy which has been identified as a major culprit for climate change (3-5).

Every year about 200 million people transit through UK's airports (6) which has resulted in demands for huge amount of energy and created an equally huge amount of carbon emission. A large airport can consume more energy than a city of 50,000 households; for example, in 2008, UK's largest airport, Heathrow Airport, consumed over 1000 GWh of energy (7) compared to an average of about 20 MWh (8) for UK's dwellings. Therefore, any little energy saving effort in the way airport terminals are built and operated can have tremendous impact.

It was surprising that given the stated importance and uniqueness of the airport terminal buildings, published studies on airport built environment energy performances are quite few. Galliers and Booth in a publication by Building Services Research and Information Association (BSRIA) carried out a physical and public's perception survey of some six public transport buildings including an airport terminal. The conclusion was that *public transport buildings have a fair way to go in order to provide the ideal environment for the travelling public (9)*. Balaras *et al* (2003) analysed, some specific measures aimed at reducing energy use without compromising comfort in Hellenic airports using thermal simulations and collected site data. By exploring various design options, it was concluded that that potential energy savings of 15-35% exist (10). Babu (2008) proffer design alternatives by varying building fabrics and HVAC configuration for Ahmedabad Airport terminal (11). Liu *et al* (2009) used CFD thermal simulations, indoor environment monitoring and thermal comfort surveys based on the PMV at Chengdu Shuangliu International Airport. The result of the study shows that 95.8% of the passengers were satisfied with their thermal environment (12). Griffith et al (2003) actually used the earliest form of EnergyPlus (Version 1.0.3) to study the influence of advanced building technologies such as optimised envelope systems and schedules for a proposed Air Rescue and Fire Fighting Administration Building at Teterboro airport and found that the results obtained compare well with those obtained using DOE-2.1E (13).

This paper discusses the indoor environment systems' comfort performance of a UK airport terminal and compares it with the standard comfort requirement for such spaces using Chartered Institute of Building Services Engineers (CIBSE) standards for indoor temperature, relative humidity and lighting levels (14) and Occupational Safety and Health Administration (OSHA) for indoor CO_2 Level (15). It also analyses the flight schedules to identify the Opportunities for implementing energy conservation strategy such as setbacks and switch offs.

2. Material and Methods

The methods comprise indoor environment system's monitoring and measurement for summer and winter to establish performance characteristics, analysis of flight schedules in summer and winter to identify opportunity for energy conservation measures and airport building computer modelling and simulation to gauge the savings in energy and carbon emission and comfort performance of a propose

energy conservation measures.

2.1. Indoor Environment Monitoring

An indoor site monitoring was conducted for winter period from 26th October to the 2nd November and for summer period 22nd August to 29th August. This site monitoring involves mounting HOBO U12 Data logger and CO_2 sensors for a week to measure temperature, relative humidity, lighting levels and CO_2 levels in four separate areas of the airport.

The places monitored include Baggage Reclaim area, a Duty-Free shop, a Departure Gate, and the Arrival Hall. The reason for the choice of these places was to focus on the airside of the terminal where passenger occupancy varies directly with flight schedules as against the landside where the structure occupancy pattern is complex and difficult to predict. Some pictures of these places are shown in Figure 1A-D; the positions of the sensors are indicated with a red arrow. The more expensive CO_2 sensors have to be hidden from view in some places since the airport is a public place.

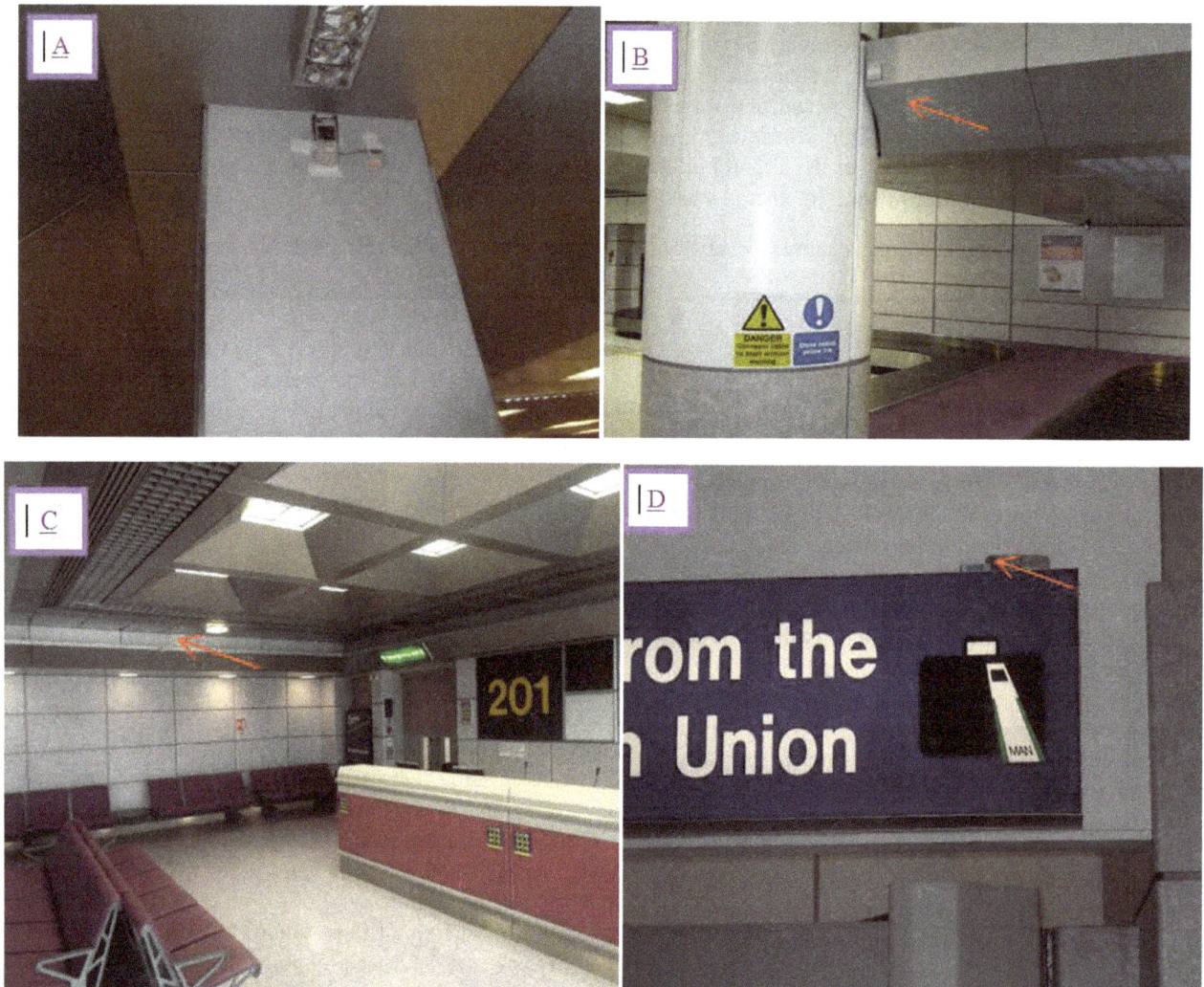

Figure 1. (A) Passport Control (B) Departure Gate (C) Baggage Reclaim (D) Arrival Hall

The indoor temperatures of the spaces were monitored and the external temperature collected from the archives of (16). External temperature influence solar heat gains, temperature of ventilation air and the convective and conductive heat exchange across the building fabrics. Therefore, when external temperature profile is compared with the indoor temperature profile it could provide clue of heating or cooling requirement to achieve the indoor comfort. It can also indicate opportunities available from the external environment to meet indoor thermal requirement either purely through passive means and/or together with active means.

2.2. Collection and Analysis of Flight Schedules

The summer and winter week's flight schedule was used for examining arriving and departing flight pattern. This was uploaded from the Chroma suite (the airport information management system) one week in advance. The average flight punctually in UK airports is about 80 % with a maximum delay time of about 10 minutes. The data collected in advance is valid.

2.3. Modelling Of Building Geometry and HVAC Systems

The case study airport is the busiest airport in the UK outside London with an annual turnover of 21 million air passengers transiting through and about 16,250 employees on site (23). It has two runways operated in two ways depending on the wind directions. Terminal 2 was recently refurbished

making it a suitable candidate for low energy refurbishing study. This terminal was constructed in 1992 on the North-West part of the airport site. It is made up of five-floor central building covering a gross floor area of about 18,000 m^2 and has two piers of four floor levels measuring about 5,400 m^2 spanning to the left and right direction of the central building. The ground and the first floor contain the arrivals hall, the third floor, the departure halls, and the fourth floor is made up of lounges, offices and the control room on the central building it mainly housed the plant rooms on the piers. The fifth floor is mainly plant rooms. So the airport building's function is already well segregated.

The terminal's hot water is served by gas boilers located in the central and eastside of the terminal. There are external air-cooled chillers located on steelwork frames in the main plant rooms. The air handling units comprises of Inlet damper, mixing box, HPHW Frost Coil, Panel Filter, Bag Filter, Carbon Filter, Cooling Coil, HPHW Re-heat Coil, Supply Fan, Extract Fan. This airport is mainly an air conditioned building. The building has no lighting control. However, the luminaires was upgraded, and the introduction of lighting control is being considered.

The first step in building modelling in DesignBuilder is in the definition of location and choice of weather data to match the location. Weather data was the hourly ASHRAE International Weather for Energy Calculation (IWEC) GBR MN6 data based on thirty years average in EnergyPlus Weather format.

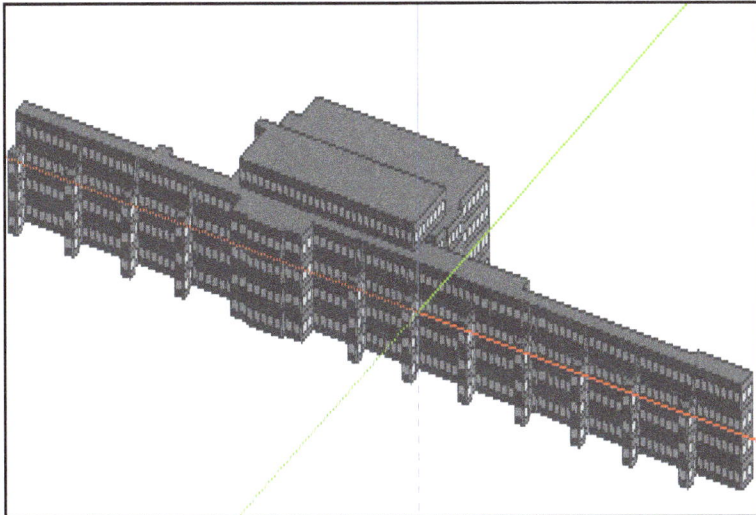

Figure 2. Thermal Model of Terminal Two

The building geometry was modelled by importing the 2D AutoCAD drawings of the building using the dxf import facility. The model was assembled by positioning blocks in the 3D space to define the external walls based on the CAD drawings. Figure 2 shows the resultant 3D geometric form of the building.

Thermal zones (internal partition walls) were defined based on the functions of the space and type of the HVAC system in the indoor space for each of the floors according to

the description obtained from Jacobs Engineering's HVAC system physical survey report and CAD drawings of terminal 2.

For this case study, there are twenty-two thermal zones in the building. However, these zones are further sub-grouped into six zone groups according to the HVAC system type. In EnergyPlus, A "zone" is different from a geometric form; it is an air volume of uniform temperature and all the heat transfer and heat storage surfaces surrounding or internal to the air volume. The building model was zoned according to passenger flow such that the areas accessible to the public were separated from the areas that were restricted to only passengers and staff. Occupancy in the restricted areas such as the Check-in, Customs, Security, passport control and baggage reclaim areas can easily be linked to arriving/departing passenger planes. However, in the public spaces such as the booking hall, some retail areas and some offices, the flow of people needs to be estimated and therefore more complicated to control.

The building façade data, lighting and opening types was chosen from the template to satisfy the Part L Building Regulation for commercial buildings in England and Wales (1990-1994) since according to the report; the building was constructed in 1992 since the details of the airport building material was not available.

The HVAC modelling was done using the recently approved Version 3 which allows access to a wide range of EnergyPlus HVAC systems component through an easy to use diagrammatic interface and satisfied compliance rating for LEED, BREEAM and Green Star. The HVAC system's specification was also based on the airport's HVAC system

survey report.

The HVAC model includes the boilers, chillers, condenser, air handling units (AHU) and the zone groups as described previously. The activity template was based on the BRE National Calculation method specifications for passenger terminal spaces contained in the DesignBuilder activity templates. This template covers occupancy profiles, internal gain data, equipment usage and plant schedules, design indoor temperature, illuminance levels and ventilation rates per person. To create the base case scenario, occupancy schedules, internal gain data and setpoints were adjusted to simulate the as measured scenario.

For the energy saving scenario, compact schedules interface was utilised to supply CIBSE thermal setpoints, lighting setpoints and air flow rates which varies with the passenger flow data. Since Terminal 2 is a jet only terminal with low cost, charter and long haul carriers. Smallest regular aircraft type is the B737-300 with 148 seats and the largest is Virgin's B747-400 with around 500 seats. This information was used to estimates the passenger number per given flight time. The flight arrival and departure data were gathered from airport's information desk.

Table 1. Summary of Parameters Used In Base and Test Case

Parameters	Base Case	Energy Case
External walls	0.45 U (W/m2 K)	0.45 U (W/m2 K)
Ground floor	0.20 U (W/m2 K)	0.20 U (W/m2 K)
Flat roof	0.35 U (W/m2 K)	0.35 U (W/m2 K)
Windows, Doors and Roof light	3.00 U (W/m2 K)	3.00 U (W/m2 K)
Occupancy	Full schedules	Varies as flight schedules but within aircraft size range
Environment Setpoints	As measured	Based on comfort standards

The summary of parameters used in base and test case was provided in Table 1.

The summer and winter week simulation dates presented in figures were chosen to reflect the monitoring period

The output of the simulation was the total electricity and gas usage in kWh combined to give the total energy usage in kWh, total carbon dioxide emission in kg of CO_2 and Fanger PMV rating.

3. Results and Discussions

3.1. Winter Case

The outside temperature varies from about 2°C on the night of the 28th to the highest day temperature of about 16°C on the 30th and 31st.

The results for indoor temperature for the monitored spaces are as shown in Figure 3A and this shows that indoor temperature range for Arrival Hall (21 – 22.5°C), Departure Gate (22 - 23°C), Duty-Free Shop (24 - 26°C) and Baggage Reclaim (20 - 22.5 °C) throughout the week. However, the CIBSE recommended temperature for arrival hall, Departure Gate, Duty-free shop is 19 – 21°C and 12 - 19°C for baggage

reclaim area.

Also, Figure 3B shows the relative humidity profile for the same spaces. The range of values for Arrival Hall, Baggage Reclaim, Departure Gate and Duty-Free Shop is 36-55%, 38-60%, 32-55% and 28-46% respectively as against the 40-70% as the CIBSE recommended values for all kinds indoor spaces. CIBSE Guide A state that a relative humidity lower than 30% is acceptable where risk of static electricity is low and above 70% where risk of microbial growth is minimal as such it is not uncommon to see practitioners quoting 20 - 80% as the acceptable range for comfort. Additionally and more important to the passenger exclusive areas of the airport, the standard stipulates that lower relative humidity is acceptable in areas of short occupancy. In this context, therefore, the relative humidity values recorded for all the indoor space except the Duty-Free Shop are acceptable. In the shops, attendants remain in the space for a long duration of time, so while it may not matter to the passenger, 28% relative humidity may be not be acceptable to the staff. However, this level was only reached briefly on a Friday afternoon, otherwise, the range has been within acceptable level for the rest of the times.

By plotting the measured indoor temperature and relative

humidity represented by the yellow shade and the CIBSE recommended setpoints for the same variables depicted with the blue shade on the psychometric chart shown in Figure 4; it can be seen that the indoor environments are warmer than

they should be compared to the standard requirements for such places However, the relative humidity in all monitored spaces are within the acceptable limits (20-80%).

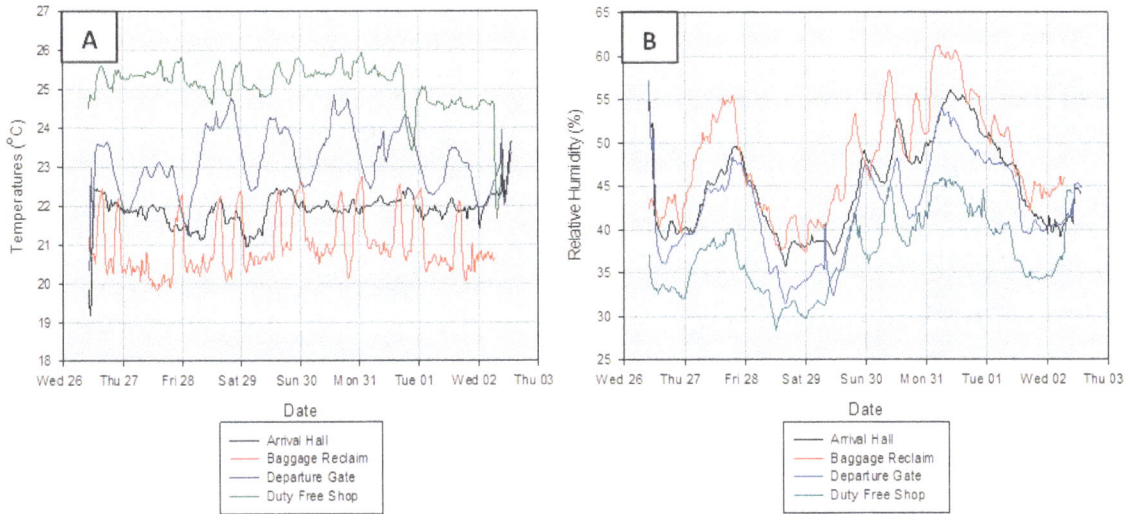

Figure 3. (A) Indoor Temperature Profile (B) Indoor RH Profile

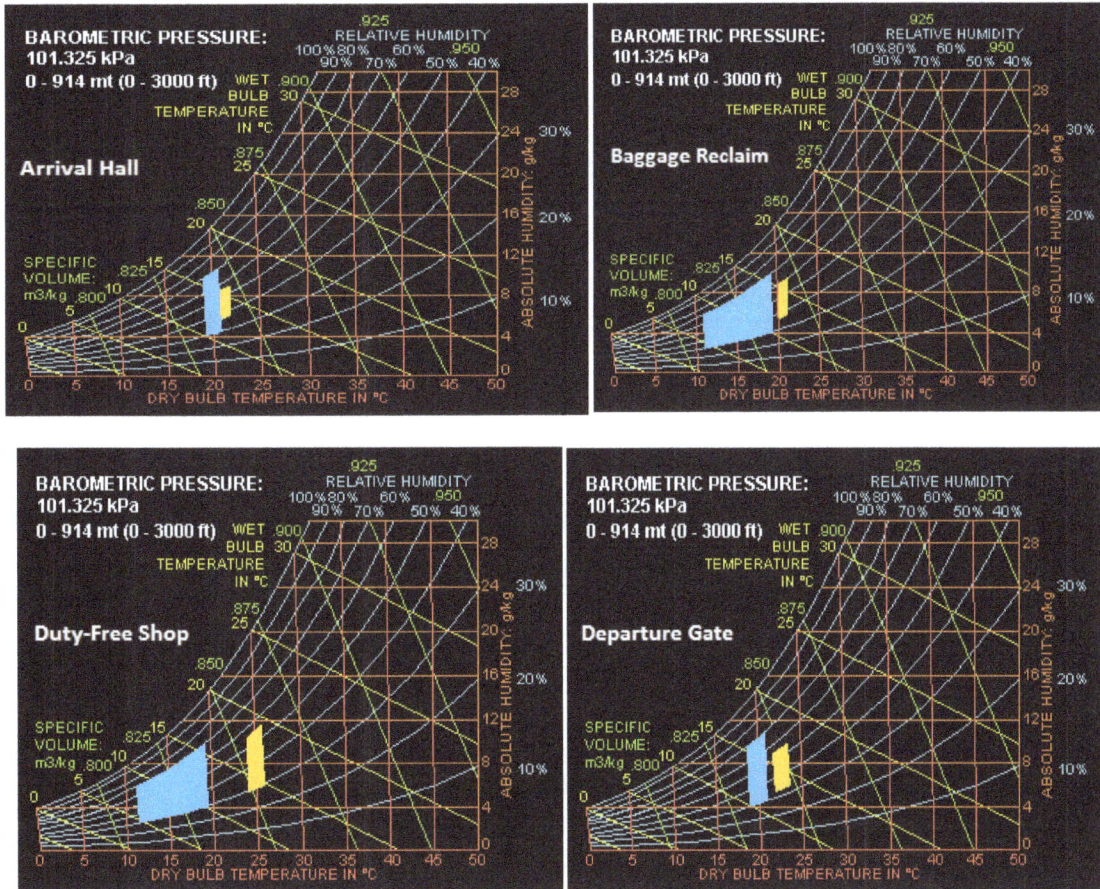

Figure 4. Measured vs. Recommended Comfort Variables

3.1.1. Indoor CO_2 Levels

CO_2 is an indicator of the amount of fresh air injected into a space to dilute pollutants and provides oxygen necessary for respiration. So, elevated CO_2 is a likely indicator of the

presence of other air pollutants and a pointer to inadequate ventilation. Although, ANSI/ASHRAE Standard 62-2007 (a very conservative standard for transient spaces) specified that an indoor concentration of not more than 700 ppm above the outdoor concentration will satisfy majority (80%) of building occupants and the National Institute for Occupational Safety and Health (NIOSH) recommends that a concentration of over 1000 ppm was a marker for inadequate ventilation. European standards however limit carbon dioxide to 3500 ppm and Occupational Safety and Health Administration (OSHA) limits carbon dioxide concentration in the workplace to 5,000 ppm for prolonged periods, and 35,000 ppm for 15 minutes (15).

The CO_2 Levels recorded in all the places monitored was less than 900 ppm during peak occupancy. Comparing this with the OSHA standards quoted above, the inference is that these spaces may have been over ventilated.

3.1.2. Indoor Illuminance Levels

As can be seen from Figure 5, the indoor illuminance level

for Arrival Hall, Baggage Reclaim, Departure Gate and Duty-Free Shop is 250-400 Lux, 310-370 Lux, 320-600 Lux and over 310 Lux respectively. These levels are higher than the recommended 200 lux (the brown line in Figure 5) for these spaces. The indoor illuminance level depends on whether the space in question is exposed to direct daylight and that is the reasons for the high illuminance spikes during the day time in the Departure Gate Area. This area is suitable for Daylighting control. During site assessment tour, it was observed that almost all the artificial lights were on even in spaces where the daylight illuminance was very high such as the Departure Gates and Departure Concourses generally. The new lighting system installed after the monitoring period, now has the capability of maximising the daylighting in terminals.

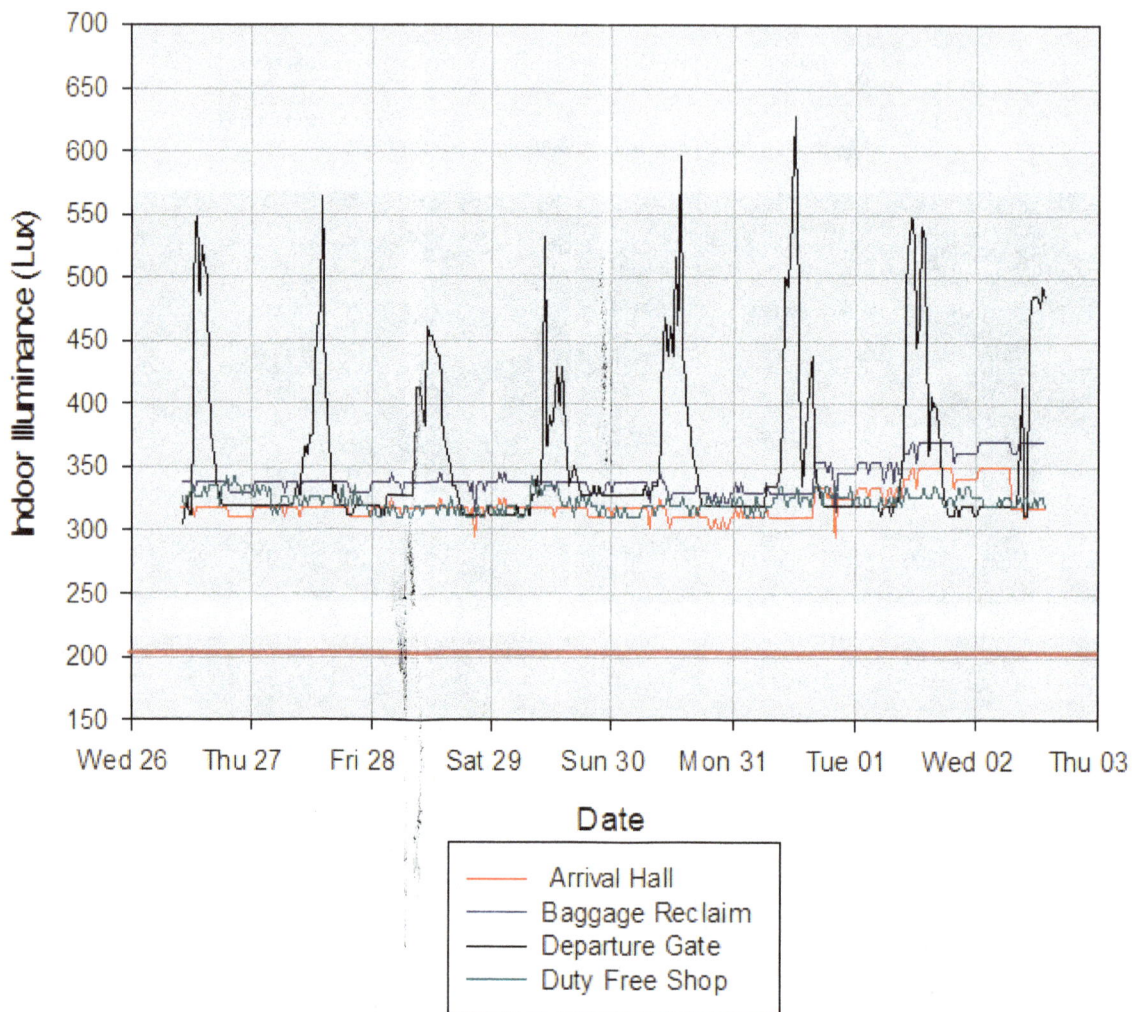

Figure 5. Indoor Illuminance

The environmental performance for winter, have clearly shown that the lighting, temperature and ventilation setpoints has exceeded the recommended CIBSE and OSHA values These excess values have led to substantial loss in energy. It can be seen from the temperature profile that there was no indication of setback operation in the space during

unoccupied times. The setback temperature during unoccupied hours will be dictated by the external temperature and occupancy. Although relative humidity level was not controlled as part of the airport's HVAC control strategy, the level recorded was about right for comfort in all the spaces monitored except for a short period in the shop.

3.2. Summer Case

External and indoor temperature were measured. The external temperature ranges from 11°C for some nights to 19°C on some days.

3.2.1. Indoor Thermal Comfort Variables

This indoor temperature profile in Figure 6A showed a week-long temperature ranges of 22-25°C for Arrival Hall, 24-26.5°C for Baggage Reclaim, 22-23°C for Departure Gate,

22.5-23.5°C Duty-Free as against the CIBSE recommended range of 21-25°C for all spaces. There was no adjustment of setpoints during unoccupied hours to reduce energy consumption in the airport terminal. So although, the recommended setpoints is the same for all the spaces, recorded temperature shows considerable variation with the Baggage Reclaim area; a deep plan space with no opening to the outside, was much warmer. However, the Departure Gate, the only space with an external wall had the lowest temperature.

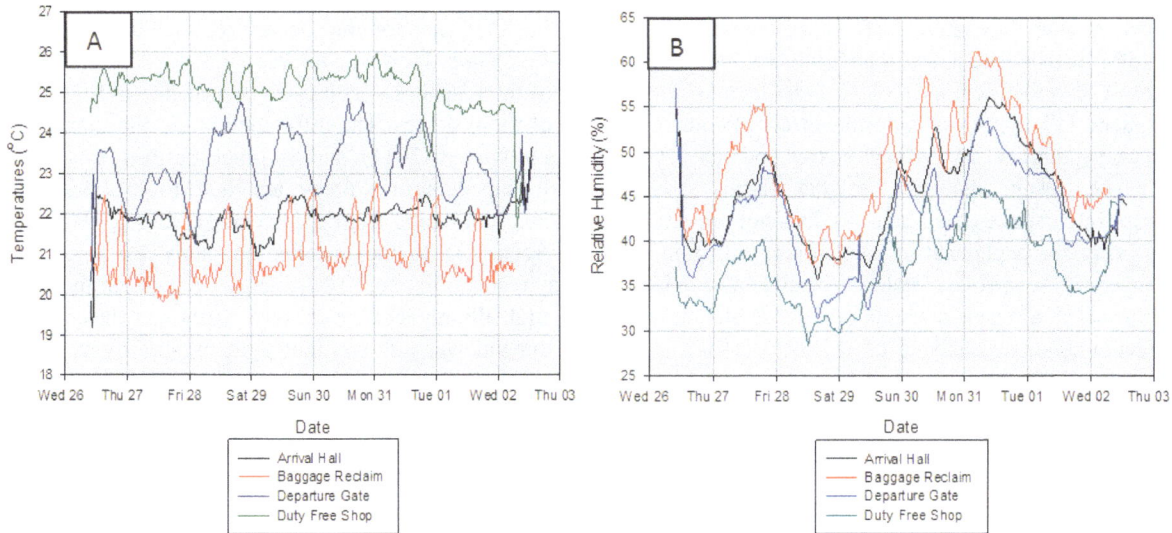

Figure 6. Summer Indoor Temperature & RH

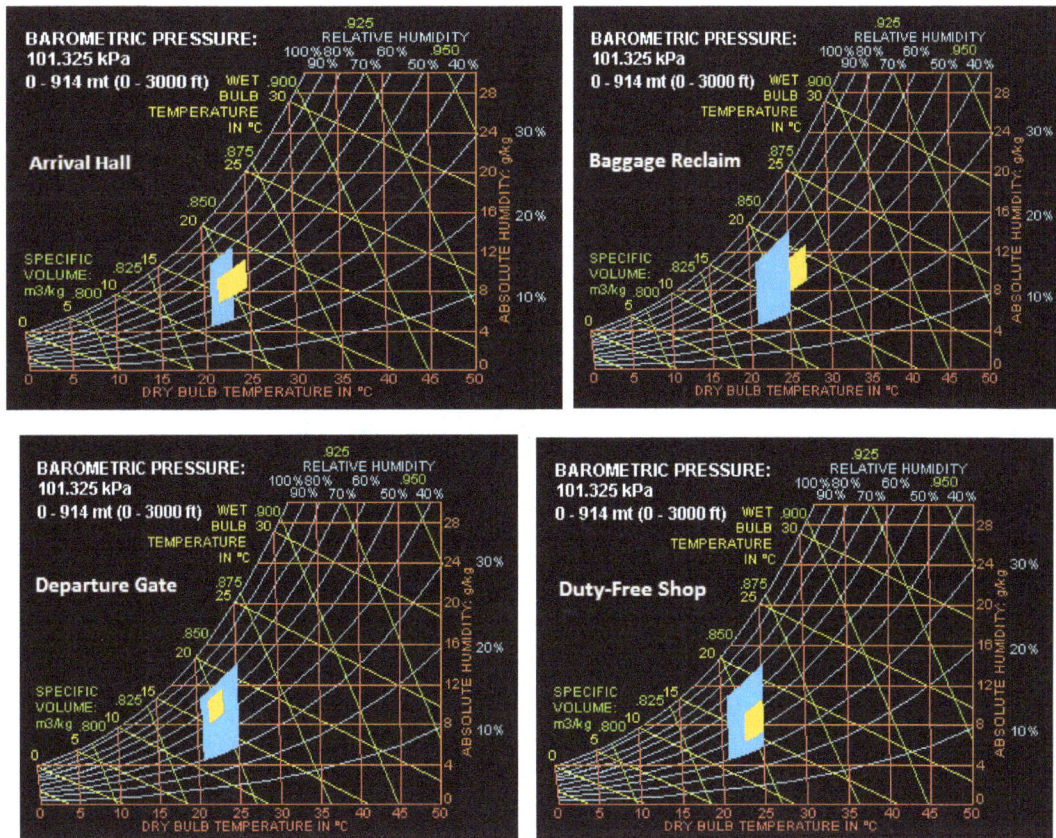

Figure 7. Measured vs. Recommended Comfort Variables

Similarly, the indoor relative humidity value for the indoor places shows considerable variation (Figure 6B). For example, the range of values measured for the Arrival Hall, Baggage Reclaim, Departure Gate and the Duty-Free Shop was 43-58%, 37-53%, 46-65% and 37-53% respectively. However, the range in all the spaces monitored was within the acceptable level for comfort.

By juxtaposing the plotted measured indoor relative humidity and temperature (Yellow shade) with the acceptable values (Blue shade) for these variables on the psychometric chart as shown in Figure 7, it can be seen also that the indoor spaces are warmer than they should be. Space temperature control for comfort usually has a dead-band (interval between higher and lower comfort setpoint) of several degrees for most indoor spaces. (17,18) investigates the effect of control deadband on acceptability of indoor space and energy consumption. The result showed that the tightly air-temperature-controlled space (dead-band 2) does not provide higher acceptability for occupants in comparison with non-tightly air-temperature-controlled spaces (deadband 4 and deadband 6). In fact in a recent revision, ASHRAE Standard 90.1 (2010) recommends a deadband of at least 5 (19). The indoor data collected for both winter and summer operation show that the HVAC is applying tight control (small area covered by yellow compare to the large area covered by the blue shade) of the variables compare to what is acceptable. Although, this is typical of many air conditioned space, it results in high energy cost (18,20).

3.2.2. Indoor CO₂ Levels

The measured CO_2 at peak occupancy is about 1150 ppm. While this does not satisfy the requirement set by ASHRAE Standard 62 and NIOSH recommendations of about 400 ppm, it still appears over-ventilated by European and OHSA Standards for transient occupancy.

3.2.3. Indoor Illuminance Levels

Also, the indoor illuminance values in Figure 8 shows a range of over 250 Lux for Arrival Hall, 300 lux for Baggage Reclaim, 250 Lux for the Departure Gate and 280 lux for Duty-Free shop. However the recommended illuminance for most of these spaces is 200 Lux (Brown line in Figure 7) for most of these spaces. The difference in the illuminance level between winter (2011) and summer periods especially in arrival and departure areas are due to upgrade of the terminals luminaires from the metal Halides to TiLite High Bay. According to the installer company, Philips, this has already resulted in about 50% energy savings but the fact that these high illuminance levels were sustained throughout the experimental week shows that there is still room for more energy saving through adjusting artificial lighting according to the occupancy pattern and daylighting (20). The monitored results from the Departure gate which mainly uses daylighting show an average daylight level of 240 lux to a daily peak of 300-1000 lux. This is more than sufficient for the requirement of this space, so, incorporating a Daylighting control in this area and similar areas within the terminals will lead to additional energy savings.

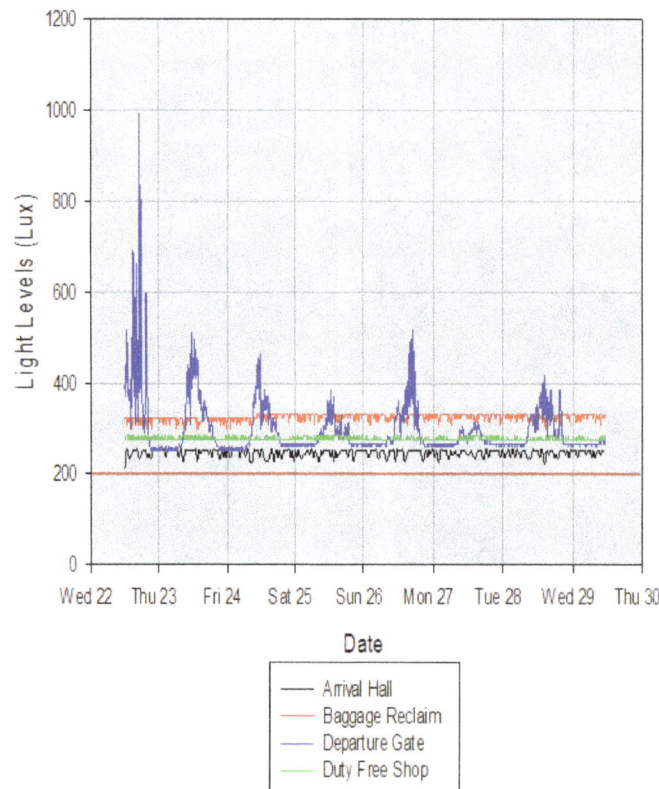

Figure 8. Summer Indoor Illuminance

From the summer and winter results it is clear that there are opportunities for reducing energy consumption within the airport terminal building. The monitoring results have demonstrated the potential of energy saving through appropriate set points for indoor air quality, thermal and visual comfort according to the occupancy pattern. Relative humidity level was generally within acceptable range; therefore this has been excluded from this study.

4. Flight Time Interval

4.1. Winter Arrival & Departure Times and Intervals between Flights

Figure 9A below shows plane arrival times plotted against the time-interval between two consecutive arrivals for the period 27th October to 3rd (7 days) November. ICAO recommended duration for arriving passenger moving from disembarkation to baggage collection was one hour, as shown

by the blue line in the figure. Aggregating the hours above the blue line, up to 50 hours per week was available for implementing energy saving strategies.

A recent survey (21) conducted in seven major UK airports (Manchester, Heathrow, Stansted, Gatwick, Luton, Edinburgh, Inverness) by Civil Aviation Authority (CAA) in 2009 shows that normal processing time for most passengers in these airports is even less than the provisions in the standard. It is 45 minutes for arrival and 1 hour for departure processing in most airports (1).

Figure 9B also shows plane departure times plotted against the time-interval between any two consecutive departures for 7 days. By using the 1 hour ICAO recommendation for the duration from presentation of passengers at first processing point to the scheduled time of flight departure; Up to 52 hours interval is available in the week for implementing energy conservation measures.

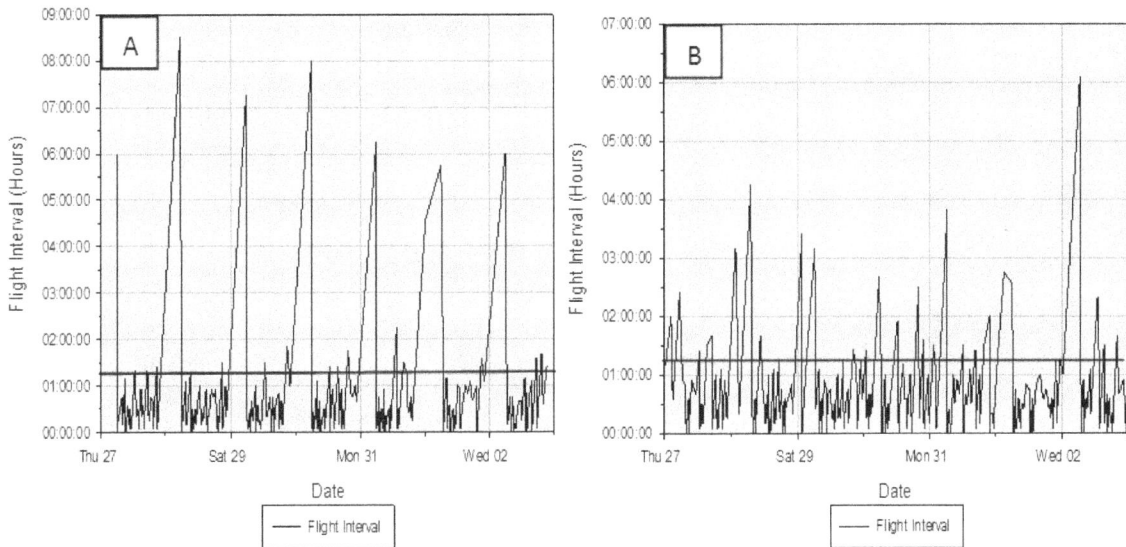

Figure 9. (A) Plane Arrival's; (B) Departure's Time Versus Arrival's Time Intervals

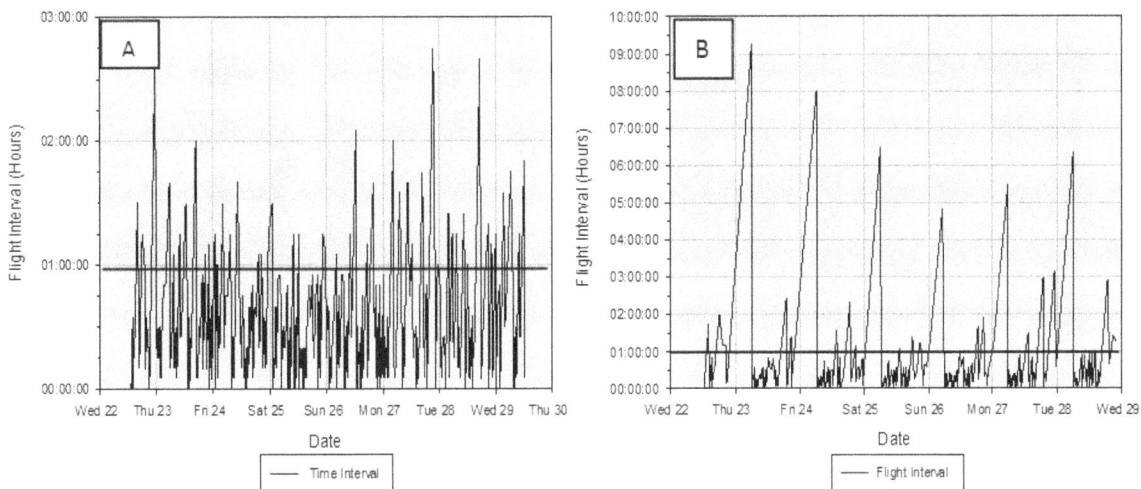

Figure 10. Plane Arrival's Time Versus Arrival's Time Intervals

4.2. Summer Arrival & Departure Times and Intervals between Flights

Similarly, Figure 10A below shows plane arrival time-interval between two consecutive arrivals for the period 22nd to 29th August. Based on the one hour clearing time, Up to 21 hours opportunity exist for the week under review to switch to energy saving mode.

Figure 10B shows departures for the period 22nd to 29th August and has Up to 50.667 hours' opportunity existed for energy conservation.

Table 2. Setback Opportunities in 7 Days Monitoring

Spaces	Winter (Hours)	Summer (Hours)
Arrival	39.05	21.50
Departure	52.00	50.67

From the winter and summer arrival and departure schedules and as summarised in Table 1, it can be seen that there are more flights in summer time than in winter period (less time interval between flights for the same number of days) and also there are more arriving than departing flights in both seasons.

A close look at the histograms in Figure 11 showing the distribution of the interval duration for the week under review, It can be seen that 75% of the time intervals is in the range of over 1 hour duration in the Winter Arrival, about 82% of the time for the Winter Departure and about 85% of the time for Summer Departure. This shows that the time available to implement energy conservation measure for duration above one hour has greater availability. The distribution in summer arrival however shows that this is a particularly busy period for the airport and so the intervals are tighter and the duration shorter (0-1 forms 70% of the range). The entire distribution shows that there are more arrivals than departure flights for both winter and summer.

What was demonstrated in this work was a week review for an airport. If all these energy conservation opportunities are extrapolated across the whole airport terminals and for a whole year, the energy saving is very significant.

This results shows the need to develop an airport environment management system capable of providing the required comfort setpoint during occupancy and implementing energy conservation measure during unoccupancy by taking into account passenger flow pattern and external (22).

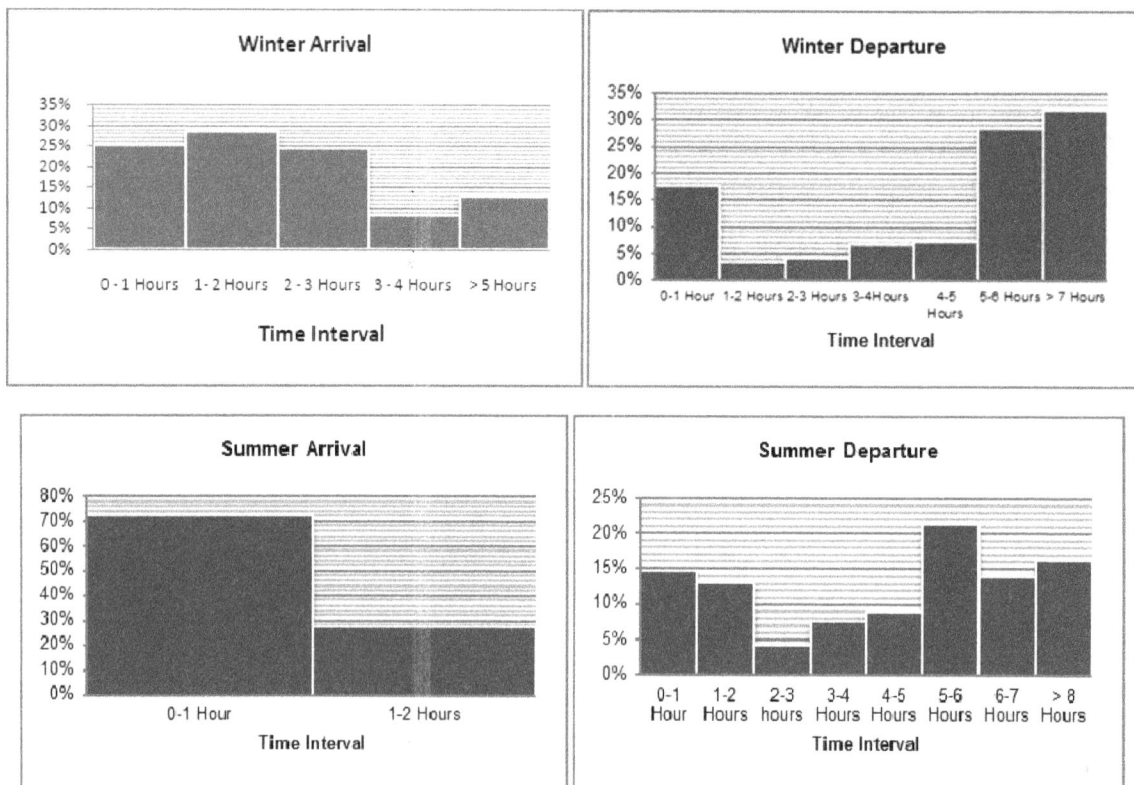

Figure 11. Distribution of Flight Interval

5. Building Simulation Results

From Figure 12 it can be seen that the energy savings of 21 to 27% was achieved for the summer case 40 to 50%

recorded for the winter time. The main reason for less energy saving during summer period is that there are more flights during summer with fewer intervals between the flights

Energy savings for both summer and winter cases are due

to:

(1) reducing the high indoor setpoints since the external temperature for both winter and summer for the period under review was less than 19 degrees, comfort setpoints can be achieved almost passively.

(2) by scheduling the system to implement setback during passenger un-occupied period in the passenger exclusive area.

Also, based on Fanger's 7 point thermal sensation scale in winter, the base case generally showed a warm environment and the energy case edged the scale towards the neutral point. Similarly in the summer case, the energy case also edges the scale father away from warm towards the neutral position. These clearly indicate that by implementing energy conservation measures, the indoor environment of the airport terminal has been made more comfortable. So these energy savings has been achieved with increase in comfort of the passengers.

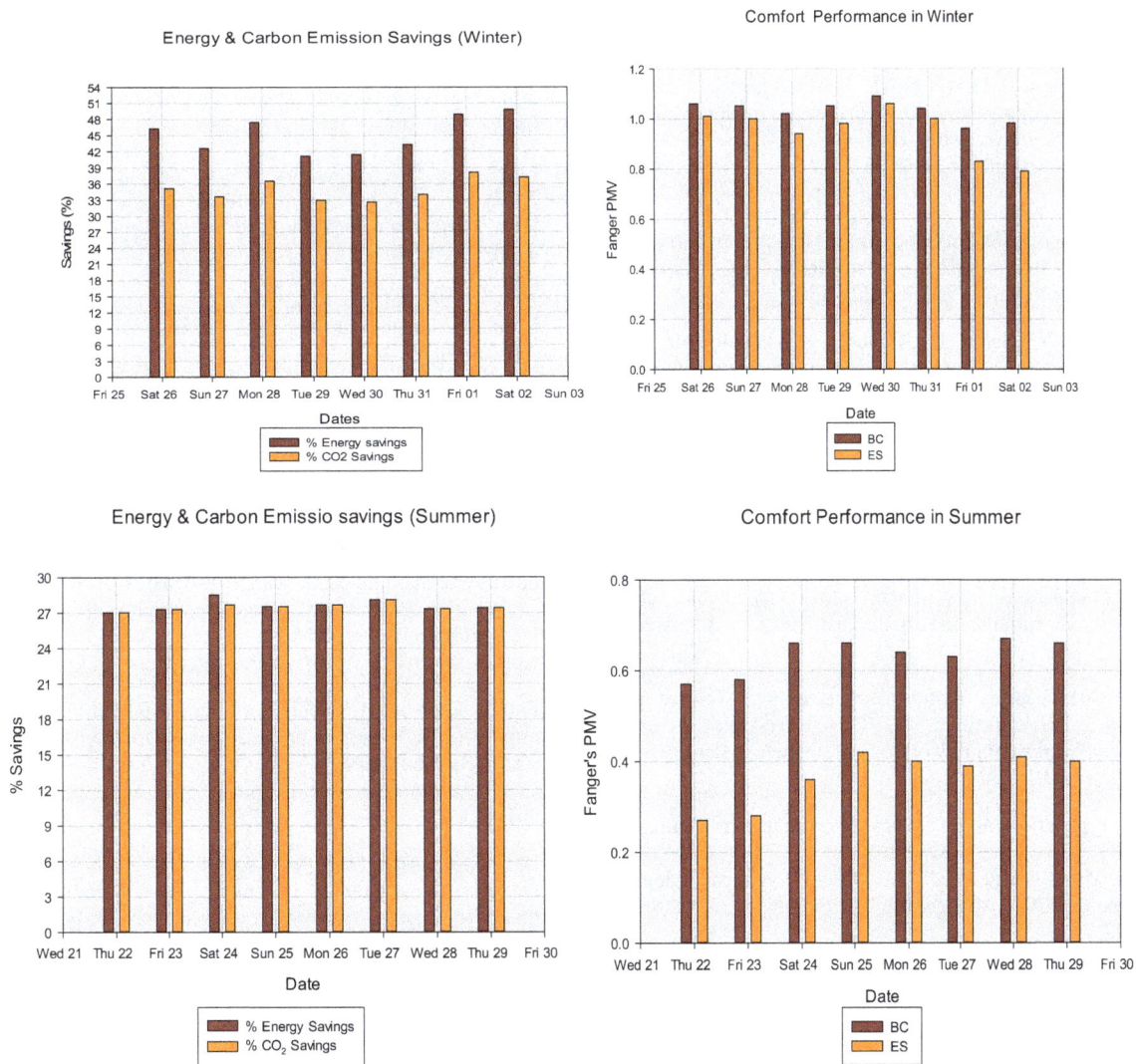

Figure 12. Energy, CO_2 emission and Comfort Rating from Energy Conservation

6. Conclusions

This paper presented the analysis of the primary data collected for both the arrival and departure indoor spaces of a UK airport during winter and summer scenarios. From the comfort variables analysed, it can be seen that that the indoor spaces temperature, lighting and ventilation rates were higher than the recommended values. However, relative humidity satisfies the comfort level though it was not being controlled. Tight deadband were also noticed in the control of temperature; a situation that will lead to higher energy consumption. Also, analysis of the flight schedules showed that there are sufficient opportunities to implement energy conservation measures especially in the passenger exclusive spaces. This Paper considers varying indoor environment comfort set points according to passenger flow through the airport and this would lead to energy saving of 20-25% while considerably improving thermal comfort of the passengers.

Acknowledgement

The authors gratefully acknowledge the support of Environment Group, Manchester Airport for facilitating the effort on data collection. We also acknowledge the financial support provided by the Engineering and Physical Sciences Research Council, UK in its Airport Sandpits Programme and Petroleum Technology Development fund (PTDF) Nigeria.

References

[1]   ICAO. International Standards and Recommended Practice, Annex 9 to the convention of International Civil Aviation Organization -Facilitation. 2005; Available at: http://dcaa.trafikstyrelsen.dk:8000/icaodocs/Annex%209%20-%20Facilitation/Facilition.pdf. Accessed 11/2014.

[2]   Worldwatch Institute. VITAL FACTS: Selected facts and story ideas from Vital Signs 2006-2007 | Worldwatch Institute. 2007; Available at: http://www.worldwatch.org/vital-facts-selected-facts-and-story-ideas-vital-signs-2006-2007. Accessed 5/20/2013, 2013.

[3]   Turnbull D, Bevan J. The impact of airport de-icing on a river: The case of the Ouseburn, Newcastle upon Tyne. Environmental pollution 1995;88(3):321-332.

[4]   Moussiopoulos N, Sahm P, Karatzas K, Papalexiou S, Karagiannidis A. Assessing the impact of the new Athens airport to urban air quality with contemporary air pollution models. Atmos Environ 1997;31(10):1497-1511.

[5]   Unal A, Hu Y, Chang ME, Talat Odman M, Russell AG. Airport related emissions and impacts on air quality: Application to the Atlanta International Airport. Atmosphere Environ 2005;39(32):5787-5798.

[6]   Aviation Foundation. Fast Facts. 2013; Available at: http://www.aviationfoundation.org.uk/Fast-Facts/. Accessed 5/20/2013, 2013.

[7]   Heathrow. Sustainability performance summary: Toward a sustainable Heathrow. 2010; Available at: http://www.refworks.com/refworks2/?r=references|MainLayout::init#. Accessed 12, 2012.

[8]   OFGEM. Typical Domestic Energy Consumption Figures. 2011; Available at: http://www.ofgem.gov.uk/Media/FactSheets/Documents1/domestic%20energy%20consump%20fig%20FS.pdf. Accessed 12/24, 2012.

[9]   Galliers S, Booth W. Quality Environments for Public Transport Buildings: Comparisons of Public Perceptions of the Environment and Physical Measuring of the Environment. CIBSE Publications 1996.

[10]   Balaras C, Dascalaki E, Gaglia A, Droutsa K. Energy conservation potential, HVAC installations and operational issues in Hellenic airports. Energy Build 2003;35(11):1105-1120.

[11]   Babu AD. A Low Energy Passenger Terminal Building for Ahmedabad 'Building Envelop as an Environment Regulator'. 25th Conference on Passive and Low Energy Architecture, Dublin 2008.

[12]   Liu J, Yu N, Lei B, Rong X, Yang L. Research on Indoor Environment for the Treminal 1 of Chengdu Shuangliu International Airport. 2009.

[13]   Griffith B, Pless S, Talbert B, Deru M, Torcellini P. Energy design analysis and evaluation of a proposed air rescue and firefighting administration building for Teterboro airport. : National Renewable Energy Laboratory Golden, CO; 2003.

[14]   CIBSE. Guide A: Environmental design. London: Chartered Institution of Building Services Engineers 2006.

[15]   OSHA, Occupational Health and Safety Administration 1990. Carbon Dioxide In Workplace Atmospheres. 1990; Available at: https://www.osha.gov/dts/sltc/methods/inorganic/id172/id172.html.

[16]   Freemeteo. Hourly weather history of Manchester Airport. 2013; Available at: http://freemeteo.com/default.asp?pid=20&gid=2643743&la=1&sid=33340&lc=1.

[17]   Arens E, Humphreys MA, de Dear R, Zhang H. Are 'class A' temperature requirements realistic or desirable? Build Environ 2010 1;45(1):4-10.

[18]   Hoyt T, Lee KH, Zhang H, Arens E, Webster T. Energy savings from extended air temperature setpoints and reductions in room air mixing. 2009.

[19]   ANSI/ASHRAE/IES Standard 90.1 (2010). Energy Standard for Buildings Except Low-Rise Residential Buildings. I-P Edition, 2010; Available at: https://www.google.co.uk/search?q=ASHRAE+90.1+%282010%29+&ie=utf-8&oe=utf-8&rls=org.mozilla:en-US:official&client=firefox-a&gws_rd=cr&ei=1qMlUvmfLeiJ0AXUuoCQAQ.

[20]   Philips. Case Study Manchester Terminal 2. 2012; Available at: http://www.lighting.philips.co.uk/pwc_li/gb_en/projects/Assets/projects/Manchester Airport T2/Case_study_Manchester_Terminal 2.pdf. Accessed 4/15/2013, 2013.

[21]   DfT. Air passenger experience: Results from CAA survey module (2009). 2010; Available at: https://www.gov.uk/government/publications/air-passenger-experience-results-from-caa-survey-module-2009.

[22]   Supervisory Control of Indoor Environment Systems to Minimise the Carbon Footprint of Airport Terminal Buildings–A Review. Sustainability in Energy and Buildings: Proceedings of the 3rd International Conference on Sustainability in Energy and Buildings (SEB´11): Springer; 2012.

[23]   Knowles M. Lighting and Lighting Controls - Manchester Airport Case Study. North-West Energy Forum 2006.

9

Investigations on Heat Loss in Solar Tower Receivers with Wind Speed Variation

Ramadan Abdiwe, Markus Haider

Institute for Energy Systems and Thermodynamics, Vienna University of Technology, Wien, Austria

Email address:

rabdiwe@ite.tuwien.ac.at (R. Abdiwe)

Abstract: The performance of the Solar Tower Receiver (STR) affects significantly the efficiency of the entire solar power generation system and minimizing the heat loss of the STR plays a dominant role in increasing its performance. Unlike the other thermal losses the convective heat loss in STR has direct relation with wind conditions. In this study a Simulation tool ANSYS® FLUENT® was used to determine the convection heat loss in both cavity and externalSTR at wind speed varies from(2) to (10) m/s. A fixed tilt angle ($\theta = 90°$) for the cavity receiver is adopted. The results show that the convection heat loss in both receivers increases with increase of wind speed. The absolute values are considerably lower in the case of the cavity with comparison to the external type. Furthermore, the radiative heat loss in the external and the cavity receivers is investigated. The results show that for the same absorbed area, the radiation loss in the cavity is lower by almost (80%) than the radiation loss in the external.

Keywords: Solar Tower Receiver, Central Receiver System, Heat loss, CFD

1. Introduction

Concentrated Solar Power technology implementation is growing fast. In 2013 , (2.136) GW are operating, (2.477) GW under construction and (10.135) GW are announced mainly in the USA followed by Spain and China and about (17) GW of CSP projects are under development worldwide [1]. Central Receiver Systems (CRS) arevery promising from the point of view of cost produced electricity [2].TheCRS is composed of the following main components: the heliostats, the receiver, the power block and thermal storage and balance of plant components allow high temperatures which lead to high efficiency of the power conversion system [3], easy integration in fossil plants for hybrid operation in a wide variety of options. It has the potential for generating electricity with high annual capacity factors from(0.4) to (0.8) through the use of thermal storage [3], and great potential for cost reduction and efficiency improvements [4]. Table (1) shows some characteristics of CRS.

Table 1. Characteristics of solar thermal central receiver systems [5] .

Typical Size	10-200 MWel
Operating Temperature	
• Rankine	565 °C
• Brayton	800 °C
Annual Capacity Factor	20 – 77%*
Peak Efficiency	16 – 23%*
Annual Net Efficiency	12 – 20%*
Available Storage technologies	Nitrate salt for molten salt receivers Ceramic bed for air receivers
Hybrid designs	Yes

Since the early1980s the tower technology has attracted worldwide a lot of interest and numerous pilot projects have been successfully tested inUSA, Spain, and France. However, the first commercial Concentrating Solar Thermal Power Plants using large heliostat field and a solar receiver placed on the top of a tower are PS10, PS20, and Gemasolar in Spain [6]. The Ivanpah and Crescent Dunes installations in the USA pass the(100) MW_{el} threshold in 2013and 2014. In the case of power towers, incident sunrays are tracked by large mirrored collectors (heliostats) which concentrate the energy flux towards radiative /convective heat exchangers, called solar receivers, where energy is transferred to a working thermal fluid. After energy collection by this solar subsystem comprised of optical concentrator, and solar receiver, the conversion of thermal energy to electricity has many similarities with the one from fossil-fueled thermal power plants [3]

As the thermal receiver plays a very important role of transferring the solar heat to the engine, and heat loss of the thermal receiver can significantly reduce the efficiency and consequently the cost effectiveness of the system, it is important to assess and subsequently improve the thermal performance of the thermal receiver [7]. Therefore, the research aiming to minimize heat losses of existing receivers and developing of new designs is of great interest.

There are different types of receivers that can be classified into four groups depending on their functionality and geometric configuration. The four groups are external receivers, volumetric receivers, the cavity receivers and the particle receivers [1]. The present study will focus only on external (outer surface) and cavity type receivers.

The essential feature of the receiver in power tower plants is to absorb the maximum amount of the concentrated solar irradiance and transfer it to the working fluid or materials heat with minimum losses. In external type, the receiver is exposed to the environment and receives the solar irradiation from the heliostats. Being exposed to the environment, the performance of the external receivers can be affected by the environment conditions such as wind speed and ambient temperature. The radiation and convection heat transfer from the hot surface temperature of the receiver to the environment will reduce its efficiency. The cavity receiver is not exposed to the environment to the same extent as external type. However, still the heat loss in cavity receiver remains an issue to be solved and quantified. To date, some publications concerning the heat loss of the STR are available. Clausing (1981) presented an analytical model which enables the estimation of convective losses from cavity receivers and his model indicated that the influence of the wind on the convective loss at normal operating conditions is minimal [8]. Stine and McDonaled (1989) stated that the convective heat loss depends on the air temperature within the cavity, the inclination of the cavity and external wind conditions [9]. Leibfried and Ortiohann (1995) reported that the radiation loss is dependent on the cavity wall temperature, the shape factors and emissivity/absorptivity of the receiver walls, while conduction is dependent on the receiver temperature and the insulation material [10]. Sendhil Kumara (2007) investigated the approximate estimation of the natural convection heat loss from actual geometry of a cavity receiver by varying the inclinations of the receiver from 0° to 90° and also investigated the effect of operating temperature [11]. Prakash M. (2008) studied the convective and radiative heat losses of a cavity receiver taking in consideration different wind speed and direction, input temperature and receiver inclination angel [12]. Shuang and Xiao (2010)obtained that the conduction and radiation in cavity receivers can readily be determined analytically, on the other hand, the determination of convection heat loss is rather difficult due to the complexity of the temperature and velocity fields in and around the cavity and usually rely on semi-empirical models [7]. Qiang and Zhifen (2011) presented that wind conditions can obviously

affect the thermal losses and the value reaches its maximum when the wind blows from the side of the receiver ($\theta = 90°$)[13]. Figure (1)shows the tilt angle in a cavity receiver

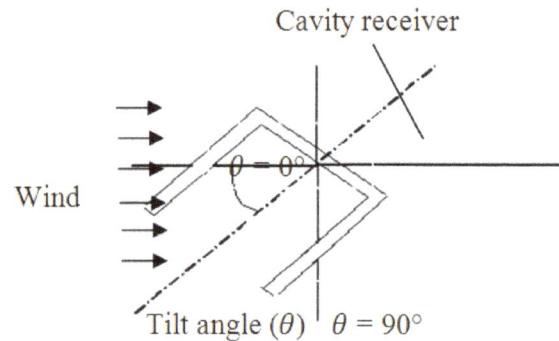

Fig. 1. A sketch illustrates the tilt angle.

The objective of this study is to investigate the convection heat loss characteristics in external and cavity receivers with a fixed tilt angle of (90°) and varying wind speed, using Computational Fluid Dynamics (CFD) simulation ANSYS® FLUENT®

2. Mathematical Model

Thermal losses in STR are mainly consists of three losses as following: conduction loss ($\dot{Q}_{loss,cond}$) ,radiation loss ($\dot{Q}_{loss,rad}$) , and convection loss ($\dot{Q}_{loss,conv}$) .The conduction heat loss occurs because of the heat-conduction through the receiver body, and it can be minimized by using insulated material with low thermal conductivity. In this model an assumption was made that all surfaces are adiabatic surfaces and therefore, heat flux passes through them is negligible. The radiation heat loss is caused by infrared radiation that emits from the receiver wall to the environment. In the other hand the convective heat loss occurs because of the convective heat exchange between the receiver walls and the air flowing along the receiver wall. The radiation and convection heat losses will be considered in this study.

2.1. Radiation Heat Loss

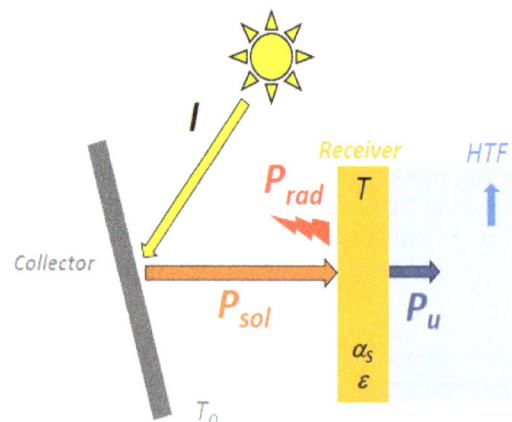

Fig. 2. Energy balance of a solar absorber [14].

The radiation loss ($\dot{Q}_{loss,rad}$) is caused by infrared radiation, which emits from the receiver walls to the environment. The radiative loss is dependent on wall temperature, the shape factors and emissivity/absorptivity of the receiver walls and independent of the inclination [9]. The figure (2) shows an energy balance of a solar absorber.

The total infrared energy can be calculated by the Stefan-Boltzmann law as:

$$E = \sigma T^4 \qquad (1)$$

$$T^4 = (T_{Abs}^4 - T_{Amb}^4) \qquad (2)$$

The solar power to the receiver ($\dot{Q}_{Sol}$) can be calculated by applying an energy balance on the solar absorber, as following:

$$\dot{Q}_{Sol} = \dot{Q}_{HTM} + \dot{Q}_{loss,rad} + \dot{Q}_{loss,conv} + \dot{Q}_{refl} \qquad (3)$$

Where

($\dot{Q}_{HTM}$) is the power transferred to the heat transfer medium.

($\dot{Q}_{refl}$) is the reflected power which happens when the incident rays of visible light that come from the heliostats are reflected at the wall of the receiver.

Rearrangement yields:

$$\dot{Q}_{HTM} = \dot{Q}_{Sol} - \dot{Q}_{loss,rad} - \dot{Q}_{loss,conv} - \dot{Q}_{refl} \qquad (4)$$

With

$$\dot{Q}_{Sol} = C \cdot A_{Abs} \cdot I \cdot \eta_{optical} \qquad (5)$$

$$\dot{Q}_{loss,rad} = \varepsilon \cdot \sigma \cdot A_{Abs} \cdot T^4 \qquad (6)$$

$$\dot{Q}_{loss,conv} = h \cdot A_{abs} \cdot (T_{Abs} - T_{Amb}) \qquad (7)$$

$$\dot{Q}_{refl} = (1 - \alpha) \cdot \dot{Q}_{Sol} \qquad (8)$$

The energy efficiency of the system can be calculated

$$\eta_{sys} = \frac{\dot{Q}_{HTM}}{\dot{Q}_{Sol}} = \frac{\dot{Q}_{Sol} - \dot{Q}_{loss,rad} - \dot{Q}_{loss,conv} - \dot{Q}_{refl}}{\dot{Q}_{Sol}} \qquad (9)$$

The convective heat loss ($\dot{Q}_{loss,conv}$) will be illustrated more in details later in this paper.

The radiative heat loss ($\dot{Q}_{loss,rad}$) in the case of the external receiver can be calculated as described above in equation (6):

$$\dot{Q}_{loss,rad,external} = \varepsilon \cdot \sigma \cdot A_{Abs} \cdot T^4 \qquad (10)$$

The radiative heat loss ($\dot{Q}_{loss,rad}$) in the case of the cavity receiver can be calculated using the following equation [15]:

$$\dot{Q}_{loss,rad,cavity} = \varepsilon_{eff} \cdot \sigma \cdot A_{Ape} \cdot T^4 \qquad (11)$$

The aperture area of the cavity receiver (A_{Ape}) in the model is equal to (10.89) m^2.

The effective emissivity (ε_{eff}) of the cavity can be derived using the following equation [15]:

$$\varepsilon_{eff} = \frac{1}{1 + (\frac{1-\varepsilon}{\varepsilon})\frac{A_{Ape}}{A_{Abs}}} \qquad (12)$$

2.2. Comparison of Radiative Losses (External VS. Cavity)

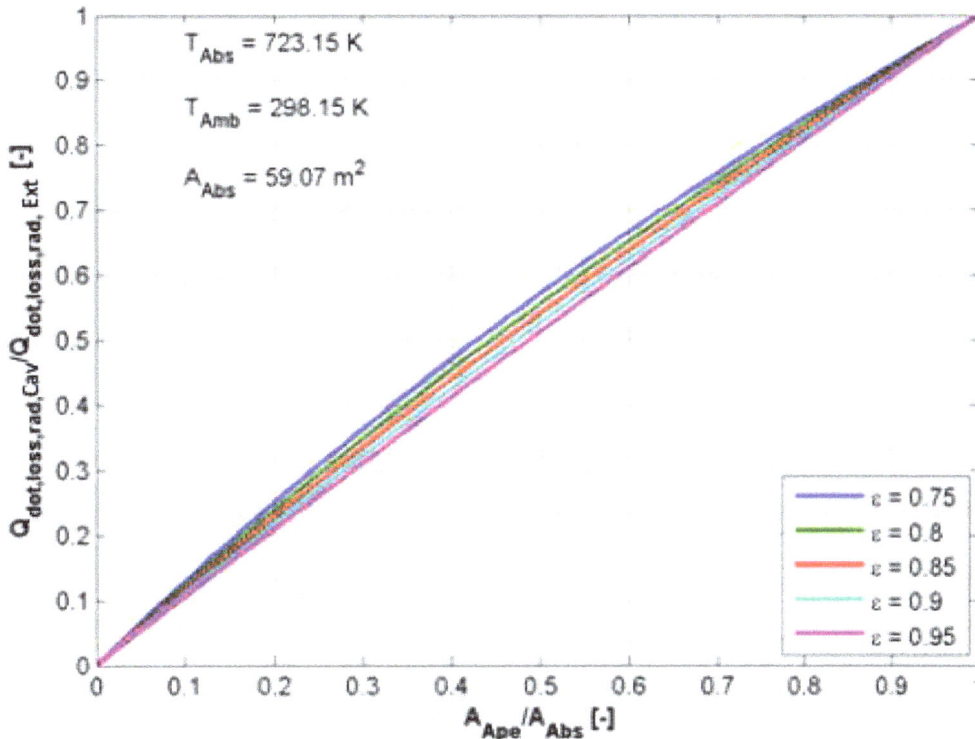

Fig. 3. Ratio of radiative losses for external and cavity type receivers versus the ratio of aperture to absorber area.

Even though the areas of the absorbers in both receivers in our model are equal $(59.07\ m^2)$; however, the radiation loss in both receivers is not the same. As described above in questions (10) and (11) the radiation lossin the cavity receiver depends on the aperture area of the cavity not on the absorbed area as the case of the external receiver. Therefore, the radiation loss associated with the external type receiver is much higher than the radiation loss associated with the cavity type receiver. Figure (3) below illustrates the ratio of radiative losses for both receivers versus the ratio of the aperture to the absorber area.

2.3. Convection Heat Loss

The convective heat loss is the only heat loss in the receiver that has direct relation with the wind conditions. Since there is an existing empirical formula to calculate the heat loss in a turbulent flow over a flat plate as shown in

equations below [15], it appears reason to do the simulation for the receiver as a flat plate with the same hot surface area, thickness, and insulation material and to compare the results with the empirical results. This allows validating the simulation model.

The average laminar Nusselt number $Nu_{1,lam}$ for a flat plate of the length l is calculated by equation (13):

$$Nu_{1,lam} = 0.664\ \sqrt{Re_1}\ \sqrt[3]{Pr} \tag{13}$$

With $Nu_{1,lam} = \frac{hl}{\lambda}$ and $Re_1 = \frac{ul}{v}$

$$Re < 10^5$$

The average turbulent Nusselt number $Nu_{1,turb}$ for a flat plate of the length l is calculated by equation(14):

$$Nu_{1,turb} = (0.037Re_1^{0.8}Pr)/\left(1 + 2.443Re_1^{-0.1}\left(Pr^{-\frac{2}{3}} - 1\right)\right) \tag{14}$$

$$5 \times 10^5 < Re < 10^7$$

$$Nu_{1,0} = \sqrt{(Nu_{1,lam}^2 + Nu_{1,turb}^2)} \tag{15}$$

Where $Nu_{1,0} = \frac{hl}{\lambda}$

$$10^1 < Re < 10^{10}$$

The convection heat transfer coefficient can be calculated by:

$$h = (Nu_{1,0}\lambda/l) \tag{16}$$

Knowing the convection heat transfer coefficient, the convection heat loss can be estimated by equation(17):

$$\dot{Q}_{loss,conv} = A_{Hot}.h.(T_{Hot} - T_{Amb}) \tag{17}$$

3. Physical Model

The physical model of the receivers is shown in the Fig. 4.There are few assumptions made in all simulation cases as following:

1. Hot surface area of all receivers is $(59.07)\ m^2$.
2. Thickness of the flat plate and cavity receivers is $(0.35)\ m$.
3. Insulation material is Rockwool.
4. Surrounding temperature is $(25)\ °C$.
5. Temperature of hot surface is $(450)\ °C$.
6. Height of the receiver is $(70)\ m$ above ground.

$$T_{Amb} = 25\ C°$$

(a)　　　　　(b)　　　　(c)

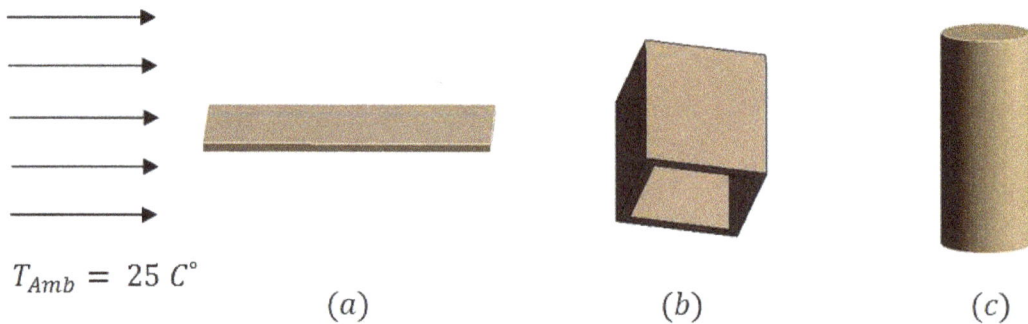

Fig. 4. *Three dimensional models of Central Receiver System.(a) Flat plate receive (b) Cavity receiver with cube geometry. (c) External receiver with cylindrical geometry.*

All receivers are facing the wind with a tilt angle of(90°). The area of the hot surfaces in all receivers is assumed to be equal and the dimensions are as following: in the flat plate (a) is $(16 \times 3.692)\ m$ as length and width respectively. In the cube receiver (b) the length of the inner cube is $(3.3)\ m$, the height is $(3.65)\ m$ and the roof area is $(3.3 \times 3.3)\ m$. In the cylindrical receiver (c) the outer diameter is $(3.3)\ m$ and

the height is$(5.7)\ m$. The imaginary wind tunnel used in the model is external type. Each one of the receivers will be centered in the tunnel in order to apply the boundary conditions such as temperature and wind speed. The cross section of the tunnel is $(50 \times 50)\ m$ with$(50)\ m$ length.

4. Velocity Distribution

Fig. 5(a, b).Show the velocity distributions in cavity and external receivers with the same wind speed of (10) m/s. It should be mentioned that the flow is from left to right. The cavity is supposed to be suspended from a cantilever type receiver. The impact of tower and cantilever on the flow field across the cavity is neglected. It is obvious that the wind speed along the hot surface inside the cavity is lower than in the wind speed around the hot surface of an external receiver. That is because of the air recirculation which reduces the heat transfer coefficient and as a result reduces the convection heat loss.

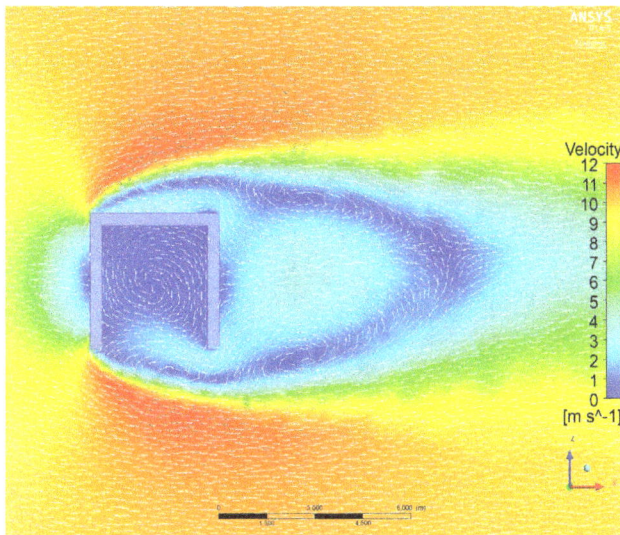

Fig. 5(a). *Velocity distribution around and in the cavity receiver.*

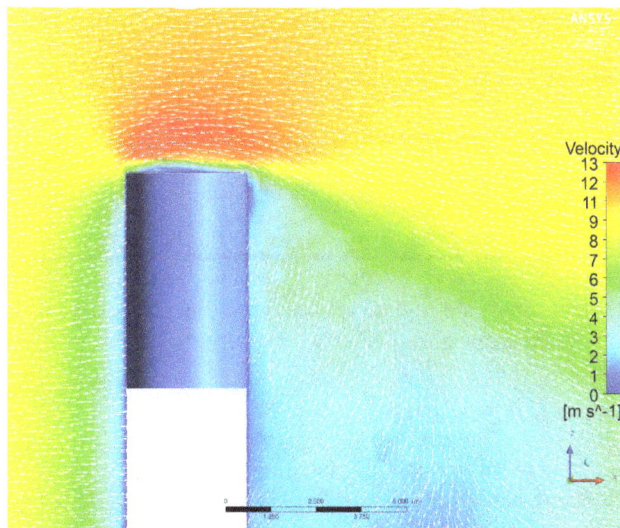

Fig. 5(b). *Velocity distribution around the external receiver.*

5. Calculation of Convective Heat Transfer Coefficient

In the flat plate receiver the convection heat transfer coefficient can be calculated theoretically by the above equations and the result shown in table (2)

Table 2. *Convective heat transfer coefficient with different wind speed.*

Wind speed (m)	Nu	h(W/m² K)	$\dot{Q}_{loss,conv}(MW)$
2	3400.71	5.58	0.1400
3	4649.02	7.62	0.1919
4	5809.09	9.53	0.2392
5	6907.98	11.33	0.2844
6	7960.60	13.06	0.3278
7	8976.33	14.72	0.3696
8	9961.58	16.34	0.4101
9	10920.96	17.91	0.4496
10	11857.97	19.45	0.4882

Figure (6) shows the calculated convective heat transfer coefficient increases approximately in early with increase of wind speed.

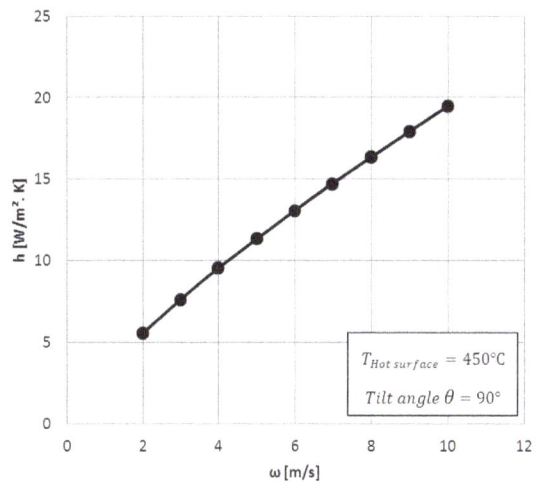

Fig. 6. *The calculated convective heat transfer coefficient of the flat plate receiver verus wind speed.*

In order to examine the accuracy of the model, the result of the analytically calculated convective heat transfer coefficient was compared to the numerical result from the model (FLUENT®) with (450) °C temperature of the hot surface, (90°) title angle and different wind speed varying from (2) to (10) m/s.Based on Figure (7) there is a good agreement between the analytical and numerical results for the convective heat coefficient.

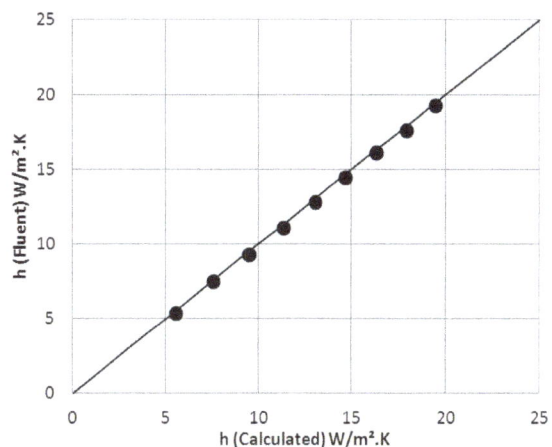

Fig. 7. *Comparison between the analytical and numerical heat transfer coefficient for the Flat Plate receiver.*

To compare the convective heat transfer coefficient values of the cavity receiver (cube geometry) and the external receiver (cylindrical geometry), it is necessary to apply the same boundary conditions on both receivers. Figure (8) shows a comparison between the convective heat transfer coefficients of all receivers with different wind speed.

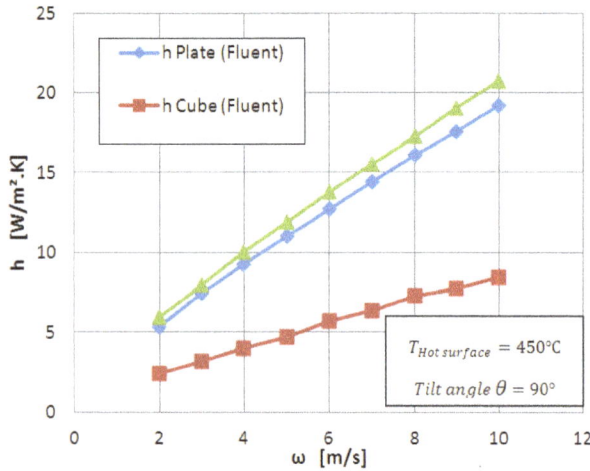

Fig. 8. The convective heat transfer coefficient of three receivers at different wind speed.

6. Results and Discussion

The convective heat loss of all receivers with different

wind speed, fixed tilt angle (90°) and fixed hot surface temperature (450) $°C$ is shown in figure (9). The convective heat lossshows for all types of receivers nearly linear dependence on wind speed. The cavity receiver is suffering the smallest convective heat loss of all the receivers examined.

For the same conditions applied on the cavity and external ("outside") receivers, the radiative heat loss in the cavity receiver reduced by almost (80) % with comparison to the radiative heat loss in the external receiver.

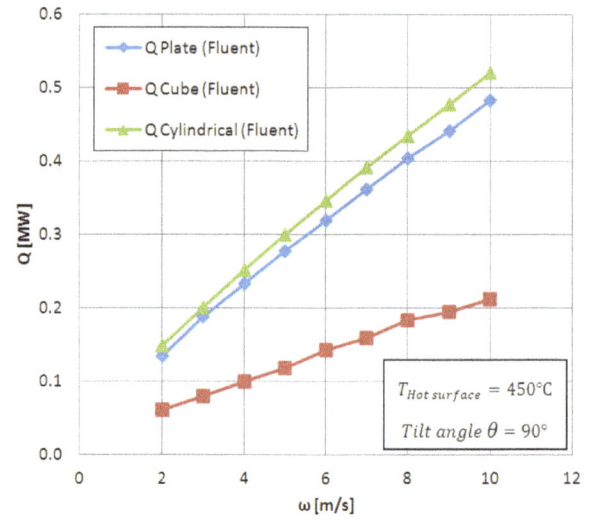

Fig. 9. The convective heat loss of all receivers at different wind speed.

Table 3. Shows the numerical result of the convective heat loss in all receivers.

Wind speed [m/s]	$\dot{Q}_{loss,conv}$ Plate [MW]($Fluent$)	$\dot{Q}_{loss,conv}$ Cube [MW]($Fluent$)	$\dot{Q}_{loss,conv}$ Cylindrical [MW]($Fluent$)
2	0.1341	0.0614	0.1485
3	0.1875	0.0801	0.2002
4	0.2327	0.0998	0.2515
5	0.2767	0.1184	0.2992
6	0.3196	0.1429	0.3455
7	0.3612	0.1589	0.3905
8	0.4044	0.1827	0.4344
9	0.4415	0.1944	0.4775
10	0.4829	0.2123	0.5200

7. Conclusions

The effects of wind speed variation on the convective heat loss of both cavity receiver and external outer surface receiver at a tilt angle ($\theta = 90°$) is studied. The following conclusions can be drawn from study:

First, the convective heat loss is increasing almost linearly with the increase of the wind speed and the highest convective heat loss is obtained at highest wind speed. Second, the convection heat loss associated with using cavity receiver is much lower than the loss associated with using external receivers. For the analyzed geometry the convective heat loss in the cavity receiver was approximately half of the convective heat loss in the external 'outside' receiver.

Furthermore, regardless of the wind speed and the tilt angle, the radiative heat loss in the cavity receiver is almost

(80)% less than the radiative heat loss in the external receiver for the same absorbed area.

Acknowledgements

The author wish to thank the institute for Energy Systems and Thermodynamics at Vienna Technical University(TU). I am also grateful to Professor. Markus Haider for the help provided in guiding the study.

Nomenclature

h : Convective heat transfer coefficient ($W/m2K$)
ω: Wind speed (m/s)
ℓ: Characteristic length (m)
ν : Kinematic viscosity ($m2/s$)

λ : Thermal conductivity $(W/m.K)$

T_{Hot}: Hot surface temperature($°C$)

T_{Amb}: Ambient temperature ($°C$)

Nu: Nusselt number

Re : Reynolds number

Pr: Prandtl number

E : Infrared energy (W)

C: Concentration factor

A_{Hot} : Hot surface area of the absorber (m^2)

A_{Abs}: Receiver's absorbed area (m^2)

A_{Ape}: Receiver's aperture area (m^2)

I: Irradiance flux (W/m^2)

ε: Emissivity

α : Absorptivity

σ: Stefan-Boltzmann constant

$\eta_{optical}$: Optical efficiency of the receiver mirrors

η_{sys}: System efficiency

References

[1] Behar, O., Khellaf, A., and Mohammedi, K.,"A review of studies on central receiver solar thermal power plants". In: Renewable and Sustainable Energy Reviews, Algeria, pp. 12-39, 2013.

[2] An Overview of CSP in Europe, North Africa and the Middle East, CSP Today, October; 2008.

[3] IRENA, 2012, "Concentrating Solar Power. Cost analysis series". In: Renewable Energy Technologies.

[4] Romero, Manuel.,Zarza, E., 2007, "Concentrating solar thermal power. Energy conversion". In: Taylor & Francis Group.

[5] Romero, M., Buck, R., Pacheco, J., "An Update on Solar Central Receiver Systems, Projects, and Technologies". In: Solar Energy Engineering, Madrid, Spain, pp. 98-108, 2002.

[6] Antonio, L., A vila-Marın., "Volumetric receivers in Solar Thermal Power Plants withCentral Receiver System technology: A review". In: Solar Energy, Madrid, Spain, pp. 891-910, 2011.

[7] Shuang-Ying Wu., Lan, Xiao., "Convection heat loss from cavity receiver in parabolic dish solar thermal power system: A review". In: Solar Energy, Chine, pp. 1342-1355, 2010.

[8] Clausing, A.M., "An analysis of convective losses from cavity solar central receivers". In: Solar Energy, USA, pp. 295-300, 1981.

[9] Stine, W.B., McDonald, C.G., 1989, "Cavity receiver convective heat loss". In:Proceedings of International Solar Energy Society Solar World Congress,Japan, pp. 1318–1322.

[10] Leibfried, U., Ortjohann, J., "Convective heat loss from upward anddownward-facingcavity solar receivers: measurements and calculations". In: Solar Energy Engineering, pp. 75–84, 1995.

[11] Sendhil Kumar, N., Reddy, K.S., "Numerical investigation of naturalconvection heat loss in modified cavity receiver for fuzzy focal solar dish concentrator". In: Solar Energy, pp. 846–855, 2007.

[12] Prakash, M., Kedare, S.B., Nayak, J.K., "Investigations on heat losses from a solar cavity receiver". In: Solar Energy, India, pp. 157–170, 2008.

[13] Qiang, Yu.,Zhifeng, Wang., "Simulation and analysis of the central cavity receiver's performanceof solar thermal power tower plant". In: Solar Energy, Chine, pp. 164-174, 2011.

[14] A. Soum-Glaude, I. Bousquet, M. Bichotte, S. Quoizola, L. Thomas, G.Flamant, 2013,"Optical characterization and modeling of coatings intended as high temperature solar selective absorbers". In: Solar PACES 2013, Las Vegas.

[15] Springer, 2010, "VDI Heat Atlas, Second Edition". Heidelberg, Germany.

10

Bio-Ethanol Yield from Selected Lignocellulosic Wastes

Ana Godson R. E. E.[*], Sokan Adeaga Adewale Allen

Department of Environmental Health Sciences, Faculty of Public Health, College of Medicine, University of Ibadan, Ibadan, Nigeria

Email address:

anagrow@yahoo.com (Ana G. R. E. E.), sokanadeaga.adewaleallen@yahoo.com (Sokan A. A. A.)

Abstract: Developing nations are experiencing energy deficit because of overdependence on fossil-based fuels. Countries such as Nigeria have abundant raw materials for biofuels, yet these have not been explored. This study was designed to evaluate the bioethanol production potentials of lignocellulosic-based wastes. The mean glucose yield and TRS obtained from the 13.1M H_2SO_4 were significantly higher than those of 9.4M and 5.6M H_2SO_4 hydrolysis. The mean glucose yield and TRS obtained from the 13.1M H_2SO_4 hydrolysis were: CP (85.1±5.7, 209.8±3.7mg/kg), YP (269.2±11.2, 541.3±7.8 mg/kg), PP (304.0±6.1, 461.2±3.6 mg/kg) and SD (343.2±4.8, 535.9±5.0 mg/kg). The 13.1M hydrolysate was used for the ethanol production and the maximum production was obtained at 48hours of fermentation, the mean ethanol yield being: CP - 160.0±15.1 mL/kg, YP -211.7±15.3 mL/kg, PP - 265.0±20.5 mL/kg and SD - 280.0±11.5 mL/kg. A linear relationship exists between the ethanol yield and fermentation time ($R^2 = 0.711$). Sawdust produced the highest glucose and ethanol yield among the substrates; hence ethanol production from sawdust should be explored and optimized.

Keywords: Bioethanol Production, Glucose Yield, Lignocellulosic Wastes, *Saccharomyces Cerevisiae*, Total Reducing Sugars (TRS)

1. Introduction

The world's energy supply is mainly dependent on non-renewable, crude oil-derived (fossil) liquid fuels, of which almost 90% are employed for energy generation and transportation. The problem of rapidly increasing population has caused many developing countries to expand their Industrial base, resulting in increased energy demands (1). It is inevitable that fossil fuels such as oil, coal and natural gas will be exhausted with time. Hence, there is need to explore the possibilities of using alternative energy source, which are as efficient as oil; ethanol fermentation is one such option (2).

Many industrialized countries are pursuing the development of expanded or new biofuels industries for the transport sector, and there is growing interest in many developing countries for similarly "modernizing" the use of biomass in their countries and developing greater access to clean liquid fuels while helping to address energy costs, energy security and global warming concerns associated with fossil fuels (3). Biofuels are considered as a replacement for fossil fuels and the answer to poverty and even the climate crisis. They are presented as being both renewable and environment friendly (4). Increasing attention is being focused on the production of biofuels as

the alternatives that will contribute to global reduction in greenhouse gas emissions (5).

In Nigeria, the use of biofuels is anticipated to make significant impact on petroleum products quality enhancement in view of the current limitations of the fossil-based fuels which have not kept pace with the increasing demand for environmentally friendly fuel. Furthermore Nigeria recently adopted an ethanol production policy with cassava as its main feedstock, in response to the global initiative (bio-fuel production), which promises a harmonious correlation with sustainable development, efficient and energy conservation. Although fuel ethanol is currently produced from sugarcane and other starch rich grains, ethanol also can be made from cellulosic materials such as wood, grass and agro-residue (6). This would in turn reduce the pressure on food security due to excessive use of food crops for bio-fuel produce and reduce dependence on imported petroleum for vehicle, ensure environmental sustainability, sound public health and create wealth and opportunities.

Ethanol production from cellulose biomass material instead of traditional feedback is known as bio-ethanol: a carbon neutral compound. Bio-ethanol is a fuel derived from renewable resources like locally grown crops and even waste product/waste paper or grass and tree trimmings etc (6). These materials contain lignocelluloses which has cellulose, hemicelluloses and lignin in its compound. The lignocellulosic structure is more resistant to decay by organism and it is not perishable like soluble sugar and starch. The complex substance may be broken down into sugars by either acid treatment at various temperatures or by enzymatic treatment (7).

Alcohol fermentation was done by using the mash of dried sweet potato with its dregs as substrate (8). In another study, cellulosic pyrolysate – containing levo – glucosan was chemically hydrolyzed and a maximum glucose yield of 17.4% was obtained through hydrolysis with 2mol/litre H_2SO_4 at 121^0C for 20minute. The total initial glucose level was maintained at 41.9g/litre by diluting the hydrolysate. The hydrolysate was neutralize with $Ca(OH)_2$ (to bring to about pH 6.0 or 10.4) and, which was completely fermented by *S.cerevisiae* and *Pichia sp.* Yz – 1. A maximum ethanol yield of 0.45/g glucose was obtained by *S. cerevisiae*(9). Another substrate, liquefied cassava starch, was used for ethanol production by co – immobilized cells of *Z. mobilis* and *S. diastaticus*. The co – immobilized cells produce 46.7 g/litre ethanol from 150 g/litre liquefied cassava starch, while the immobilized cells of yeast *S. diastaticus* alone produced 37.5g/litre ethanol. Thus, co-immobilized cells of *S.diastaticus* and *Z.mobilis* produced a high ethanol concentration as compared to the immobilized cells of *S. diastaticus* during batch fermentation of liquefied cassava starch (10).

For direct and efficient ethanol production from cellulosic materials, a novel cellulose – degrading yeast strain was developed by genetically modifying two cellulolytic enzymes on the surface of *S.cerevisiae*. This could grow in a synthetic medium containing glucan as the sole source of carbon and could directly ferment 45g of glucan per litre to produce 16.5g of ethanol per litre within 50 hours. Thus, 0.48g of ethanol was produced per gram of carbohydrate utilized, which corresponded to 93.3% of the theoretical yield. This result indicates that efficient and simultaneous sacharification and fermentation of cellulose to ethanol was carried out by recombinant yeast cells displaying cellulolytic enzymes (11).

Alfenore *et al.,* (2002) described a nutritional strategy that allowed *S.cerevisiae* to produce a final ethanol litre of 19% (V/V) ethanol in 45hours in a fed – batch culture at 30^0C. This performance was achieved by implementing exponential feeding of vitamins throughout the fermentation process. A maximum instantaneous productivity of 9.5g/litre/hour was reached in the best fermentation. These performances resulted from improvements in growth, ethanol production rate, and concentration of viable cells in response to the nutritional strategy (12).

In other studies, (13) introduced new genes into a cyanobacterium in order to create a novel pathway for fixed carbon utilization, which results in the synthesis of ethanol. The coding sequences of the PDC and ADH II from the bacterium *Z. mobilis* were cloned into the shuttle vector pCB4 and were then used to transform the cyanobacterium *Synechococcus* sp strain PCC 7942. The PDC and ADH genes were expressed at high levels, as demonstrated by Western blotting and enzyme activity analyses. The transformed cyanobacterium synthesized ethanol, which diffused from cells into the culture medium. As cyanobacteria have simple growth requirements and use light, CO_2, and inorganic elements efficiently, production of ethanol by cyanobacteria is a potential system for bioconversion of solar energy and CO_2 into a valuable resource. Metabolic engineering of *Z. mobilis* strains was tried to maximize the ethanol production from mixtures of hexose and pentose sugars through the application of metabolic flux control techniques (14).

Currently about 90% of the world ethanol is produced from food substances such sugar cane and other starch grains. This process may lead to global food crisis while achieving energy security. Hitherto, little attention had been paid by researchers and policy makers in energy sector to the viability of lignocellulosic based wastes in ethanol production. Hence in our study we explored bioethanol production from selected Lignocellulosic wastes using *Saccharomyces cerevisiae* as the ethanologenic organisms.

2. Materials and Methods

2.1. Sample Source

The different lignocellulosic wastes utilized in this study were collected from the following sources in Ibadan: The Cassava Peels (CP) was obtained from International Institute of Tropical Agriculture (IITA). The Institute has a Cassava Processing Plant (CPP) where large quantity of Cassava Peels (CP) is generated. Yam Peels (YP) was collected in Abadina Quarters (AQ) of the University of Ibadan (UI). Plantain Peels (PP) was obtained from the Ajose Building Canteen (ABC) which is located within the University College Hospital (UCH) a sub campus of UI. The Sawdust (SD) was obtained from the Bodija Timber Processing Centre (BTPC) in Ibadan.

2.2. Biomass Sampling

Sampling of the Biomass

A representative sample of each biomass was obtained from the parent substrates. From each heap of biomass wastes, a grab sample was collected into a polythene bag ready for physical and biochemical characterization.

EXPERIMENTAL FLOW CHART

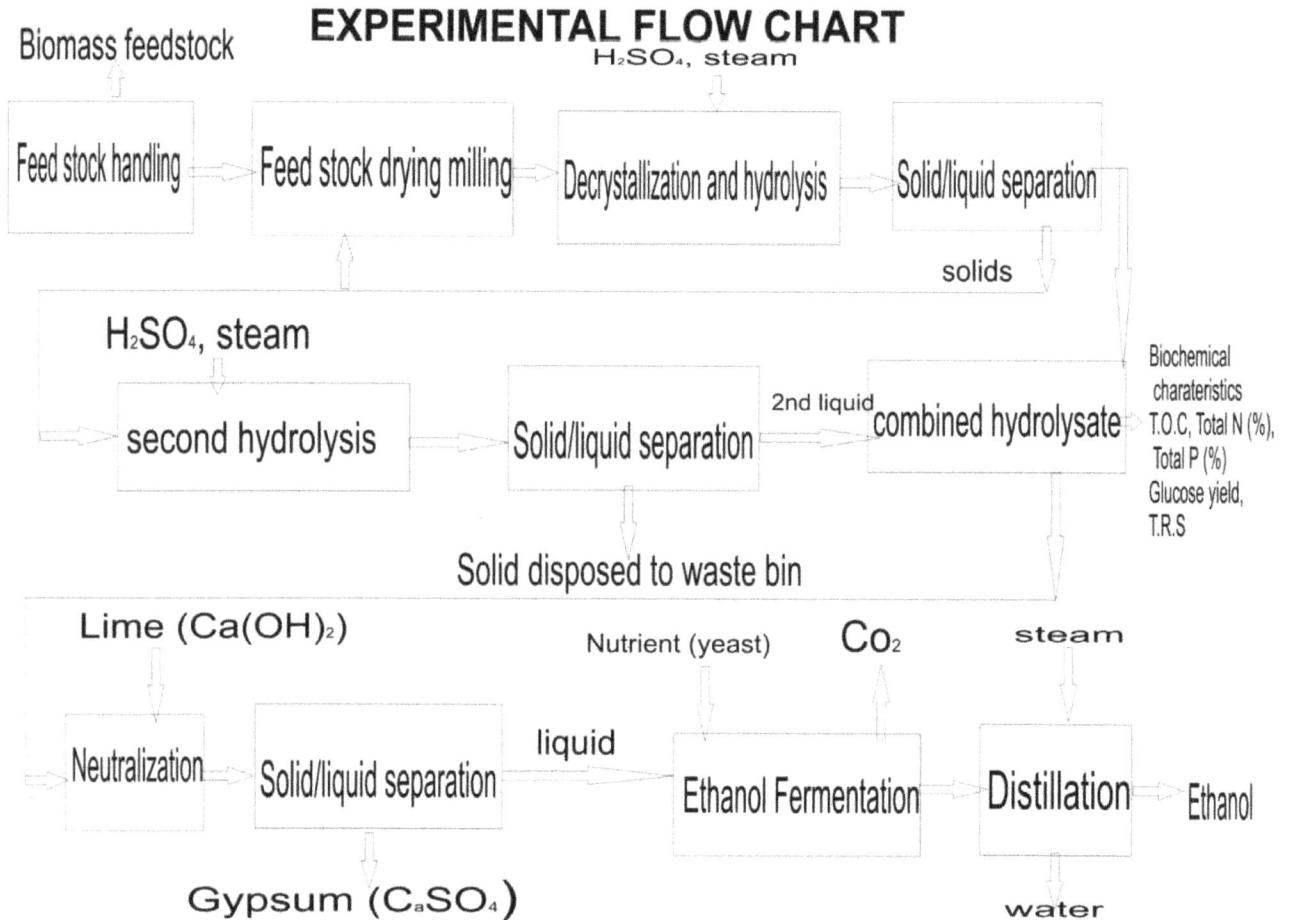

Figure 1. Flow chart illustrating the various stages involve in the bio-ethanol production process (Farone and Cozen, 1996).

2.3. Pre-treatment of Biomass

The various samples were sun-dried for about 3-5 days to reduce the moisture content to about 10%. The dried samples were pulverized to a size of about 15mm. This allowed for a large surface of the substances to facilitate chemical hydrolysis.

2.4. Acid Hydrolysis

Twenty grammes (20g) each of the powdery biomass was hydrolyzed separately with 100ml (1:5w/v) of various concentration of H_2SO_4 of 5.6M, 9.4M and 13.1M in a two stage hydrolysis. In the first hydrolysis, the mixture of acid and biomass was heated to 100^0C for 60 mins to hydrolyze the lignocelluloses. This resulted in the formation of a thick gel, which was pressed on a sieve to obtain an acid-sugar stream. The solids remaining after the first hydrolysis was again hydrolyzed with H_2SO_4 at 100^0C for 50min. The resulting gel was again pressed to obtain a second acid-sugar stream. The stream from the two hydrolysis steps was combined. The mixed hydrolysates were analyzed for glucose and total reducing sugar (TRS) to determine which of the acid hydrolysis gave best yield of glucose and TRS.

Equation of the reaction:

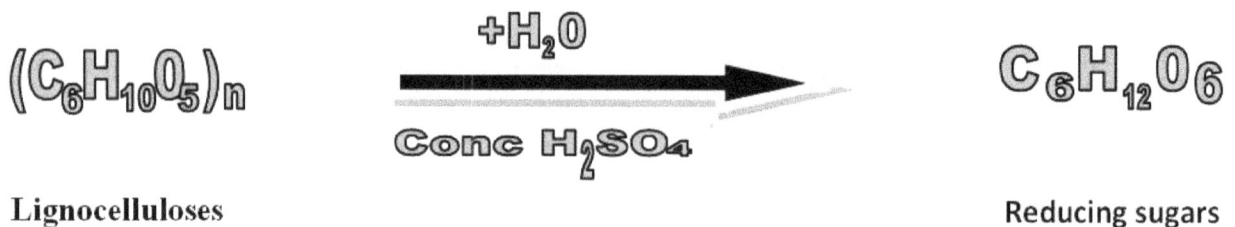

$$(C_6H_{10}O_5)_n \xrightarrow[\text{Conc } H_2SO_4]{+H_2O} C_6H_{12}O_6$$

Lignocelluloses   **Reducing sugars**

The left over solid which is lignin, the most recalcitrant to degradation out of the 3 component of lignocelluloses material (lignin, hemicelluloses and cellulose) was discarded.

2.5. Glucose Yield and TRS Determination

The AOAC method (15) was employed in the determination of glucose yield and TRS. The glucose yield in the hydrolystate was determined by using the ferric cyanide method while the total reducing sugar content was determined quantitatively by using the Phenol-sulphuric acid method as outlined by Dubois *et al*, (1956). The amount of reducing sugar released was colorimetrically determined using UV spectrophotometer at a wavelength of 420 nm. A calibration curve was obtained using D- glucose as standard.

Equation of the reaction

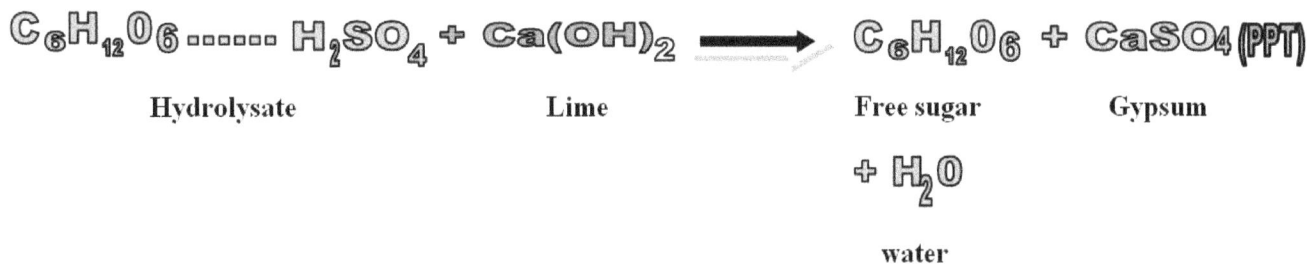

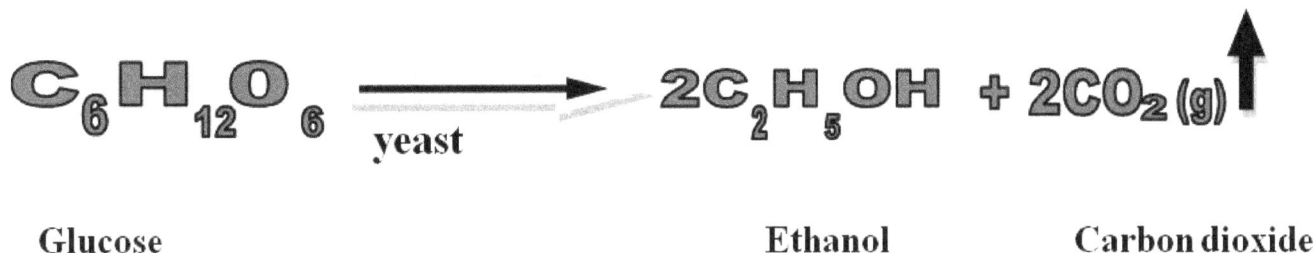

$$C_6H_{12}O_6 \cdots\cdots H_2SO_4 + Ca(OH)_2 \longrightarrow C_6H_{12}O_6 + CaSO_4 (PPT)$$

Hydrolysate   Lime   Free sugar   Gypsum

$$+ H_2O$$

water

2.7. Glucose Fermentation

The 13.1M H_2SO_4 gave the best yield of glucose and TRS hence was fermented with *Saccharomyces cerevisiae* in a 250ml fermenter at 30^0C for 72 hrs to ensure maximum

Equation of the reaction

$$C_6H_{12}O_6 \xrightarrow{\text{yeast}} 2C_2H_5OH + 2CO_2 (g)\uparrow$$

Glucose   Ethanol   Carbon dioxide

2.8. Bio-Ethanol Yield Determination

The ethanol yield (v/w) was determined by using the AOAC methods (15). The ethanol was distilled from the sample and collected in an acid solution of potassium dichromate where it is oxidized by acetic acid at 60^0C. The residue dichromate was determined by back titration with ferrous sulphate in a strong acid solution using feroin indicator (1, 10-phenathroline ferrous sulphate complex).

2.9. Statistical Analysis

All data was summarized using descriptive statistics such as proportions, mean and standard deviation. The result obtain from the biochemical analysis were subjected to one – way ANOVA at 5% level of Significance. A Simple Linear Regression Model was used to indicate the relationship between the ethanol yield of the substrates and the

2.6. Neutralization Process

The sugar-acid stream/hydrolysate obtained from the acid hydrolysis was neutralized by adding lime [Ca(OH)₂], which forms a gypsum precipitate. The CaSO₄ was removed by filtration using a Whatman No1 filter paper and then discarded. The filtrate which is a free sugar stream was tested for the presence of reducing sugar using Fehling solution before subjected to ethanol fermentation.

ethanol production. Samples were taken from the fermenting broths every 24hours to test for the presence of ethanol and ethanol yield determination.

fermentation time.

3. Results

3.1. Levels of Glucose Yield

Table 1 shows the levels of Glucose Yields (mg/kg) of the substrate hydrolysates at different acid concentrations of 5.6M, 9.4M and 13.1M respectively. In each of the acid concentration hydrolysis, the SD hydrolysates gave the highest mean glucose yield followed by PP, then YP and CP hydrolysate being the least. The mean glucose yield of the substrate hydrolysates were significantly different from each other for each of the acid concentration hydrolysis (p<0.05). The mean glucose yield obtained from the 13.1M H2SO4 were significantly higher than those obtained from the 9.4M and 5.6M H2SO4 hydrolysis (p<0.05).

Table 1. *Levels of Glucose Yield from Substrate Hydrolysates at different acid concentrations.*

Sample Description	5.6M H_2SO_4				9.4M H_2SO_4				13.1M H_2SO_4			
	1st	2nd	3rd	Mean ±S.D	1st	2nd	3rd	Mean ± S.D	1st	2nd	3rd	Mean ±S.D
SD Hydrolysate	286.0	280.6	290.5	285.7±5.0	300.0	292.5	309.5	300.7±8.6	343.0	338.5	348.0	343.2±4.8
PP Hydrolysate	257.0	249.5	260.0	255.7±5.4	275.0	273.8	285.5	278.1±6.5	305.0	297.5	309.6	304.0±6.1
YP Hydrolysate	230.0	228.0	235.0	231.0±3.6	240.0	235.0	245.0	240.0±5.0	271.5	257.0	279.0	269.2±11.2
CP Hydrolysate	65.0	41.0	45.5	50.5±12.8	70.0	69.5	75.0	71.5±3.0	84.0	80.0	91.2	85.1±5.7

* The hydrolystate obtained from the 13.1M hydrolysis gave the highest yield or mean value of Glucose Yield throughout the three (3) trials than the 5.6M and 9.4M hydrolysis. Hence it was used for the ethanol production.

3.2. Levels of Total Reducing Sugars

From Table 2, the mean Total Reducing Sugars (TRS) increased as the concentration of the acid increased and viceversa. Among the substrates, YP hydrolysates recorded the highest mean TRS (mg/kg) at different acid concentrations while the least mean TRS was found in CP hydrolysates. The mean TRS obtained from the 13.1M H_2SO_4 were significantly higher than those obtained from the 9.4M and 5.6M H_2SO_4 hydrolysis ($p<0.05$). At 13.1M hydrolysis, mean TRS of PP was significantly higher than those of CP, YP and SD ($p<0.05$). Hence the 13.1M hydrolysate was used for ethanol production, since glucose and reducing sugars serve as a precursor for ethanol production.

Table 2. *Levels of Total Reducing Sugars (TRS) from Substrate Hydrolysates at different acid concentrations.*

Sample Description	5.6M H_2SO_4				9.4M H_2SO_4				13.1M H_2SO_4			
	1st	2nd	3rd	Mean ±S.D	1st	2nd	3rd	Mean ±S.D	1st	2nd	3rd	Mean ±S.D
SD Hydrolysate	375.0	367.0	381.5	374.5±7.3	450.0	448.8	460.8	453.2±6.6	537.5	530.3	540.0	535.9±5.0
PP Hydrolysate	315.0	309.5	319.5	314.7±5.1	395.0	391.3	403.0	396.4±6.0	460.5	458.0	465.0	461.2±3.6
YP Hydrolysate	390.0	381.3	395	388.8±6.9	460.0	455.6	465.0	460.2±4.7	544.5	532.5	547.0	541.3±7.5
CP Hydrolysate	91.5	89.0	95	91.8±3.0	125.0	117.5	127.0	123.2±5.0	209.0	206.5	213.8	209.8±3.7

* The hydrolystate obtained from the 13.1M hydrolysis gave the highest yield or mean value of Total Reducing Sugars (TRS) throughout the three (3) trials than the 5.6M and 9.4M hydrolysis. Hence it was used for the ethanol production.

3.3. Ethanol Yield

Figure 2 - 3 shows the ethanol production of the fermenting broths of the various substances every 24 hours. The mean ethanol yields at 24 hours of fermentation were: CP (123.3 ± 11.1mL/kg), YP (172.0 ± 17.5ml/kg), PP (217.7 ± 13.5 mL/kg) and SD (240.3±14.0mL/kg) ($p<0.05$) respectively. The maximum ethanol production was obtained at 48 hours, the mean ethanol yield being: CP - 160.0±15.1mL/kg, YP – 211.7±15.3mL/kg, PP – 265.0±2.0mL/kg and SD – 280.0±11.5mL/kg. Mean ethanol yield at 48 hours of fermentation were significantly different from those obtained at 24 hours. A simple linear Regression established a linear relationship between the ethanol yield of the substrates and the time of fermentation ($R^2=0.711$) as shown in Figure 4.

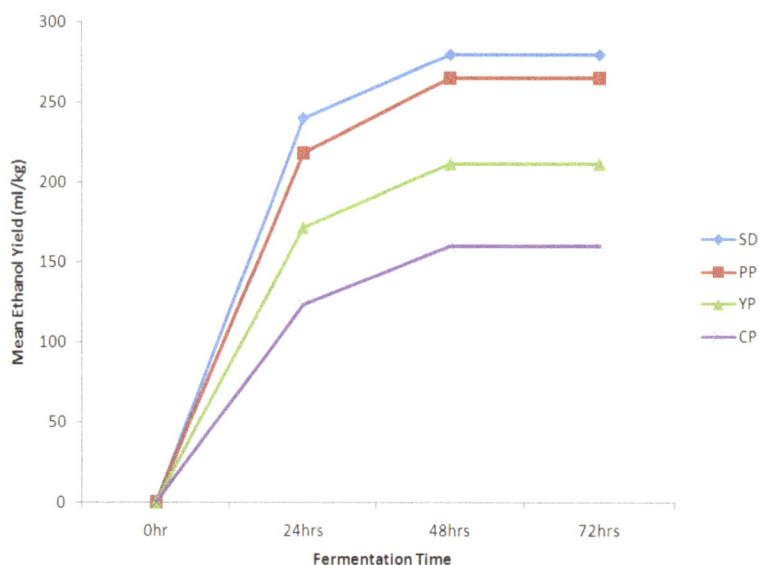

Figure 2. *Mean Values of Ethanol Yield of the Substrates at various Fermentation Time.*

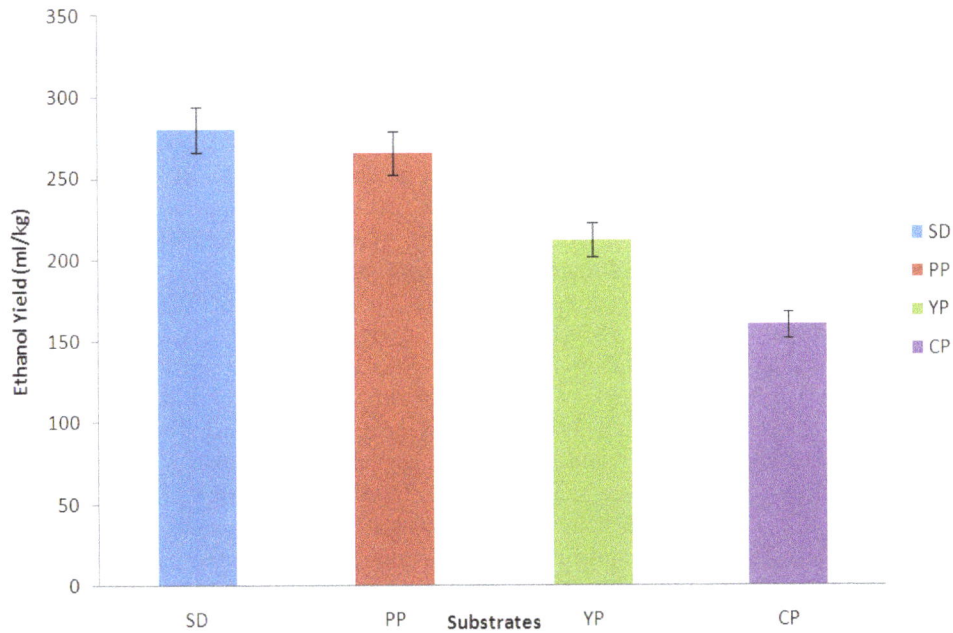

Figure 3. Maximum Ethanol Yield (ml/kg) obtained from the various Fermenting Broths at 48hrs of Fermentation.

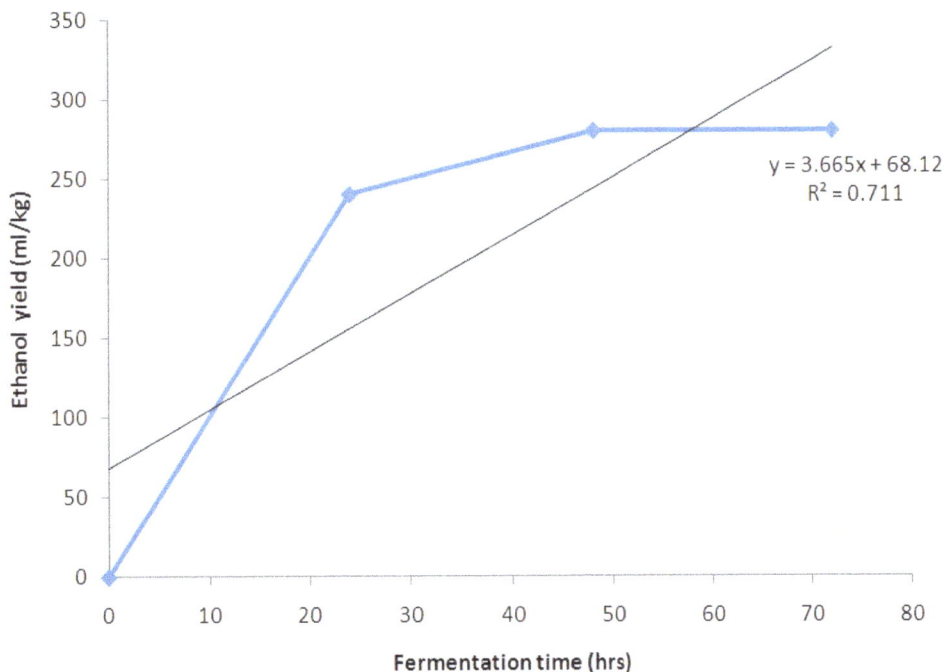

$$y = 3.665x + 68.12$$
$$R^2 = 0.711$$

Figure 4. Shows the Simple Linear Regression curve between the Ethanol Yield and Fermentation Time.

4. Discussion

With increased population growth there is a corresponding demand for energy resources especially for non-renewable forms. This over dependence has result in the depletion of the resource base and gross degradation of the environment. This has led to the search for alternative and renewable energy sources (16). In the present investigation, we explored the production of bio-ethanol from selected Lignocellulosic wastes commonly found within Nigeria's south-western region.

In this study, the optimization of sugar production from cellulose hydrolysis under different acid strengths was assessed. The result showed that hydrolysis at 13.1M (70%) provided the maximum sugar content in the substrate. This agrees with the concentrated acid technology of using 70% conc. H_2SO_4 for sugar production from cellulosic material developed by (17). At over 70% H_2SO_4 concentration, a lot of charming or dehydrating reactions occurred to a varying degree. Similar result at a higher acid concentration was reported by (18) on cassava granted waste (CGW) biomass at 120^0C for 30mins and using a high concentration of H_2SO_4 (1-5M) hydrolysis was achieved but with excessive charring or dehydrating reactions. Other chemical reactions reported in previous studies include the formation of furfural from xylose. Furfural was reported to inhibit activities of some glycolytic enzymes particularly dehydrohygenase in *S.cerevisiae* for ethanol production (19).

The finding of this study revealed that hydrolysis at 13.1M H_2SO_4 gave the best glucose and TRS yield for all the substrates when steam at 100°C for 60mins and 50mins respectively. Jeffries and Lee (1999) also reported auto-hydrolysis (steam explosion) as an effective pretreatment method for lignocelluloses materials for hydrolysis (20). In fact, (21) reported an increasing glucose concentration in hydrolysate as the severity of steam explosion increases.

Among the substrates, the highest glucose yield was obtained from sawdust. The high amount of glucose yield in sawdust is due to the lignocelluloses content of hard and softwood stem as reported by (22 – 23) from which it is produced. Ojumu *et al.*, (2003) also reported that sawdust obtained from the tree *Triplochiton scleroxylon* contained 69.5 – 80% cellulose and hemicelluloses and 25 – 30% lignin. The high cellulose content of SD is responsible for its high mean glucose yield; since cellulose is a homogenous polymer of glucose (24). Badmus (2002) also produced glucose from palm tree trunk using auto hydrolysis prior to acid hydrolysis (25). The lowest amount of glucose yield and TRS found in CP can be attributed to its containing cellulose and hemicelluloses at levels of 24.99% and 6.67% (w/w) respectively as reported by (26). This agrees with previous study done by (26) who reported that the maximum reducing sugar of 6.09% (w/v) was recovered from cassava waste after pretreatment with 0.6M H_2SO_4 at 120°C for 30mins. At concentration of H_2SO_4 higher than 0.6M, the reducing sugar was lower than 6.09%.

The high mean TRS found in YP, SD and PP may be attributed to the high amount of hemicelluloses content. Hemicelluloses are macromolecules often polymers of pentoses (xylose and arabinose), hexoses (mostly mannose) and a number of sugar acids. Hemicelluloses are particularly industrial interest since they are readily available bulk source of xylose from which xylitol and furfural can be derived (27 - 28).

Ethanol produced from cellulosic biomass materials instead of traditional feedstock is known as bioethanol: a carbon-neutral compound. The traditional process of ethanol production is through fermentation of sugars with a species of yeast called *Saccharomyces cerevisiae*. However, the changing needs, energy demands, and technological advances to overcome the general limitations in yeast-based ethanologenic fermentations have led to an exploration of different methods using a broad range of substrates and novel organisms, indigenous or genetically modified. New technologies are being developed that convert the fibrous portion of plant material to bioethanol. These feedstock materials are abundant and inexpensive (29).

In the present study, the Simultaneous Saccharification and Co-Fermentation (SSCF) was employed which involves the fermentation of both six-carbon hexoses (glucose, mannose,

and galactose) and five–carbon pentoses (xylose and arabinose) sugars to ethanol. This is in line with several authors who reported that the SSCF is superior to the Simultaneous Saccharification and Fermentation (SSF) technology in terms of cost effectiveness, better yields, and shorter processing time (6, 30). A complete conversion of glucose and xylose mixture was obtained by a respiratory deficient mutant of *S.diastaticus* co-cultivated with *Pichia stipilis* in continuous culture (31).

Of all the substrates, SD gave the highest ethanol yield. This may be attributed to its high glucose yield and TRS, since glucose is a precursor for ethanol production. According to (32), the total sugar content is important for the ethanol yield; a key economic parameter depending upon the sugar content. The maximum ethanol production was obtained at 48hours of fermentation for all the substrates and after which the level remained constant. This outcome corroborates previous findings in which different substrates were used to assess the efficiency of the strain *klebsiella oxytoka* viz., mixed office paper (33 – 35) and sugar beet pulp (36). The best strains of the transformants converted 10% glucose and 10% cellobiose into 44-45g/litre of ethanol within 48 hours. Integrating cellulose components like extracellular endoglucanase can reduce the ethanol production costs (37). When a comparative study was done, in which galacturonic acid-rich sugar beet pulp was fermented, K011 produced significantly higher quantities of ethanol production due to *E.coli* K011 affinity for the substrate. Dien *et al.*, (1998) developed a novel hexose and pentose utilizing the ethanologenic *E.coli* strain FBR3 by incorporating the plasmid pL01297. An ethanol yield of 4.38% - 4.66% (w/v) with 90-91% theoretical conversion in 70-80 hours was achieved (38).

Mixing has an important role in fermentation. The influence of mixing (from 100-110 rpm [revolutions per minute]) on the performance of *Z.mobilis* anaerobic continuous culture was studied. It was found that the biomass yield and ethanol productivity were improved at higher stirring intensities along with a decrease in the by-product formation. Vigorous mixing led to a better coupling between catabolism and anabolism (39). In another study (40) used immobilized *S. cerevisiae* cells and found that the maximum fermentation capacity of the system was at 30°C and was relatively pH – sensitive. A packed column reactor was used to test this biocatalyst's operational sensitivity to key fermentation variables. Results of this study as well as characteristics of the polymer, prepared by an epoxy resin and di-amino polyethylene oxide polymerization establish the suitability of this method for ethanol production.

Although several microorganisms, including Clostridium sp., have been considered as ethanologenic microbes, the yeast *S. cerevisiae* and the facultative bacteria *Z. mobilis* are better candidates for the industrial alcohol production (41). The feedstock typically account for more than one third of the production costs, thus maximizing the ethanol yield is imperative. A high ethanol yield means using strains of bacteria that can produce fewer side products and metabolize all major sugars, which typically include glucose, xylose, arabinose, and mannose (42).

5. Conclusions

The purpose of this study was to determine the ethanol yielding capacity of some selected lignocellulosic based-wastes. The results show that sawdust produced the highest glucose and ethanol yield among the substrates. Bioconversion offers a cheap and safe method of not only disposing the agricultural residues, but also it has the potential to convert lignocellulosic waste into usable forms such as reducing sugars that could be used for ethanol production. Hence the conversion of lignocellulosic "wastes" into biofuels such as ethanol will help reduce environmental pollution, contribute toward the mitigation of greenhouse gases emissions and serve as a sustainable solid waste management strategy.

Acknowledgements

The technical input of Mr. Yomi of the Institute of Agricultural Research and Training (IART), Ibadan and Drs. Bolaji of the Department of Environmental Health Sciences and Adepoju of the Department of Human Nutrition, University of Ibadan is highly appreciated.

References

[1]   Tripetchkul S, Hillary ZD, and Ishizaki A. (1998) strategies for improving ethanol Production using *Z. Mobilis*. Recent Research Developments in Agricultural and Biological Chemistry 2:41-55.

[2]   Nomura M, Bin T, and Nakao S. I, (2002). Selective ethanol extraction from fermentation broth using a silicate membrane. Separation purifying Technology. 27:59-66.

[3]   Green, Harvey (2006) Wood: Craft, Culture, History Penguin Books, New York, Page 403. ISB 978-1-101-2-0185-5.

[4]   Bassey, N. (2010). *Oil Politics: Nigeria's Unacceptable Biofuels Policy*. Retrieved from http://234next.com/csp/cms/sites/Next/Money/5643461-183/story.csp, on August 15, 2011.

[5]   Oniemola, P.K. & Sanusi, G. (2009). *The Nigerian Bio-Fuel Policy and Incentives (2007): A Need to Follow the Brazilian Pathway*. International Association for Energy Economics.

[6]   Lynd, Lee R. Cushman, Janet, H., Nicholas, Roberta J., Wyman, Charles E. (2003) Fuel Ethanol from cellulosic Biomass: Science, New Series, Vol 251 No. 4999, pp1318-1323.

[7]   Shilo H, and Neimo, L. (1975). The Structure and properties of Cellulose. In Proc. Symp on Enzymatic Hydrolysis of Cellulose, Aulanka, Finland, (ed. M. Baily, T. M Enaria and M, Linko), pp 9-21 SITRA.

[8]   Yu B, Zhang F, Zheng Y, Wang P. (1994). Alcohol fermentation from the mash dried sweet potato with its dregs using co-immobilized yeast. Process Biochemistry 31 (1): 1-6.

[9]   Yu B, Zhang H. (2002). Pretreatment of cellulose pyrosylate for ethanol production by *Saccharomyces cerevisiae, Pischia sp.* YZ-1 and *Zymomonas mobilis*. Biomass and Bioenergy 24: 257 – 267.

[10] Amutha R and Gunasekaran P. (2001). Production of ethanol from liquefied cassava starch using co- immobilized cells of *Zymomonas mobilis* and *Saccharomyces diastaticus*. Journal of Biosciences and Bioengineering 92(6): 560-564.

[11] Fujita Y, Takahashi S, Ueda M, Tanaka A, Okada H, Morikawa Y, Kaqaguchi T, Arai M, Fukuda H, Kondo A. (2002). Direct and efficient production of ethanol from cellulosic material with a yeast strain displaying cellulolytic enzymes. Applied and environmental microbiology 68 (10): 5136 – 5141.

[12] Alfenore S, Jouve CM, Guillout SE, Uribelarrea JL, Goma G, and Benbalis L. (2002). Improving ethanol production and viability of *Saccharomyces cerevisiae* by a vitamin feeding strategy during feed-batch process. In Applied Microbiology Biotechnology 60: 67 – 72.

[13] Deng M D and Coleman J R, (1999). Ethanol synthesis by genetic engineering in *Cyanobacteria*. Applied and Environmental Microbiology. 65(2): 523 – 528.

[14] Jirku V. (1999). Whole cell immobilization as a means of enhancing ethanol tolerance. Journal of Industrial Microbiology and Biotechnology 22:147 – 151.

[15] Association of Official Analytical Chemists (A.O.A.C, 1984, 1990, 1998). Official Methods of Analysis". Association of Official Analytical Chemists (1984) Official Methods of Analysis, 14.023. A.O.A.C.

[16] Agarwal A.K (2005) Biofuels: In wealth from waste; trends and technologies. (ed Lal and Reedy), 2nd Edition. The Energy and Resources Instituted (TERI) Press. ISBNSI-7993-067-X.

[17] Farone W. A and Cuzens J. E. (1996a). Method of Producing sugars using strong acid hydrolysis of cellulosic and hemicellulosic materials. (US Patent No. 5562777) USA: Arkenol, Inc.

[18] Agu, R. C., Amadife, A.E., Ude, C. M., Onya, A., Ogu, E. O., Okafor, M. and Zejiofor,E (1997) "Combined heat treatment and acid hydrolysis of cassava grate waste (CGW). Biomass for ethanol production" waste management. 17(1), 91-96.

[19] Banerjee, N., Bhatnagar, R. and Viswanathan, L. (1981). "Inhibition of glycolysis by furfural in Saccharomyces cerevisiae". European Journal of Applied Microbiology and Biotechnology. 11:226- 228.

[20] Jeffries, T.W. and Y.Y.Lee, (1999). Feedstocks new supplies and Processing. Applied Biochem. Biotechnol. 34:77-79.

[21] Bousssaid, A., J. Robinson, Y. Cai, D.J Gregg and J.N. Saddler, (1999). Fermentability of the Hemicelluloses – derived sugars from steam – exploded softwood (Douglas fir). Biotechnol. Bioeng., 64: 284-289.

[22] Betts WB, Dart R.K, Ball A.S. Pedlar S. L. (1991). Biosynthesis and Structure of Lignocelluloses and Synthetic Materials, Springer-Verlag, Berlin, Germany, pp. 139-155.

[23] Sun Y, Cheng J. (2002). Hydrolysis of Lignocellulosic Material from Ethanol Production: A review Biores. Technol. 83:1-11.

[24] Ojumu, T.V. B.E. Attah – Daniel. E. Betiku and B.O. Solomon, (2003). Auto – hydrolysisof lignocellulosics under extremely low sulphuric acid and high temperature conditionsin batch reactor. Biotechnol. Bioprocess Eng., 8:291-293.

[25] Badmus, M.A.O., (2002). Auto hydrolysis Production of glucose from palm tree trunk. Nig. J.Ind. Syst. Stud., 1: 1-4.

[26] Teerapatr S, Lerdluk K and La-aied S (2006). Approach of cassava Waste Pre-treatments for Fuel Ethanol production in Thailand. Biotechnology Department, Thailand Institute of Scientific and Technological Research (TISTR), 35 19003, Techno polis, Klong 5,Klong Luang, Pathumthani 12120, Thailand.

[27] Roberto J. C, Mussato S, J., Rodriguez RCLB (2003) Dilute-acid hydrolysis for optimization of xylose recovery from rice straw in a semi-pilot reactor. Indust. Crops Prod 17:171-176.

[28] Parajo JC. Dominquez HD, Dominquez JM (1998). Biotechnological production of xylitol. Part 1: interest of xylitol and fundamentals of biosynthesis. Biores, Technol 65:191-201.

[29] Licht, F.O (2006). "World ethanol markets: The outlook to 2015." Tunbridge Wells Agra Europe special report UK.

[30] Chahal DS (1992). Bioconversion of polysaccharides of lignocelluloses and simultaneous degradation of lignin. In Kennedy et al. (eds) Lignocellulosics: Science, Technology, Development and Use. Ellis Horwood Limited, England, pp, 83 – 93.

[31] Delgenes J.P, Laplace J.M, Moletta R, Navarro J.M, Moletta R, Navarro J.M. (1996). Comparative study of separated fermentations and co-fermentation process to produce ethanol from hard wood derieved hydrolysate. Biomass and Bioenergy 11 (4): 353 – 360.

[32] Kadam K. L, Forest L H., Jacobson W. A., (2000). Rice Straw as a lignocellulosic resource collection, processing, transportation and environmental aspects. Biomass, Bioenergy, 2000. 8:369-389.

[33] Brooks T. A. and Ingram L. O. (1995) Conversion of mixed waste office paper to ethanol by genetically engineered *Kebsilla Oxytoka* strain P2. Biotechnology Progress 11:619-625.

[34] Doran JB, Aldrich HC, and Ingram LO. (1994). Saccharification and fermentation of Sugarcane bagasse by *Klebsiella oxytoca* P2 containing chromosomally integrated genes encoding the *Zymomonas mobilis* ethanol pathway. Biotechnology and Bioengineering 44:240-247.

[35] Moniruzzaman M, Dien BS, Ferrer B, Hespell RB, Dale BE, Ingram LO, Bothast RJ. (1996). Ethanol production from AFEX pretreated corn fiber by recombinant bacteria. Biotechnology Letters 18:985 – 990.

[36] Doran J. B., (ripe), Sutton M, Foster B. (2000). Fermentations of pectin-rich biomass withrecombinant bacteria to produce fuel ethanol. Applied Biochemistry and Biotechnology. 84-86:141-152.

[37] Dien B. S, Cotta M. A., and Jefferies T. W. (2003). Bacteria Engineered for Fuel Ethanol Production: Current Status: Applied Microbiology and Biotechnology 63(3):258-266.

[38] Dien BS, Hopspell RB, Wyckoff HA, Bothast RJ. (1998). Fermentation of hexose and pentose sugars using a novel ethanologenic *Escherichia coli* strain. Enzyme and Microbial Technology 23:366-371.

[39] Toma M, Kalnenieks U, Berzins A, Vigants A, Rikmains M, Viesturs U.(2002). Theeffect of mixing on glucose fermentation by Z.mobilis continuous culture Process Biochemistry:1-4.

[40] Jirku V. (1999). Whole cell immobilization as a means of enhancing ethanol tolerance. Journal of Industrial Microbiology and Biotechnology 22: 147 – 151.

[41] Bothast RJ, Nichols NN, and Dien BS. (1999). Fermentations with new recombinant organisms. Biotechnology Progress 15:867-875.

[42] Wiselogel A, Tyson S, and Johnson D. (1996). Biomass feedstock resources and composition. Handbook on Bioethanol: production and utilization, edited by CE Wyman [Applied Energy Technology Series], pp.105 – 118.

11

Evaluating Green Public Procurement Practices

Patrick Boampong-Ohemeng[1], Simonov Kusi-Sarpong[2, *], Adam Sandow Saani[3], Martin Agyemang[4]

[1]Graduate School, Ghana Technology University College/Coventry University, Kumasi, Ghana
[2]School of Management Science and Engineering, Dalian University of Technology, Dalian, PR of China
[3]Procurement Department, Tamale College of Education, Tamale, Ghana
[4]School of Business Management, Dalian University of Technology, Dalian, PR of China

Email address:
pajuxl@yahoo.com (P. Boampong-Ohemeng), simonov2002@yahoo.com (S. Kusi-Sarpong),
saaniadamsandow@yahoo.com (A. S. Saani), martinon463@yahoo.com (M. Agyemang)

Abstract: The paper introduces and develops a green public procurement practices (GPPP) analytical framework for the Ghanaian Public Sector involving six major practices including Acquisition and Material Specification Planning, Environmental Requirements, Green Purchasing, Strategic Supplier Partnership, Green Information & Communication Technology, and Employee Training. The study involved five Polytechnics in Ghana selected using random sampling and ten public procurement professionals from the five selected Polytechnics using a hybrid of convenient and purposeful sampling techniques. The data for the study was obtained using questionnaires and interviews techniques. A grey-based DEMATEL technique, a multi-criteria decision-making (MCDM) tool,was utilized to help identify the most influential major practice and important sub-practice in terms of the overall goal of achieving green economy. The MCDM tool revealed that 'Strategic Supplier Partnership' is the most influential major practice whilst 'Review of Material needs to include Green Procurement Requirements' was identified as the most important sub-practice. For the first time, a MCDM tool is utilized to identify the most influential and important GPPP for implementation in Ghanaian public sector. Finally, the paper will allow public sector procurement professionals to make thoughtful decisions on products and service purchases with a focus on environmental and societal consequences to achieve sustainable development.

Keywords: Green Public Procurement Practices, Grey-Based DEMATEL, Polytechnic Institutions, Ghana

1. Introduction

In the recent decades, green procurement activities have taken the central part of global warming discussion [1], [2], [3]. Green Procurement is defined as the approach to procurement in which decisions on environmental impacts play crucial roles in purchasing decisions [4]. On the other hand, green procurement can also be described as the selection of products or services which minimize environmental waste [5]. The green procurement discussion have centered more on environmental effects of what are been procured [5], [6], [11]. Most organizations require assessment of products or services to ascertain the environmental consequences of the entire products lifecycle which include raw materials, manufacturing, transporting, handling and storing, their use and disposal [5].

A number of aspects of the traditional procurement process may be adjusted to ensure green procurement is implemented [4],[5] The procedures include the identification and setting of environmentally friendly requirements, tendering, selection of contractors and purchasing products or services[4],[6]. Unlike the traditional procurement where departments and employees were required to submit purchase orders as well as requests on papers, green procurement will require them on electronic media to avoid or minimize paper usage. This suggests that company-wide environmental resource planning system may be needed for communication or information flow [4], [7], [8], [12], [13].

The environmental issue has gradually become a part of the corporate tradition [9]. The polytechnics are becoming more dynamic and innovative so new courses and projects are being introduced. New technologies are coming up and competing among themselves. They need to adapt to the modern trend of technological changes, hence, have to deal with repeated processes of procuring new equipment and materials for their operations. Zhu et al (2007) [10] pointed out that procurement is frequent and that optimizing the procurement decisions to reduce relative cost is a critical cost strategy and development. It is therefore imperative for Polytechnic Institutions to identify the most effective green procurement practices for their operations. Effective green procurement practices will enhance the environmental performance of the Polytechnic Institutionsand contribute to their sustainable development.

The objective of this paper is to introduce and develop a green public procurement practices (GPPP) analytical framework for the Ghanaian Public Sector. The paper adopts a combined literature review and brainstorming among public procurement professionals to identify a potential GPPP framework. After identifying the potential GPPP framework, the framework will be subjected to initial review by some selected public procurement professionals to achieve a framework. The framework will furtherbe applied to the ten polytechnics in Ghana using the grey-based DEMATEL technique, a multi-criteria decision-making (MCDM) tool.

The rest of the paper is organized as follows. Section 2 presents the literature review. Section 3 discusses our research methodology. Section 4 elaborates on the data presentation and discussions. General conclusions and recommendations for future research are finally presented in section 5.

2. Literature Review

In order to understand the concept and identify the green public procurement implementation practices, it was necessary to explore and combine the literature under environmental management and public procurement. The literature reviewed depicted that, there was significant literature on Green Public Procurement (GPP) Practices in certain parts of Asia, Europe, and America; yet, there were significant numbers of nations that do not practice green procurement with Ghana not an exception.

The literature review revealed six (6) major green public procurement implementation practices and twenty three (23) sub-practices. The six major implementation practices adopted for this study are; Acquisition and Material Specification Planning (AMSP), Environmental Requirements(ER), Green Purchasing (GP), Strategic Supplier Partnership (SSP), Green Information Communication Technology (GICT), and Employee Training (ET) with the sub-practices put under their respective major practices. Table 2.1 below shows the Green Public Procurement Implementation framework.

We now overview the green public procurement practices framework using literature.

2.1. Acquisition and Material Specification Planning (AMSP)

Generally for companies to procure materials or

productsrequire a planning stage where the procurement professionals decide on the necessity and specification of the product to purchase. This is also described as the product assessment stage. The professionals assess the material to ascertain the environmental impact of the product [4], [14]. Unlike the traditional indiscriminate purchasing habit of some organizations, at the AMSP stage of green procurement, the professionals establish the need for the product. They also determine the material specifications-which are the required characteristics thus, what should be included in the product including the content, energy efficiency [6], [8], [11], [17], [18]; recyclability, packaging [11], [17], [19], [20], [21]; delivery [18], [20], disposability[16], [18], [21], [31], [37].

That is, a thorough study of the environmental impact of the product to be purchased is undertaken [4], [15]. The professionals would like to assess the entire life cycle of the product ranging from theraw material extraction, manufacturing, packaging, transport, storage and utilization and disposal compared with the other similar available products[14],[15]. However, some organizations havedevelopedprocurement policy that include green requirement [17]. The green policy usually include purchase of less packaging products, products which use bio-materials, facilities which bear environmental friendly certifications or suppliers who are committed to environmental stewardship[4], [37]. The objective of the green policy is to enable the organizations to create a green procurement guide which provide green information [14], [15]. Such policies enable procurement experts to make required green decisions on products to be purchased [14]. [15], [17], [19], [21], [37].

One critical factor which can evolve inappropriate criteria and specification is lack of training on green acquisition and material specification planning of procurement [17], [18]. In a study of 25 member states, the report of [18] revealed that only two: Sweden and Germany added clear green specifications in just over 60%. Majority of the 25 states perceived that they were practicing green procurement but in reality they were not, due to inappropriate specifications. It should be noted that inappropriate and unclear specification can affect characteristics of the products acquired [18].

To buttress it, in 2007, the government of Japan took the initiative to promulgate a law that promoted the reduction of GHG in Japan state [20].

It is normal for organizations to assess their performance in regular periods and revise their operations to meet the current trend of technology and innovations. This may necessitate the procurement of new materials or new equipment [18], [20], [31], [37].

2.2. Environmental Requirements(ER)

The whole concept of green procurement is focused on protecting the environment from further pollution and if possible re-habilitates or improves upon it to make the earth healthy to live in. It is therefore appropriate for organizations to incorporate environmental requirements in their procurement criteria [19], [20].

Many institutions and organizations have adopted various

initiatives to promote the green environment. One of such initiatives is the use of mass transport [21].

Meanwhile studies have shown that certain organizations incorporate environmental management system in their operations to effectively coordinate and manage the entire environmental requirement [11],[15],[17],[18]]37].

A study undertaken by [15] reported that European Union (EU) public authority has implemented EMS in their green procurement activities. EMS follows a management cycle that requires coherency between executive priorities, procedures and reporting mechanisms in the system [15], [18], [31],[37]. It should be noted that without coherency in EMS,green procurement implementation is difficult to practice. Studies reveal that, certain Japanese companies practice green procurement by successfully integrating EMS at all levels in the company. Such companies include Ikea, Fujitsu and Japanese Travel Bereau[20],[21].

2.3. Green Purchasing (GP)

Green purchasing can be defined as all purchasing decisions and award of contracts that are based on environmental friendly factors as well as other factors such as price and quality [19]. However Green purchasing can also be defined as the environmental desirable purchasing such as eco-friendly products [11]. It involves the decision to buy products and services that are free from environmental toxins and hazards [11], [21], [35].

Green purchasing does not only contribute positively to the environmental protection, but it also creates avenue for

awareness [15], [3], [11], [32], [33]. Thus the environmental protection does not lie in the care of environmental protection agencies only, but it also lies in the care of procurement managers [3],[11[35].It suggests that green procurement is an essential concept that should be embraced by all organizations and institutions in order to sustain the environment and maintain healthy life [11], [35],[31],[37], [43].

The European Union(EU) and the International Standard Organization(ISO) are some of the institutions that ensure manufacturing companies meet certain international green requirements[13],[22], [31],[37], [43].

One easiest way for procurement experts to identify green products is the eco-labeling [13], [19],[22],[31] [34],[37], [43]. Products labeled "ISO"," Blue Angel", "Austrian Tree", "Nordic Swam" and "EU flower" are a few of the famous reliable eco-labels[13],[31],[37]. Studies have shown that eco-labeled products are more environmental friendly compared with the similar or alternative options [13]. [23], [33].

There is an EU tool which enables organizations to perform proper green management audit [13], [19], [24].

Studies have shown that in 1993, a number of recycling organizations evolved in USA which aimed at improving waste management. Such organizations included Industrial Materials Exchange (IMEX), Gifts In Kind America, California Integrated Waste Management Board and California Material Exchange (CALMAX) among others [12], [37], [43].

Table 2.1. Green Public Procurement Practices (GPPP) Implementation Framework.

Pos	Major Practices	Sub-Practices	Literature
1	Acquisition and Material Specification Planning(AMSP)	Energy efficient and power standby devices requirement (AMSP1)	USA DoD1 (2004);Wang and Wu(2004);ECFESOPEC2(2001);UNSD3(2012); OOFEC4(2013)
		Recyclable products and materials requirements(AMSP2)	UNSD(2012);US EPARV5(1999);OOFEC(2013); USA DoD(2004)
		Meeting green procurement requirements prior to contracting(AMSP3)	UNSD(2012);Plas et al.(2000);Wang and Wu(2004);USEPARV(1999)
		Review of material needs to include green procurement requirement(AMSP4)	USADoD(2004);ECFEOSPEC(2001);UNSD(2012);OOFEC(2013)
2	Environmental Requirements(ER)	Environment Management System(ER1)	US EPARV(1999);OOFEC(2013);USA DoD(2004)
		Bio-based products(ER2)	USA DoD(2004);ECFESOPEC(2001); UNSD(2012)
		Environmental friendly and safe disposal products(ER3)	ECFESOPEC(2001);Plas et al(2000);Mine & Galle(1997)
3	Green Purchasing(GP)	Purchasing of eco-labelled products(GP1)	USA DoD(2004); ECFESOPEC(2001); UNSD(2012);Chien& and Shih(2007); OOFPEC(2013);Panasonic6(2014)
		Purchasing of recyclable products(GP2)	Ho et al(2010); Panasonic (2014); USA DoD(2004)
		Green supplier selection using environmental information(GP3)	Chien& Shih(2007); OOFPEC(2013); USA DoD(2004)
		Green supplier performance assessment(GP4)	US EPAR V(1999); Min & Galle(1997)

---

1 US Department of Defense

2 European Consultative Forum on Environment, Sustainable Develop and Office of official Publications of European Communities

3 UN Secretary of Defense

4 Office of Official Publication of European Communities

5 US Environmental Protection Agency, Region V

6Panasonic Green Procurement

Pos	Major Practices	Sub-Practices	Literature
4	Strategic Supplier Partnership(SSP)	Integration of information system with key suppliers(SSP1)	Ho et al(2010);OOFEC(2013);Panasonic(2014)
		Regular supplier audit to ascertain meeting green requirements(SSP2)	US EPAR V(1999);USA DoD(2004)
		Green information sharing with strategic supplier partners(SSP3)	Ho et al(2010);USA DoD(2004)
		Supplier location and/or delivery method(SSP4)	Ho et al(2010)
5	Green Information Communication Technology(GICT)	Teleworking(GICT1)	Driscoll(2010);Naveen et al(2009)
		Use of e-Ordering system(GICT2)	Driscoll(2010); Neupane et al(2004)
		Use of company-wide ERP system(GICT3)	Driscoll(2010); Neupane et al(2004)
		Use of energy efficient computing(GICT4)	Driscoll(2010);Naveen et al(2009); USA DoD(2004)
		Intelligent network system(GICT5)	Boucher et al(2001); Dugan et al(2002)
6	Employee Training(ET)	Initial awareness training on green procurement systems(ET1)	Wang and Wu(2004)
		Refresher training on green procurement goals(ET2)	Wang and Wu(2004);DoD(2004); Bouwer et al; (2006);Plas and Erdmenger(2002);Bouwer et al(2006)
		Annual training on green procurement systems(ET3)	USA DoD(2004);Wang and Wu(2004);

2.4. Strategic Supplier Partnership (SSP)

In order for an organization to establish effective environmental system, there must be effective integration with the various organizations involved in a project. This enables experts to undertake proper environmental impact assessment [26]. Research shows that proper environmental impact assessment depends on effective communication between all experts and organizations involved in a business or project [26].

It is therefore imperative to integrate key supplier and customer information systems for free information flow between the organizations. However the field interview indicated that integration of information system with key supplier cannot be applied in public procurement where in the view of transparency and fairness, tender is made open.Meanwhile, studies show that public institutions usually have external auditor who regularly audit the records to determine transparency and maintain corruption free organizational setting [15]. For instance, EU developed a legal framework for an "Environmental Management and Audit Scheme (EMAS) to monitor green public procurement [15].

However Art and Faith-Ell, (2010) [26] believed that if suppliers are made to implement EMS and are monitored, it will enable efficient green environmental auditing. They argued that the EMS will enable the contracting organizations to regularly evaluate contractors green environmental performance to ascertain their level of compliance to international standards.

2.5. Green Information Communication Technology (GICT)

Organizations which practice green procurement require new data and information about the environment, the impact of the various products and the effects so that they can introduce green innovations into their operations [11], [27], [26]. GICT is the concept of practicing green procurement via the application of information communication technology.

Proper management of GICT reduces the carbon (VI) oxide,

CO_2 and carbon (II) oxide, CO emissions. In addition, it builds transparency, accountability and eliminates redundancy in public procurement and maximizes energy efficiency [8], [29].

Studies revealed that the Republic of Bangladesh developed e-Government Procurement (e-GP) portal in 2006 which was manned by Central Procurement department. It is reported that the portal increased transparency, value for money, competition among bidders, quality, and accountability and reduced barriers in their procurement [29] Further studies indicated that countries which have instituted GICT and are benefiting from it include Czech Republic, Denmark, Germany, Greece, Hong Kong, Hungary, Japan, Korea (South), Malaysia, New Zealand, Netherlands, Philippine, Thailand, Turkey, UK, USA and VietNam[29].

Following advancement in technology, tele-working has become one method of implementing green [14]. Works that can be performed electronically include order processing; invoice processing, contract processing and management [14], [31].

2.6. Employee Training (ET)

For green procurement to take strong root, the various stakeholders must understand the entire concept [13], [14], [18]. One important stakeholder is the employees who put the green concept into real practice. The incorporation of the environmental factors into the criteria for purchasing requires parallel steps to train and motivate the employees [13].In fact training on green procurement is an integral part of green implementation practice in any organization [13], [14],[44].

Generally green training is categorized into two levels, including: (1) initial general introduction (this level is what this study refers to as initial awareness stage), (2) the detailed green procurement training. Some organizations separate the detailed green procurement training into (a) refresher green training and (b) the annual green training[13], [18].

3. Research Methodology

In order to achieve the set objective of the research, this study proposed and utilized a grey-based DEMATEL model. The methodology involves a two-phase data collection process: the first-phase focused on a simple YES/NO questionnaireaiming to validateand conceptualize the selected green public procurement criteria and indictors from literature relevant to the study for effective implementation. The second-phase uses real field study approach facilitated by questionnaire to seek public procurementprofessionalsopinion through pair-wise comparison influence matrices to measure the complex influence among the criteria and indicatorsand validate the proposed grey-based DEMATEL model [5], [8], [11], [18], [19],[20],[26],[32],[33],[37],[43].

3.1. Overview of Grey Theory

The grey system theory was developed by Deng in 1982 to deal with problems with uncertainty or systems with a lot of setbacks and imperfect or incomplete information. [38],[39],[40] .

Table 3.1. *The Linguistic variables and grey numbers for Criteria weighting.*

Linguistic Terms	Linguistics Variables	Grey Numbers
No Influence	N	(0,0,)
Very Low Influence	VL	(0,0.25)
Low Influence	L	(0.25,0.50)
High Influence	H	(0.50,0.75)
Very High Influence	VH	(0.75,1.00)

3.2. Some Basic Grey Mathematical Definitions are Presented Below

Let us define x_{ij}^p as the grey number for a expert (decision maker p) evaluation of the influence of factor$i$on factor$j$.Let l and r be the lower and the upper limits respectively, then the grey number for the lower limit is xl_{ij}^p and that of the upper limit is xr_{ij}^p respectively.

Transformation of the grey numbers into crisp numbers (scores) is necessary. We therefore adopt the modified-CFCS (Converting Fuzzy data into Crisp Scores) defuzzification method for this operation.The modified-CFCS method involves three key steps to convert grey numbers into crisp numbers and is given below:

Step 1: Normalization:

$$xr_{ij}^p = \left(r_{ij}^p - minl_{ij}^p\right)/\Delta_{min}^{max} \quad (1)$$

$$xl_{ij}^p = \left(l_{ij}^p - minl_{ij}^p\right)/\Delta_{min}^{max} \quad (2)$$

$$\Delta_{min}^{max} = maxr_{ij}^p - minl_{ij}^p$$

Step 2: Compute total normalize crisp values:

$$x_{ij}^p = \left[xl_{ij}^p\left(1 - xl_{ij}^p\right) + \left(xr_{ij}^p \times xr_{ij}^p\right)\right]/\left[1 - xl_{ij}^p + xr_{ij}^p\right] \quad (3)$$

Step 3: Compute crisp values:

$$z_{ij}^p = minl_{ij}^p + (x_{ij}^p \times \Delta_{min}^{max}) \quad (4)$$

Step 4: Integrate crisp data matrices:

$$z_{ij} = \frac{1}{h}\left(z_{ij}^1 + z_{ij}^2 + \cdots + z_{ij}^h\right) \quad (5)$$

3.3. Overview of DEMATEL (Decision Making Trial and Evaluation Laboratory)

DEMATEL is a structured analysis tool developed at the Geneva Research Centre of the Battelle Memorial Institute [27], [45], to help determine the causality of criteria in relatively small sample size setting [46]. DEMATEL helpsto effectively evaluate complex relationship amongst implementation criteria and support management strategic decision using the influence relationships shown on digraphsor in some cases uses matrices. In our case, DEMATEL is used to evaluate the green public procurement implementation practices and sub-practices for both influence relationship (major practices) and importance rankings (sub-practices). DEMATEL method follows four basic step-wise processes [27].

Step1: Develop the initial pair-wise direct-relation matrix and find the average matrix

Given that there are h experts available to solve a complex interdependent problem and there are n practices to be considered, if each expert gives n× n non-negative feedback matrix,x^p with $1 \le p \le h$. This implies thatx_1, $x_2 \ldots x_h$ are the feedback matrices for each of the h experts, each element of x^p, is an integer denoted by x_{ij}^p. $x^p = \left[x_{ij}^p\right] n \times n$ .The diagonal elements of each feedback matrixx^p, are all set to zero. We can then compute the n×n average matrix A by averaging the h experts score matrices. The (i,j)element of average matrix A is denoted by:

$$\left[a_{ij}\right] n \times n = \sum_{p=1}^h \left[x_{ij}^p\right] n \times n \quad (6)$$

The average matrix$\left[a_{ij}\right]n \times n$is called the direct-influence matrix which indicates the initial direct effect of row practice i on the column practice j.

Step 2: Normalize initial direct-relation matrix D
Normalize initial direct-relation matrix is obtained using Eqs. (7) and (8) below:

$$s = max\left\{max \sum_{j=1}^n a_{ij}, max \sum_{i=1}^n a_{ij}\right\} \quad (7)$$

$$D = \frac{A}{s} \quad (8)$$

Step 3: Determine the total direct-relation matrix
The total relation matrix T is determined by using Eq. (9) below

$$T = D^1 + D^2 + \cdots + D^p = D(I - D)^{-1} \quad (9)$$

Where I is identity matrix
Step 4a: Calculate the sums of rows and columns of matrix T

In the total-influence matrix T, the sum of all rows and the sum of all columns are represented by vectors r and$c$ respectively.

$$T = [t_{ij}]n \times n \ , \mathrm{i}, \mathrm{j} = 1, 2, 3 \dots \mathrm{n}$$

$$r_i = \sum_{j=1}^{n} t_{ij} \ \forall_i \qquad (10)$$

$$c_j = \sum_{i=1}^{n} t_{ij} \ \forall_j \qquad (11)$$

Step 4b: Determine the overall importance/prominence (P_i) and net cause/effect (E_i)

$$P_i = (r_i + c_j / i = j) \qquad (12)$$

$$E_i = (r_i - c_j / i = j) \qquad (13)$$

The greater the P_i value, the greater the influence of the practice i with respect to the overall relationship. If $E_i > 0$, then practice i is a net cause for other practices. If $E_i < 0$, practice i receives the net effect [47].

Step 4c: Set a threshold value to filter the minor relationship amongst the criteria from the total-relational matrix and plot digraph

A threshold is set to filter the minor relationships from within the total-relation matrix. The relationships above the agreed threshold and P_i and E_i values are combined to plot a directional graph (digraph) to display the cause and effects relationships among the practices [41], [42].

3.4. Proposed Grey-Based DEMATEL Methodology

This is an integration of the grey theory and the DEMATEL methodology. The computational steps involved with the proposed methodology are as follows:

Step 1: Obtain the initial linguistic rating direct-relation matrix from all experts/decision-makers using the linguistic variables shown on Table 3.1 column 2.

Step 2: Reassign the initial linguistic direct-relation matrix for all experts/decision-makers with equivalent grey numbers from Table 3.1 column 3.

Step 3: De-grey the grey initial direct-relation matrix to achieve crisp values/data following Eqs. (1) – (4) for all experts/decision-makers

Step 4: Integrate all experts/decision-makers crisp datamatrices into a single crisp data matrix using Eq. (5)

Step 5: Normalize the aggregated crisp initial direct-relation matrix using Eqs. (7) & (8).

Step 6: Determine the total direct-relation matrix using Eq. (9)

Step 7: Determine the overall importance/prominence and net cause/effect using expressions (10) and (11).

Step 8: Set or agree on a threshold value to filter the minor criteria relationships from within the total-relation matrix and plot the digraph.

3.5. The Study Setting

This study focused on Green Public procurement with special recognition to the ten Polytechnics in Ghana. The Polytechnics in Ghana are becoming more dynamic and innovative so new courses and projects are being introduced. New technologies are coming up and competing among themselves. They need to adapt to the modern trend of

technological changes, hence, have to deal with repeated processes of procuring new equipment and materials for their operations. The pressing need for polytechnic institutions to greening their operations is becoming increasingly important. This has motivated us to initiate this study to investigate and attempt to identify the most important criteria and indictors that can contribute to greening the Ghanaian polytechnic institutions operations.

3.6. Population and Sample and Sample Technique

3.6.1. Population

In order to identify the major and sub practices of green public procurement implementation practices, all the ten Polytechnic institutions in Ghana were considered as the entire population.

3.6.2. Sample and Sample Techniques

Due to the nature of the study, random sampling was used to sample five of the Polytechnic Institutions. The sampled Polytechnic Institutions included Kumasi Polytechnic, Koforidua Polytechnic, Sunyani Polytechnic, Tamale Polytechnic and Wa Polytechnic.

Furthermore, due to the availability of the experts, a hybrid of convenient and purposive sampling was adopted to select the sample from the five Polytechnic Institutions. Ten public procurement professionals from the five Polytechnics inGhana, who availed themselves and had the requisite capacity, were selected for their inputs in the framework development and respond to the interview and the questionnaire.

3.7. Linguistic-Based Questionnaire Design and Pilot Testing

The authors administered two-phases of questionnaires in the study. Each of the questionnaires had five parts- the demography of the respondents, the cover letter, introduction which provided detailed explanations of how the questions would be answered, followed by the questions and definition of abbreviations. The introduction section of the second questionnaire had a sample question to guide the procurement professional when answering the main questions.

3.8. Data Collection

The authors partially self-administered the questionnaires and through e-mails. Series of follow-ups were done after thequestionnaires weredistributed to ascertain if the procurement professionals encountered any challenge when completing the questionnaire.

3.9. Validity and Reliability

3.9.1. Validity

The questionnaire was developed based on the literature review. This was to ensure that the procurement professionals know and understandwhat they were responding to. Further explanations were given to the procurement professionals to ensure that they provided valid responses.

3.9.2. Reliability

Self-administered questionnaire and interviews were conducted in order to minimize completely data collection biases.

3.10. Data Analysis

The data collected were methodologically analyzed using the proposed grey-based DEMATEL model aided by Microsoft Office Excel and Matlab.

4. Data Presentation and Discussions

4.1. Application of the Proposed Grey-Based DEMATEL Methodology to a Real Case

This study adopted literature review and procurement professionals input to arrive at the theoretical framework. This theoreticalframework was then evaluated with the aid of the proposed grey-based DEMATEL model with inputs from the public procurement professionals. The application of the proposed model was completed in two-stages. The first-stage dealt with the major practices whist the second-stage dealt with the sub-practices.

Stage-1: Major Practices Application

Theproposed methodology procedure is as follows:

Step 1: Obtain the initial direct-relation matrix.

Table 4.1 below is the direct-relation matrix completed by expert-1.

Table 4.1. Manager-1 Direct-relation matrix for major practices.

Major practices	AMSP	ER	GP	SSP	GICT	ET
AMSP	0	*H*	VH	VH	VH	VH
ER	H	0	H	VH	H	VH
GP	H	VH	0	VH	VH	VH
SSP	H	H	VH	0	H	VH
GICT	H	L	VH	H	0	VH
ET	VH	VH	VH	VH	VH	0

Step 2: Assign linguistic initial direct-relation matrix with grey numbers.

To address the vagueness in the responses from the experts, we used expert1's response as an illustration. Let us consider element $a_{12} = H$ from expert 1 response from Table 4.1 and assign the linguistic variable *H* with the grey value $(0.50, 0.75)$ at the intersection of AMSP and ER. The rest of the linguistics variable grey assignments are shown in Table 4.2 below.

Table 4.2. Expert-1Direct-relation matrix for major practices reassigned with grey numbers.

Major practices	AMSP	ER	GP	SSP	GICT	ET
AMSP	0	(0.50, 0.75)	(0.75, 1.00)	(0.75, 1.00)	(0.75, 1.00)	(0.75, 1.00)
ER	(0.50, 0.75)	0	(0.50, 0.75)	(0.75, 1.00)	(0.50, 0.75)	(0.75, 1.00)
GP	(0.50, 0.75)	(0.75, 1.00)	0	(0.75, 1.00)	(0.75, 1.00)	(0.75, 1.00)
SSP	(0.50, 0.75)	(0.50, 0.75)	(0.75, 1.00)	0	(0.50, 0.75)	(0.75, 1.00)
GICT	(0.50, 0.75)	(0.25,0.50)	(0.75, 1.00)	(0.50, 0.75)	0	(0.75, 1.00)
ET	(0.75, 1.00)	(0.75, 1.00)	(0.75, 1.00)	(0.75, 1.00)	(0.75, 1.00)	0

Step 3: De-grey the grey initial direct-relation matrices

To covert the grey rating/response of the experts into crisp values requires the need to de-greyusing Eqs. (1) - (4). Again, considering the grey valuesat $a_{12} = (0.50, 0.75)$ for expert 1 in Table 4.2, we followed the following steps:

Step 3.1: Normalize grey:

$$\Delta_{min}^{max} = (1 - 0.25) = 0.75$$

$$xr_{ij}^{1} = \frac{(0.75 - 0.25)}{0.75} = 0.6667$$

$$xl_{ij}^{1} = \frac{(0.5 - 0.25)}{0.75} = 0.3333$$

Step 3.2: Compute total normalize crisp values:

Eq. (3) is used to compute the normalized crisp values. The normalized crisp value for exper-1 is computed as shown below.

$$x_{ij}^{1} = \frac{[0.3333(1 - 0.3333) + (0.6667 \times 0.6667)]}{[1 - 0.3333 + 0.6667]} = 0.5$$

Step 3.3: Compute crisp values:

The total crisp value of manager-1 is computed using Eq. (4). The total crisp value of expert-1 is as shown below.

$$z_{ij}^{1} = 0.25 + (0.5 \times 0.75) = 0.625$$

Step 4: Aggregate crisp data matrices of all experts:

The aggregate crisp data matrixwas obtained by using Eq. (5). Table 4.4 below depicts the aggregated crisp matrix.

Table 4.3. Total Crisp data of expert-1.

Major practices	AMSP	ER	GP	SSP	GICT	ET
AMSP	0	0.625	0.938	0.938	0.938	0.938
ER	0.625	0	0.625	0.938	0.625	0.938
GP	0.625	0.938	0	0.938	0.938	0.938
SSP	0.625	0.625	0.938	0	0.625	0.938
GICT	0.625	0.313	0.938	0.625	0	0.938
ET	0.938	0.938	0.938	0.938	0.938	0

Table 4.4. *Aggregated crisp matrix for all experts.*

Major practices	AMSP	ER	GP	SSP	GICT	ET
AMSP	0	0.628	0.569	0.713	0.644	0.869
ER	0.641	0	0.716	0.478	0.550	0.719
GP	0.641	0.944	0	0.644	0.569	0.794
SSP	0.484	0.638	0.794	0	0.563	0.791
GICT	0.566	0.409	0.869	0.488	0	0.869
ET	0.794	0.794	0.719	0.566	0.706	0

Step 5: Normalize initial crisp direct-relation matrix.
Table 4.5 depicts the normalized initial crisp direct-relation matrix of all the experts

Table 4.5. *Normalized initial direct-relation matrix of all the experts.*

Major practices	AMSP	ER	GP	SSP	GICT	ET
AMSP	0	0.175	0.158	0.198	0.179	0.242
ER	0.178	0	0.199	0.133	0.153	0.200
GP	0.178	0.263	0	0.179	0.158	0.221
SSP	0.135	0.178	0.221	0	0.157	0.220
GICT	0.158	0.114	0.242	0.138	0	0.242
ET	0.221	0.221	0.200	0.158	0.197	0

Step 6: Determine the total direct-relation matrix.
Table 4.6 displays the total direct-relation matrix

Table 4.6. *Total direct-relation matrix.*

Major practices	AMSP	ER	GP	SSP	GICT	ET	R
AMSP	2.336	2.671	2.766	2.320	2.408	3.047	15.559
ER	2.308	2.329	2.591	2.107	2.214	2.797	14.346
GP	2.577	2.831	2.728	2.388	2.478	3.140	16.141
SSP	2.373	2.586	2.717	2.077	2.310	2.929	14.992
GICT	2.364	2.513	2.701	2.174	2.150	2.913	14.815
ET	2.602	2.795	2.890	2.369	2.501	2.955	16.112
C	14.559	15.725	16.393	13.436	14.062	17.780	91.954

Step 7: Determine the overall importance/prominence and net cause/effect of the practice (criteria).

The overall importance of the practices and sub-practices are determined based on the final weights after adding the row sums to the column sums and the greater the better. Additionally, the net cause/effect is also determined based on the final weights after subtracting the column sums from the row sums and the greatest the better. This is further put into two main groups. Cause group with positive final weights and effect with negative final weights. Table 4.7 below depicts the weights of all the major practices.

Table 4.7. *Relative Weights(R+C) and net cause-effect(R-C) values of practices.*

	AMSP	ER	GP	SSP	GICT	ET
R	15.559	14.346	16.141	14.992	14.815	16.112
C	14.559	15.725	16.393	13.436	14.062	17.780
R+C	30.118	30.071	32.534	28.428	28.877	33.892
R-C	1.000	-1.379	-0.252	1.556	0.753	-1.668

Step 8: Set a threshold value to eliminate some of the minor effect elements from the matrix. To minimize the complexity of the digraph for a more manageable relationship, we set a threshold to helpfilter and only display relationships above the set threshold considered most influential. Table 4.8 depicts the values above the threshold values.

Table 4.8. *Shows the values above the threshold value.*

Major practices	AMSP	ER	GP	SSP	GICT	ET
AMSP	2.336	*2.671	*2.766	2.320	2.408	*3.047
ER	2.308	2.329	*2.591	2.107	2.214	*2.797
GP	*2.577	*2.831	*2.728	2.388	2.478	*3.140
SSP	2.373	*2.586	*2.717	2.077	2.310	*2.929
GICT	2.364	2.513	*2.701	2.174	2.150	*2.913
ET	*2.602	*2.795	*2.890	2.369	2.501	*2.955

Threshold = 2.554, Values above threshold value*

The relative importance and the net cause/effect weights combined with the major practices influence relationships above set threshold dataset were used to plot a directional graph (digraph). This digraph is shown in Figure 4.1 below.

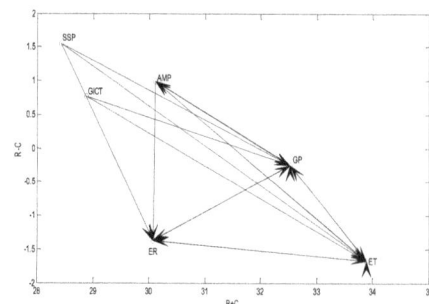

Figure 4.1. *Diagraph of the relationship among the major GPPIP.*

Stage -2: Sub-Practices Application.

The sub-practices application and computation followed similar procedure as the major practices but stopped at step 7. In step 7 however, we only computed the importance/prominence of the sub-practices since we only needed to rank the sub-practices based on the relative importance weights.

Table 4.9 depicts the sum of rows and sum columns. It also shows the addition of the sum of rows and sum of columns to achieve the relative importance weights or prominence of the sub-practices. Furthermore, the sub-practices are ranked in descending order using the relative importance weights as shown in Table 4.10.

Table 4.9. Relative importance/prominence Weights(R+C) of sub- practices.

	AMSP1	AMSP2	AMSP3	AMSP4	.	ET1	ET2	ET3
R	6.007	6.305	6.618	7.199	.	6.142	7.237	6.612
C	6.358	7.017	6.886	7.080	.	6.323	7.012	6.597
R+C	12.365	13.321	13.504	14.278	.	12.465	14.249	13.208

Table 4.10. The ranking of the sub-practices.

Practice	Weight	Rank	Practice	Weight	Rank
AMSP4	14.278	1	GP4	13.129	12
ET2	14.249	2	SSP3	12.948	13
ER3	14.190	3	SSP1	12.911	14
SSP2	14.180	4	ET1	12.465	15
ER1	13.938	5	AMSP1	12.365	16
GP1	13.796	6	GICT2	12.337	17
GP3	13.625	7	GICT4	12.251	18
AMSP3	13.504	8	GICT5	12.062	19
GP2	13.355	9	GICT1	12.017	20
AMSP2	13.321	10	GICT3	11.841	21
ET3	13.208	11	ER2	11.357	22

4.2. Discussion and Analysis

This section will consider the analysis of the major practices followed by the sub-practices.

4.2.1. Part 1: Discussion and Analysis of Major Practices

This part presents the cause-effect relationship analysis from digraph followed by the analysis of the results of the major practices.

4.2.2. The Cause-Effect Relationship (Diagraph) Among the Major GPPIP

Figure 4.1 (digraph)shows that SSP influences ER, ET and GP whilst GICT influences ER, GP. AMSP influences ER, ET and GP whilst ER influences ET and GP. Close observation of the digraph shows that GP also influences ER, ET, and AMSP whereas ET influences ER, GP, AMSP and itself. Table 4.9 below shows the summary of the cause-effect relationship among the major GPPIPs.

Table 4.11. Summary of the relationship among major GPPIPs.

Pos	Cause Practices	Influencing Practices
1	SSP	ER,ET,GP
2	GICT	ET,GP
3	AMSP	ER,ET,GP
4	ER	ET,GP
5	GP	ER,ET,AMSP
6	ET	ER,ET,GP,AMSP

The major GPPIPs were divided into:

• Cause category
• Effect (Influenced) category as shown in Table 4.11 above

Table 4.12. Shows the cause and effect groupings for major practice.

Cause group	R-C value	effect group	R-C value
SSP	1.556	ET	-1.668
AMSP	1.000	ER	-1.379
GICT	0.753	GP	-0.252

4.2.3. Analysis of the Results of Major GPPIPs

The values of R-C indicate the net cause and effect of one practice on another practice. All the positive values of the major GPPIP were categorized as cause group. These practices influence other practices more than they are being influenced. The highest positive value of R-C means the practice has the most influence (impact) on others. The results on Table 4.7 depicts that SSP, AMSP and GICT have positive R-C values of 1.55, 0.99, and 0.76 respectively, hence considered the cause practices. However SSP (Strategic Supplier Partnership) has the highest positive value of 1.55, hence considered as the practice with the most direct influence on the others. Thisimplies that SSP is the most influential practice for management consideration. This could mean that, in a system of interdependent six GPPIPs identified, if SSP is improved, the entire system may improve significantly. It further suggests that, if SSP is well resourced, the entire system benefits accordingly.

The negative values of R-C mean that, the practice is being influenced (affected) by others more than they influenced others. The results on Table 4.7 indicates that ET, ER and GP have negative R-C values of -1.67, -1.37, -0.25 respectively, hence considered as the effect practices (refer to Table 4.12). ET (employee training) has the highest negative value of-1.67 hence considered the most directly affected practice by the other practices. Considering the relative importance of the practices, ET was identified as the practices with the highest relative importance weights of 33.89 whilst SSP observed as the practice with the least relative importance weight of 28.43 (refer from Table 4.7).

From the results, employee training is considered the most important major practices whilst strategic supplier partnership is considered the least important practices. On the contrary, employee training is categorized under the effect group and considered the highest effected practice within the effect group whilst strategic supplier partnership is categorized under the cause group and considered the highest caused practice within the cause group. This may mean that, because the concept of green public procurement is relatively new to the Ghanaian public sector in general and the polytechnic in particular, there is a nature tendency for the polytechnic to encounter enormous difficulties during the green concept program implementation hence in the absence of employee training, the program is bound to achieve worsened results or may end up unsuccessful as such the experts considered employee training as the topmost important practices. It may also mean that, strategic supplier partnership may have been developed already from the traditional public procurement strategy given it the highest causation as a result their importance may be less needed for this implementation requiring the strong need of the less mature GPPIP practices such as employee training.

Therefore, for the program to survive and exceedingly succeed, requires the need for strong employee training (ET), as well as educating the general polytechnic staffs on the new procurement strategy. This finding is in line with [19]argument that, for an organization to strongly incorporate environmental factors into the criteria for purchasing and become successful, requires parallel steps to train and motivate their employees. It is therefore clear from this study that SSP and ET are the two key significant practices that may drive the success of the green public procurement implementation in the Ghanaian polytechnic institutions.

4.2.4. Analysis of Sub-Practices

The GPPIP that has the highest weight is considered the most important practice. Table 4.10 indicates that AMSP4 is the most important practice with the highest weight of 14.28. On the other hand, ER2 is considered the least important with sub-practice since it has the smallest weight of 11.36. Thus, AMSP4 and ER2 take the first and the twenty-second positions respectively on the ranking as shown on Table 4.10. Review of material needs to include green procurement requirement (AMSP4) is considered the most important sub-practices amongst others. This may sound right since the whole procurement strategy commence with establishment of the requirements for the needs and that if at this stage the procurement staffs together with the internal customers/users incorporate the green concept before communicating the needs to the supplier, it will be a good start. Bio-based products (ER2) are considered the least important sub-practices probably because the green concept is still in the fancy in the country, public sector and polytechnic. This advanced approach to addressing the environmental issues will need time for the polytechnic to adopt and mature to a certain level before moving to the advance stage. Taking up such practice may be less beneficial,less attractive and expensive.

5. General Conclusions and Recommendations for Future Research

This paper introduced and developed an analytic framework for green public procurement implementation in

the public sector of Ghana using a combination of literature review and public procurement professional opinions. Additionally, we assessed complex relationships among the practices and the most important sub-practice in terms of the overall goal of achieving green economy.

Based on the six major practices and twenty two sub-practices, the study proposed and applied the grey-based DEMATEL methodology to analyze and prioritize the practices. The study identified that the most influential major practice was SSP and the most vulnerable or affected major practice was ET. However the ET had the highest weight to indicate that it was the most prominent practice. Among the sub-practices, the most prominent was identified as AMSP4, and ER2 was the least prominent.

The study findings support some prior researches. Earlier researches suggest that Employees training is crucial for management of green procurement implementation [14], [18], [19], [26]. In order for an organization to establish effective environmental system, there must be effective integration with the various organizations involved in a project [26].

5.1. Managerial Implications

The following important managerial implications are obtained from the proposed methodology with its application to green public procurement implementation practices. The proposed grey-based DEMATEL methodology provided general decision making framework for green public procurement implementation practices (GPPIPs). Thus, procurement managers can plan their activities along the relative importance of the practices. Again managers confronted with green procurement resources challenge can apply the proposed grey-based DEMATEL methodology to identify which practices influence others the most using the diagraph. The diagraph provides relationship on which GPPIPs should be emphasized and resourced to improve the system. State agents can also use information in the study to examine public procurement policies. Lastly, the results of the study cannot be a generalization of GPPIPs. It is therefore imperative for managers to perform critical analysis since the results of this study were based only on experts in five Polytechnic Institutions.

5.2. Further Research

In terms of further research, researchers can adopt our proposed analytical frameworkand grey-based methodology to conduct to conduct further similar studies in the developing nation and try to compare the results.Other researchers might study green public procurement implementation practices in other sectors of the economy using the proposed methodology.Further extension of the methodologyto include ANP, GRA or even fuzzy-based DEMATEL may be an interesting tool to further investigate the level of maturity of the green concept in the developing countries using our proposed analytical framework will be an interesting area for future research.

References

[1] Walker, Helen, and Stephen Brammer. "The relationship between sustainable procurement and e-procurement in the public sector." *International Journal of Production Economics* 140, no. 1,pp 256-268,2012.

[2] Sarkis, Joseph, and MaurryTamarkin. "Real options analysis for "green trading": the case of greenhouse gases." *The Engineering Economist* 50, no. 3 pp273-294, 2005.

[3] Chien, M. K., and Li-Hsing Shih. "An empirical study of the implementation of green supply chain management practices in the electrical and electronic industry and their relation to organizational performances." (2007).

[4] Mary M.(2013),*what is Green Procurement?,* Wise GEEK, Follow @wiseGEEKhttp://repository.lib.polyu.edu.hk/jspui/bitstream/10397/626/1/RFID4DSS_v5.pdfretrieved 29 Jan 2014

[5] Office for Official Publications of the European Communities. *Managing Natura 2000 Sites: The Provisitions of Article 6 of the'Habitats' Directive 92/43/EEC.* Office for Official Publications of the European Communities, 2000.

[6] Tripathi, Shiv, and Gerald Petro. "Evolving Green Procurement and Sustainable Supply Chain Practices in the Organizations: A Framework to Align Functional Strategy Implementation to Organization's Corporate Social Responsibility (CSR) Objectives." *Management Convergence* 1, no. 1 pp 24-32, 2011.

[7] Christopher, Martin. "The agile supply chain: competing in volatile markets." *Industrial marketing management* 29, no. 1 pp 37-44, 2000

[8] Naveen Chilamkurti,SheraliZeadally, FrankMentiplay. *Green Networking for Major Components of Information Communication Technology System*s,Hindawi Publishing Corporation, EURASIP Journal onWireless Communications and Networking Volume 2009,pp 1-7, 2009

[9] Madu, Christian N., ChuhuaKuei, and Ifeanyi E. Madu. "A hierarchic metric approach for integration of green issues in manufacturing: a paper recycling application." *Journal of environmental management* 64, no. 3 pp 261-272,2002.

[10] Zhu, Qinghua, Jo Crotty, and Joseph Sarkis. "A cross-country empirical comparison of environmental supply chain management practices in the automotive industry." *Asian Business & Management* 7, no. 4 pp 467-488, 2008.

[11] Tripathi, Shiv, and Gerald Petro. "Evolving Green Procurement and Sustainable Supply Chain Practices in the Organizations: A Framework to Align Functional Strategy Implementation to Organization's Corporate Social Responsibility (CSR) Objectives." *Management Convergence* 1, no. 1 pp 24-32, 2011.

[12] Min, H., & Galle, W. P. Green purchasing strategies: trends and implications. *Journal of Supply Chain Management, 33*(3), pp 10-17, 1997.

[13] Plas, Géraldine, ChristophErdmenger, and BögleWerbeagentur. "Green Purchasing Good Practice Guide." , 2000.

[14] Driscoll, T. *Green Procurement Practices in the London Borough of Croydon* (Doctoral dissertation, WORCESTER POLYTECHNIC INSTITUTE), 2010.

[15] Plas, G., &Erdmenger, C.*Green purchasing good practice guide: How local authorities spend their budgets responsibly.* International Council for Local Environmental Initiatives, 2000

[16] Ângela Denise da Cunha Lemos and Antonio Giacomucci. Green procurement activities: some environmental indicators and practical actions taken by industry and tourism, *Int. J. Environment and Sustainable Development, Vol. 1, No. 1, pp 59-71, 2002*

[17] Wang, X. H., & Wu, W. A REVIEW OF ENVIRONMENTAL MANAGEMENT SYSTEMS IN GLOBAL DEFENCE SECTORS. *American Journal of Environmental Sciences, 9*(2). 2013.

[18] Bouwer M, Jonk M, Berman T, Bersani R, Lusser H, Nappa V, Nissinen A, ParikkaK,Szuppinger P and Viganò C, *Green Public Procurement in Europe – Conclusions and recommendations, Virage Milieu & Management bv, KorteSpaarne 31, 2011 AJ Haarlem, the Netherlands.pp 1-42,2006.*

[19] Plas, G., &Erdmenger, C.*Green purchasing good practice guide: How local authorities spend their budgets responsibly.* International Council for Local Environmental Initiatives, 2000

[20] Ho, L. W., Dickinson, N. M., & Chan, G., Green procurement in the Asian public sector and the Hong Kong private sector. In *Natural resources forum*, Blackwell Publishing Ltd,. Vol. 34, No. 1, pp. 24-38, 2010.

[21] Walters, C. S., & Headquarters, U. S.,2012 Sustainability Plan Public Version, 2012

[22] Heritage, C., Departmental Sustainable Development Strategy Report 2011-2012, 2012

[23] Lemos, A. D. D. C., &Giacomucci, A. Green procurement activities: some environmental indicators and practical actions taken by industry and tourism. *International Journal of Environment and Sustainable Development, 1*(1), pp 59-72 2002.

[24] Ângela Denise da Cunha Lemos and Antonio Giacomucci. Green procurement activities: some environmental indicators and practical actions taken by industry and tourism, *Int. J. Environment and Sustainable Development, Vol. 1, No. 1, pp 59-71, 2002.*

[25] Panasonic, *Supplier Partnership*http://panasonic.net/procurement/green/retrieved 28 Jan 2014

[26] Arts, J., & Faith-Ell, C., Environmental impact assessment in green procurement and partnering contracts looking for environmental performance beyond EIA. In *30th Annual Meeting of the International Association for Impact Assessment April* (pp. 6-11) ,2010.

[27] Fontela, E., & Gabus, A. (1976). The DEMATEL observer. Geneva, Switzerland.

[28] Carberry, J., "Using Environmental Knowledge Systems at DuPont," in *Information Systems and the Environment,* Washington, DC: National Academy of Engineering, pp. 81-86, 2001.

[29] Neupane, A., Soar, J., Vaidya, K., & Yong, J., Role of public e-procurement technology to reduce corruption in government procurement. In *Proceedings of the 5th International Public Procurement Conference (IPPC5)*, Public Procurement Research Center , pp. 304-334, (2012. August).

[30] Hull, A. (2008). Policy integration: what will it take to achieve more sustainable transport solutions in cities?.*Transport Policy*, *15*(2), 94-103.

[31] European Consultative Forum on the Environment, Sustainable Development, & Office for Official Publications of the European Communities.,*Towards a new environment action programme for Europe: a contribution to the co-decision process by the European Consultative Forum on the Environment and Sustainable Development*. Office for Official Publications of the European Communities, 2001.

[32] Authority, G. L., The Mayor's Annual Report 2007/08, 2008

[33] Authority, G. L., Ahluwalia, A., Hutchinson, D., Lester, A., Sadler, L., Vowles, D., ... &Cheetham, N., Cleaning London's Air The Mayor's Air Quality Strategy. *ISBN*, *1*(85261), 403, 2002.

[34] Delmas, M. A., & Grant, L. E.,Eco-Labeling Strategies and Price-Premium The Wine Industry Puzzle. *Business & Society*, *53*(1), pp 6-44, 2014.

[35] Green, K., Morton, B., & New, S.,.. Green purchasing and supply policies: do they improve companies' environmental performance?.*Supply Chain Management: An International Journal*, *3*(2), pp 89-95, 1998.

[36] Green, K., Morton, B., & New, S. Green purchasing and supply policies: do they improve companies' environmental performance?.*Supply Chain Management: An International Journal*, *3*(2), pp 89-95, 1998.

[37] USA Director of Defense Momorandum for the Assistant Secretary of the Army: Acquisition, Technology and Logistics: Establishment of DoD Green Procurement Program (2004) http://www.partnersinprojectgreen.com/files/GreenPurchasing Guide.pdf retrieved 29 Jan 2014

[38] Moran, J., Granada, E., Míguez, J. L., &Porteiro, J., Use of grey relational analysis to assess and optimize small biomass boilers. *Fuel Processing Technology*, *87*(2), pp 123-127, 2006.

[39] Hsu, Y. T., Yeh, J., & Chang, H., Grey relational analysis for image compression. *The Journal of Grey System*, *12*(2), pp 131-138, 2000.

[40] Wu, H. H., A comparative study of using grey relational analysis in multiple attribute decision making problems. *Quality Engineering*, *15*(2), pp 209-217, 2002.

[41] Yang, W. T., Liu, W. H., & Liu, H. H., Evaluating Influential Factors in Event Quality Using DEMATEL Method. *International Journal of Trade, Economics & Finance*, *4*(3), 2013.

[42] Shieh, J. I., Wu, H. H., & Huang, K. K., A DEMATEL method in identifying key success factors of hospital service quality. *Knowledge-Based Systems*, *23*(3),pp277-282, 2010.

[43] Odeen, P. A., & Howard, W. G., *The Defense Science Board 1998 Summer Study Task Force on DoD Logistics Transformation. Volume 1: Final Report*. DEFENSE SCIENCE BOARD Fields, C., &Haver, R. (2008). *Challenges to Military Operations in Support of US National Interests. Volume 1: Executive Summary (Defense Science Board 2007 Summer Study)*. DEFENSE SCIENCE BOARD WASHINGTON DC.WASHINGTON DC., 1998

[44] United States. Environmental Protection Agency. Region V. (1999). Purchasing Strategies to Prevent Waste and Save Money. Source Reduction Forum of the National Recycling Coalition.

[45] Gabus, A., & Fontela, E. (1973). Perceptions of the world problematique: Communication procedure, communicating with those bearing collective responsibility. Geneva, Switzerland.

[46] Fu, X., Zhu, Q., & Sarkis, J. (2012). Evaluating green supplier development programs at a telecommunications systems provider. *International Journal of Production Economics*, 140(1): 357–367

[47] Tzeng, G. H., Chiang, C. H., & Li, C. W. (2007). Evaluating intertwined effects in e-learning programs: A novel hybrid MCDM model based on factor analysis and DEMATEL. *Expert systems with Applications*, *32*(4), 1028-1044.

12

The Role of Universities in Developing a Sustainable Economic Model Based on Solidarity

Matarazzo-Neuberger Waverli Maia, Alves Luiz Roberto, Bernardes Marco Aurélio

Sustainability Center at the Methodist University of São Paulo, Methodist University, São Bernardo do Campo, Brazil

Email address:

waverli.neuberger@metodista.br (Matarazzo-Neuberger W. M.), luiz.alves@metodista.br (Alves L. R.),
marco.bernardes@metodista.br (Bernardes M. A.)

Abstract: The Methodist University of São Paulo, a confessional institution with three campuses in São Bernardo do Campo, in the São Paulo Metropolitan Region, which has 1,600 employees and 25,000 undergraduate students and postgraduate students, adopted in 2008 Sustainability as a transversal theme in its Institutional Educational Project. In 2009, started the Methodist Sustainable Program with the goal to mainstream sustainability in all undergraduate programs, in a co-creative process that included staff, professors and directors. Theory U and other similar methodologies were used in order to connect with new perspectives and attitudes that could arouse a co-creation field and a process able to reveal the connections of all expertises and knowledge produced with the sustainability area. The expectation is that this methodology continues to expand, involving university sectors and participants and oriented by educational purposes, research and community outreach. The expansion of the program led to the development of Actions Research Projects as reported in the present work, designed to consolidate a network of regional solidarity economy. The experiment enrolled 12 professors and lecturers, 20 students and 11 social enterprises linked with Solidarity Economy Network of the populous neighborhood of Montanhão in São Bernardo do Campo, Brazil. The development of the project and the results accomplished created concrete actions of the University that benefit peripheral areas of the Metropolitan Region of São Paulo and could be used for promoting emergence and organization of individual and collective enterprises in other six nearby cities, that together with São Bernardo do Campo forms the Great ABC region, home to 2,500,000 inhabitants. Collective training of entrepreneurs in technology, human reasoning and political citizenship in the urban world were the axis of project. The results obtained can engage other groups and nearby cities, strengthening the social consciousness and regional identity.

Keywords: Solidarity Economy, Regional Identify, Theory U, Peripheral Areas, Education for Sustainability

1. Introduction

We live in a period of crisis. You can name it: the energy crisis, the water crisis, the food crisis, the security crisis, the leadership crisis, the health care crisis, the educational crisis, the climate crisis (Scharmer, 2010:3). The sustainability agenda of the twenty-first century is a huge example of this. Our major challenges requires that we stabilize climate, become much more efficient in the use of fossil fuels, make a rapid transition to solar-based technologies, stabilize and then reduce population, reverse the loss of forests, protect biodiversity, conserve soil, rebuild rural areas, clean up toxic messes and sharply reduce poverty (Orr, 2004: 39).

What is the role of universities in bringing new solutions and creating a sustainable future? How universities could help to do that in Metropolitan Areas densely urbanized? Those questions, for sure, need to be addressed as part of universities challenges and possibilities.

During the 20th century Latin America metropolitan areas were inflated by the impact of migrant workers at the service of the unfolding industrial revolution, specially reread in the eighties by a globalizing perspective. Poor urban planned areas, high demographic density and all sort of environmental issues were the consequence. Civic organizations also had to face the socio-economic and environmental restructuring consequences of the globalizing capitalism. Despite its economic and social contradictions, the best translation of the city as a place to live includes the ability to include, distribute and share, that is, to try being truly the *polis*. All authors quoted by Andrusz, Harloe & Szelényi (1996), who deal with post-socialist cities in Europe, recognize this phenomenon. Already in its introduction (1996, p. 2) we find

> … The cities of capitalism and socialism are conditioned by their respective modes of economic planning, organization of social classes and political structures. The socio-spatial organization of cities, their policies and forms of administration, their dwelling and their markets, their models of social interaction are directly linked to the larger purposes of the socialist and capitalist order.

Considering the development of Brazilian cities Compans (1999, p. 91-113), discusses the paradigms attributed to so-called global cities: a) decentralization, decomposition processes and re-centralization of control functions, b) proclivity for the advanced tertiary sector, especially information processing, c) coordination of the global with the local. To the extent that cities become the *nodes* of the great connection, the solution to the problems is, roughly speaking, the implementation of new forms of social agreement between actors and protagonists of the city, of course in favor of integration in networks. The values of local society, certainly including their cultures, would also compound the globalizing project and at the same time would be a marketing procedure and a training of new human resources. In the analysis of the author, the planners of global cities do not give answers to the socio-cultural sense of the territory, as well as to the peripheral reality of the cities or to the macro-economic movements. The territories and common people, the regulations of markets and the production and distributions processes aiming public good, all remain outside the global showcase. Therefore, planners-sellers, such as the Catalan Jordi Borja, take part in some of opportunistic projects of the cities. Kurz (1993, pp. 07-41) had glimpsed the turning off of the city-market lights, specially the urban illusion of being a lucrative showcase and node, simply because the agglomerations are inserted in world economy chains and their technological and communication supports.

It is possible to infer from this that medium and large cities should be understood as unsustainable, not only by the contradictions of their socio-environmental reality, but also by their lack of definition as to their courses, the meaning of their ground and their history, their management modes and other variables that ask about their past and future as a place for collective life. In the same direction, Sassen (in a TV interview, 2006) moves forward in the reflection on the errors resulting from the exacerbated privatization, under the presumption that the State would be the problem, or that individuals-citizens would be the obstacles to development and progress. What concerns her, after the known studies on global cities, is the loss of rights of the person, and the immediate demand of new and more direct forms of social representation, capable of symbolizing the overcoming of privatization. These lines are also followed by socio-economic thoughts, with communicational and cultural perspectives. Muniz Sodré (2002, p. 257) has a clear proposal to rethink the mediatized society, that is, a society strongly influenced by media communication, the global city:

> … It is a matter of going scientifically against the expansive movement of reduction of the vital experience by the exponential growth of the technological framework of the world, whitewashed by the indifferent commercialism of the organizations that control the media. This implies to include the horizon of revitalization of the democratic experience in the communicational thought starting from the 'ordinary', that is, from the ability of ethico-political articulation of regional and popular organizations.

It is clear therefore that it is in the city that occur the great wrangles, which are denser when symbolic. Although often verging on chaos, such disputes can build solidarity.

The public debate and crisis response we witness nowadays continue to be framed by the same old categories of thought that got us into this whole mess in the first place. The crisis we face is first and foremost one of mind, perception and values; hence, it is a challenge to those institutions presumed to shape minds, perceptions and values, it is an educational challenge. More of the same kind of education can only make things worse. (Orr, 2004: 27). To paraphrase Albert Einstein's famous observation, "The significant problems we have cannot be solved by the same type of thinking that created them." That, however, is exactly what we are busy trying to do (Scharmer, 2010: 2).

As Sterling (2001: 12) states, it is a change of mind on which change towards sustainability depends. It is the difference of thinking that stands between a sustainable or chaotic future. The qualities, depth and extent of learning that take place globally in the next ten to twenty years will determine which path is taken: moving towards or further away from sustainability. Or, according Morin's words:

> I would indicate the option of a civilization policy that would revive solidarity, would force the retreat of selfishness and, still more deeply, would reform society and our lives. In fact our society is in crisis. At this point where it and we arrived, the material well-being did not necessarily lead to a mental well-being, as is witnessed by the unbridled consumption of drugs, anxiolytics, antidepressants and sleeping pills. The economic development did not lead to a moral development. The application of the calculation, the use of the chronometer, the hyper-specialization, the compartmentalization in work, in businesses, in the administrations and, finally, in our lives, often has caused the degradation of solidarity, widespread bureaucracy, loss of initiative, fear of responsibility. (Edgar Morin, 25 April 2007 in Le Monde, as cited by Marco Aurelio Weissheimer, Carta Maior).

The aim of this paper is to describe how we are addressing at our university this education challenge, mainstreaming sustainability through the curricula of the all undergraduate programs and to discuss a first community interactive result of this program.

2. The Methodist University Sustainable Program (MUSP)

Colleges and universities are critical loci for the change sustainability requires. They are learning centers for new ideas and for change. They guide other sectors and have the potential to serve as societal models. Although they are unique in many ways, their innovations are largely replicable by other institutions. And their mission and responsibilities for defining educational scope give them a reason to update and reevaluate in a way that invites institutional learning and openness to sustainable thinking (Edelstein 2004: 271). Universities also transcend boundaries in space and time. With their senior professors and junior students, they also connect society's elders with its youth. With their interdisciplinary studies and many learned associations, they connect across intellectual and geographic boundaries and are thus participants across space (M'Gonigle and Starke 2006: 13).

Sustainability provides universities an opportunity to confront its core values, practices, entrenched pedagogies, the way it uses resources and its relationship with the broader community (Wals and Jickling 2002: 230). It can be a catalyst for institutional change and for a transition towards new ways of knowing, introducing teachers, students and administrators alike to a new pedagogical world that opens up promising avenues for both institutional and individual practice (Wals et al. 2004: 348).

Methodist University of São Paulo is located in São Bernardo do Campo, in the Metropolitan Region of Sao Paulo. It has 26,564 undergraduate students distributed in the knowledge areas of humanities, communication, business, science and technology, totaling 51 undergraduate programs. In addition it runs 43 specialization and six MSc and PhD programs, totaling 1,000 students. The institution has 1,713 employees, 537 professors and lecturers and 94 interns. The university has great regional influence and is recognized as the third-best private university of São Paulo state.

The Methodist University Sustainable Program (MUSP) began to take form at 2008, during the discussions of the new Political-Pedagogic Project (PPP) of the university for the period of 2008 to 2013. This document describes the main policies adopted by the university. Matarazzo-Neuberger, one of the authors, suggested the creation of a Sustainability Axis. The meetings to renew PPP resulted in the adoption of sustainability as a new core value in addition to two others already existent, namely, Christianity and Common Good. The PPP was approved and presented to the academic community by the end of 2008. Sustainability has been recognized since then as a central value in the institution, and has opened a new and strong scenario for change.

The initial plan of MUSP defined two major structuring components: (1) an Educational Program, designed to introduce sustainability in a transversal way through all the undergraduate courses; and (2) Structural Diagnosis of three environmental resources: water, energy and greenhouse gases emissions, in order to offer subsidies to plan the reduction of the ecological footprint of the university. This initial plan provided enough space for incorporation of other ideas coming from university members. The MUSP was launched in early April 2009.

The structural diagnosis was conducted through June to December 2009. Data for the last three years of water and energy consumption was collected and analyzed. The report was presented to the Sustainability Committee and gave birth to a series of initiatives dedicated to reduce the university footprint.

The elaboration of the educational program began in May and June 2009, when meetings with directors and course coordinators of all faculties happened. The objective was to identify which modules/disciplines in every undergraduate program could begin to introduce sustainability. A fundamental principle of the MUSP was that sustainability should not be treated as one more discipline within current curricula, but should be transversally mainstreamed into curricula in order to really be effective. The mind map methodology was used in these meetings to identify in each faculty themes, areas and ways in which sustainability could be interweaved.

A group of professors and lecturers was identified for implementing sustainability in the curricula of each program and was invited to join the Education for Sustainability Leadership Program in Higher Education (FLESES, in Portuguese).

Opening the door for emerging new leaders was viewed as a *sine quo non* condition for the development of the program. Chase and Rowland (2004: 104) demonstrated that the success of higher education sustainability projects was based on decentralized leadership structure, avoiding a small group becoming 'the' environmental organization on campus and allowing other groups to emerge and play critical roles. Keeping its focus on education for sustainability through faculty development could give rise to a university-wide commitment that was one of the main objectives of MUSP.

FLESES was designed as a modular program. Module I was named 'Sustainable Futures: A collective creation'. FLESES Module II was devoted to developing procedures and techniques to work with sustainability issues with students and Module III was related to the systematization and sharing of the results obtained by each undergraduate program in the implementation of sustainability within its curricula. The objectives were to allow the understanding of the relations between sustainability and future scenarios, to reflect on the relationship of professional practice and planetary sustainability in the context of the careers offered in the university, to delineate the role envisioned for university students in the future, to introduce new forms of learning and teaching and to stimulate the rising of a learning community devoted to sustainability.

While designing the modules, sustainability was considered an emergent quality arising from sets of relationships in a system, whether viewed at the macro or micro scale. As Sterling (2004: 55) states, sustainability is likely to arise depending upon the degree to which our attention shifts from 'things' to relationships, and from a segregated and dualistic view of the world towards an integrative and participative perspective. This involves more than a simple and dualistic environmentalism and indicates, instead, the need for 'whole system thinking'. The activities, reflections, text readings and videos chosen to be part of the modules immersed the participants in this kind of vision, offering a glimpse of how we can create the opportunity for people to imagine and work towards life-centered forms of development (Clover 2002: 167).

The Earth Charter was adopted as a guideline for FLESES, because it represents an important contribution for a holistic and integrated vision of the social and environmental problems of humanity. 148 professors and lecturers representing all undergraduate courses of the university and 20 administrative managers attended FLESES. The main objective of this initiative was to establish a real sustainability academy—a learning community that will be responsible for introducing sustainability transversally in university curricula, develop research and engage community and stakeholders aiming to improve the local environment. Our goal for the next two years is to offer FLESES to all professors and lecturers of the university.

One of the main objectives of an education for sustainability program should be to create space for social learning that includes spaces for alternative paths of development; for new ways of thinking, valuing and doing; for participation minimally distorted by power relations; for pluralism, diversity and minority perspectives; for deep consensus, but also for respectful disagreement and differences; for autonomous and deviant thinking; for self-determination; and finally, for contextual differences (Wals and Corcoran 2004: 224). The design of the Educational Program of MUSP allowed the creation of these spaces and the program was built bottom-up, considering the points of view and experience already embodied in the academy.

By the first semester of 2010, sustainability was introduced in the curricula of almost all undergraduate courses. Mitigation measures and monitoring plans to reduce the university's footprint were also implemented.

3. The ABC Region: Social, Economic and Environmental Scenarios

The ABC region represents the southeastern micro-region of the metropolitan region of São Paulo and covers an area of 841 km² occupied by 2,500,000 inhabitants distributed in seven cities. Every square kilometer of the micro-region is occupied by almost 4,000 people. Located on the transit route between the coast and the highlands, the region was discovered as a place to assist the development project of São Paulo, just after the stream of immigrants began in 1877 and the establishment of the railway line by the British, at the same time, connected São Paulo to the seaport of Santos. In 1920 its population was 25,215 inhabitants and today this region is a significant part of the nearly 18,5 million

inhabitants of São Paulo metropolis. Until the thirties the immigrants, especially Italians, Spaniards, Slavs, and later the Japanese, constituted the basis of the professional and cultural basis of its population. The explosion of the industry-based capitalist project in the aftermath of World War II draws thousands of Brazilians from the poorest regions of the country during the fifties when began the industrial pole of companies formed by chemical, petrochemical, automotive parts, electro-mechanical branches, and the automobile industries, with a globalizing trend and a Taylor-Ford style of administration. This pole replaces the old style industries established in early twentieth century, during the first industrial revolution and represented by furniture and textile companies, most based on cooperative ways. The socio-economic development associated with the building of new democratic experiment in the region would be unthinkable without the concurrence of the cultures of work, of renovation of public powers, and learning of the economic and financial forces, which since 1989 have organized themselves in forums, chambers and working groups. Today, the region has about five thousand industries and more than thirty thousand commercial outlets and service providers, while 35.34% of jobs are still available in the large transformation industries. The services sector already employs 40.04%, according to 2008 data (Reis,2008, pp. 133-135). The ABC region has a strong capacity to rebuild itself, which could be observed during the capitalist restructuring of the end of eighties and nineties, when at least 20% of the workforce of about 1,200,000 citizens, lost their jobs and were unemployed. This number has been reduced to less than a half in the last ten years, although one should not forget the informality and precarious work took the place of the late job positions.

Another characteristic of the region is that 56% of its total area of 841 km² is still covered by Atlantic Rainforest. Conservation International included this rainforest as one of the 25 hotspots of the world, locals with high diversity, high endemism and high degree of threat, that are considered priorities for preservation and conservation of biodiversity all over the world. This organization also ranks this forest among of the top five of these hotspots (Mittermeier et al., 1999). Despite most of this area is protected and should be preserved in order to keep springs, forests and help to regulate the climate, all the area suffers real estate speculation and invasions of people claiming for housing. The total area covered by this forest in the ABC region is under federal, state and local law protection, although this not guarantees its real conservation. State/local parks and natural reserves concentrate environmental studies efforts, but knowledge scarcity is still a major concern. There is a great demand to environmental restoration of disturbed areas and enrichment of secondary forests. This forested area also shelter Billings reservoir and 80% of its watershed. This reservoir, an artificial lake constructed in 1925 with energetic purposes, has strategic interest for water production to the Metropolitan Region of São Paulo. With 127 square kilometers of area and a perimeter of 900 kilometers, it stores a maximum volume of 1.2 10^8 cubic meters of water. (Matarazzo-Neuberger, 1994) a essential and scares resource at São Paulo Metropolitan Region.

The regional per capita income remains twice the national income. The gross domestic product of the region corresponds to 2.43% of Brazilian GDP and 7.21% of the Sao Paulo State. The illiterate rate of the region does not go above 8%, comparing to the country rate of 17% . Technological and functional illiteracy, however, are more problematic given the demands of the new industries and services, interested in keeping the third consumer market position of the country. The data on the informal economy are similar to those of the state capital: 55% of the persons are engaged in the productive process.

From the perspective of its economic and cultural history, the region crossed three cycles and lives today the fourth historic-cultural period. The first cycle can be understood as being a passage, since it was the road for goods and natural products for the first cycle of commercial exchange of the eighteenth and nineteenth centuries. The second cycle embodies the construction of the industrial identity, coupled with the political autonomy of the various cities, completed in the fifties of last century. It is during this cycle that is consolidated the symbolization of the region as a restless, and claimant space and cradle of social and political movements, which would greatly help to ensure the re-democratization of Brazil in the seventies and the eighties. The third cycle was that of an acute conflict between capital and labor, finished in the early' 90s, which signaled intense socio-urban losses while helping Brazil to construct meanings for citizenship and social participation, accumulating values for the creation of the new consensus of the micro-regional society, already inserted in the fourth cycle in an accelerated process of globalization and known consequences.

Sader (1988) noted that the projects and practices of these organized people of the periphery, are a signal of victory over the physical and symbolic disintegration, the acquisition of rights amidst critical stress, the inter-communication of segments to produce the condition of being a person within the associativity of workers and the use of the wealth of the cities to promote a better income distribution. It is a project of humanity, in which the gap between knowing and doing diminishes. Thus, these cultures organized around work, present since the beginning of the previous century, proofs and examples of social organization, creation of parties, associativity of residents and neighborhoods to demand public services infrastructures, labor and credit cooperatives (such as the *società di mutuo soccorso*, the old peoples' banks), and union's claims. It can be understood therefore that the social basis historically organized guarantees some regional organity and promotes new managerial ventures. Perhaps it can be of less interest the fact that three of the seven cities in the region are among the best fifty to live in Brazil than discover in these cultures defined by work a process fully favorable to management developed in a direct way by the different sectors of society.

4. Globalize the Regional Intelligence

Santos (1998, pp. 19-20) shows that "the (trans nationalized) territory reaffirms itself by the place. (…) It is wise, however, to remember that thanks to the miracles allowed by science, technology and information, the forces that create fragmentation may in other circumstances, serve their opposite." This micro-region, with large contingents of workers connected by multiple origins, coped with the senses of the periphery and overcame them. This makes it difficult to accept new and crueler forms of living in the economic, political, and cultural periphery. It is not without reason that in this regional social micro-physic, the coordinating institution of the Regional Literacy Movement is the Steelworkers Union. From 1997 to 2008 about 70,000 people overcame illiteracy. Today the municipalities are developing new educational policy for youth and adults from the new forms of technological illiteracy, with promising experiences, specially provided by the proximity of regional universities.

To achieve strategic objectives and deepen themes opened by the breakdown of those supply chains derived from the industrial policy of fifties, the civil society in the region and the public administration of the seven cities created new institutions and new methods of social articulation: strictly speaking, a new policy agenda. Based on European and American experiences, discussed at international seminars held in Brazil and abroad, were created during the period of 1991 to 1998 three institutions that produce, transmit and negotiate information and services: The Consorcio Intermunicipal das Bacias do Tamanduateí e Billings (Inter-cities Consortium of the Tamanduateí and Billings Watersheds), commonly called Regional Consortium, the Câmara Regional do Grande ABC (Great ABC Regional Chamber) and the Agencia de Desenvolvimento Econômico (Economic Development Agency). The Consortium, formed by the seven mayors and their advisers, specialized in regional policies priorities, is the place where the initial studies about programs and policies are generated. Launched in 1991, its major concerns have been sustainable development, final disposal of waste, revitalization of the productive chains, creation of an infrastructure for business tourism and ecological tourism, prioritization of children and adolescents at risk and the combat against several forms of illiteracy. The Regional Chamber, made up of mayors, advisers, state representatives, civil state servants, and representatives of local civil society, began its activities in March 1997. It sought to organize priorities, expand studies by means of ten working groups, approve 31 basic demands of the region and negotiate agreements and processes of implementation of policies and actions, decided by consensus, with the state government and with the Union. The Board was constituted symbolically: in addition to politicians and local managers, the creation agreement (12.03.1997) was endorsed by five members of the Citizens' Forum, five representatives of the productive arrangements and five union officials nominated by their peers. The main agreements, accompanied by representatives of society, prioritized: the implementation of a local technology hub, production chains increase of the competitiveness, creation of new drainage systems for rain waters and industrial effluents, establishment of a regional hospital, amplify the offer of popular housing, establish a federal university, provide workers qualification for new jobs, revitalization of industrial processes in still productive plants – like furniture companies –, improvement of public transportation, development of opportunities for the first job, and strengthening of the movement in favor of impoverished children and adolescents.

5. Marketing and Engagement

The Economic Development Agency is the product of the previous institutions. It is both a database and tool for regional marketing. It concentrates socio-economic information, develops research, supports and fosters the development of companies with a view to sustainable development. It is a public-private organization (51% private, 49% public), while the Regional Chamber is apolitical instrument of partnership between the powers and the Consortium an official core for the generation of projects and identification of needs in the various public policies. Of fundamental importance for the creation of the Chamber and the Agency was the formation of the Citizenship Forum, an exclusive organ of the civil society that begun in 1994 by dozens of associations, schools, unions and service clubs and groups. It acted as an ombudsman during the whole process, encouraging, criticizing and analyzing the actions of the Consortium, the Chamber, and the Agency.

As it is possible to see, the permitted agenda of the regional microphysics meant a necessary moment of political consciousness, an act of identification with the movement of managing the diversified public good in the region. Following that, international experiences contributed to the discussion of cases and the discovery of new projects. The best working relationships were established with the Ruhr Valley, Detroit, Great Leipzig, Rotterdam, Lombardy, and the German area of Baden-Württemberg, plus the social inclusion projects in Latin America, linked by the Mercocities project. The presence of former Mayor Celso Augusto Daniel was instrumental for the knowledge, debate and presentation of regional proposals. Bulletins of unions and trade associations analyzed continuously the issues of the regional crisis and at least one hundred thesis and dissertations of the major Brazilian universities cataloged by the Laboratory on Regionalism and Management (University of Sao Caetano do Sul) and by the Celso Daniel Chair on Cities' Management (Methodist University of São Paulo) attested to the importance of the micro-regional debate. In truth, it is necessary to say that two groups of institutions are not yet fully involved in this mode of management: universities and city councils. The first ones, despite the analysis and criticism, only recently began to evaluate the new regional dynamics and the importance of critical mass for the strengthening of projects agreed upon, unlike what occurs, for example, in Lombardy and Baden-Württemberg,

especially in agencies and observatories. For their part, the legislative chambers, with exceptions, still prefer a traditional way, even archaic, of doing politics. The new political actors overcame, then, the traditional social representations, built upon roles and functions sanctioned by the old politics. The representations that are directly created are engendered in the society that best felt the challenges, whether in the emptying of the executive power, whether at the work in the neighborhood or in the production and distribution that were lived on the factory floor. The legislative representations resemble those whom the poet called "farmers of the air"1, since their domain spaces are unconnected to what the poet also called "sense of world." Contrary to what waited the political history, the popular vote, when transports these people to the legislative palaces, also transvests them. Their look upon the city becomes wrapped up in their own myths of interest, sometimes collective, but really myths of usurped communities. In contrast, what should be highlighted in this new process of dialogue, which creates communication and new social mediations, is that the stigma of exclusion – that is known in immigration and migratory processes, that is combated in movements for the urbanization of slums, that is widely publicized by the strong regional unions – became transparent and was made public in all its breadth what led to projects and practices to overcome them.

6. The Montanhão Community Project: An Example of Sustainable Practice Leading to Results and Community Change

The Montanhão Community Project was designed by professors and students from Management and Services Faculty two years ago and received the official name of Services and Management Network for a Solidarity Community. Solidarity economy is defined as a set of economic activities, like production, distribution, consumption and credit, organized and conducted collectively by workers (SENAES: 11, 2006). Economy solidarity researches as a consensus assert that community of workers that adopt this kind of economy should do it with a collectively, communitarian and solidary character, be their proper managers and not follow the established society rules for labor.

Methodist University of São Paulo already had a partnership with Associação de Resgate Humano e Cidadania Padre Léo Comissari, a local league community that is responsible to promote a Solidarity Economy Network in the Montanhao neighbourhood. At the begining of the last decade they already supported one hundred informal small enterprises.

The main goals for this Project were:
1. To develop entrepreneurs plans, which values the balance between personal and collective interests, since they

---

1 The expression was coined by the Brazilian poet Carlos Drummond de Andrade.

belong to a network of Solidarity Economy and must have common goals.

2. To offer technical support for the network enterprises.

3. To develop negotiator skills for the entrepreneurs of the network that are members of the Municipality Forum of Solidarity Economy.

4. To support and participate in meetings and trainings promoted by community league, developing skills linked with sustainability, planning and designing, best practices, human resources management, conflict solutions and solidarity economy.

5. To link other universities areas and programs demanded by the league and the developing enterprises.

A diagnosis of the Montanhão community related to cultural aspects, enterprises profile of the network and their results was conducted in the very beginning. One of the discoveries of this diagnosis was that they didn't have a proper way to share their achievements and experiences. Community meetings named Listener University were suggested introducing deep discussions about solidarity economy and sustainability, and causing reflections on the kind of relationship and practices they already established between them and their suppliers and clients. This process was the basis to shape the future university group interventions at this community.

During the Listener University Meetings they discussed the results they achieved, their hopes and fears, labor laws, human resources conflicts and management, negotiation skills and how the enterprises linked to the network should work and accept the local social coin named "Comissari", how they could support and help each other, in order to attract more entrepreneurs to the network and how they could work collectively and support their needs as a community. All those themes are related to sustainability in the way we developed it at MUSP looking to the social, economic and environmental issues and creating a vision that brings together a systemic view and different way of living.

Those are some of the results already achieved:

1. Developing a methodology to work with the network and the social incubator of the Montanhão league: all the information, methodologies and experiences were summarized and available in seven papers published on our digital magazine REGES - Revista Eletrônica da Faculdade de Gestão e Serviços [link] and at XXXV ENANPAD 2011 meeting: [link].

2. Publishing the book: Work, Solidarity Economy and Social Development: The case of the Montanhão Community Network on Solidarity Economy in press.

3. Three new enterprises were established that hired 8 persons.

4. Promoting the Seminar: Incubation methodology: challenges and paths to students, professors, entrepreneurs, and government representatives of the nearby cities of Diadema, Santo Andre, Osasco e Maua, in September, 2010.

5. Developing skills and practices demanded by the Montanhão community, also related to Padre Leo league.

6. Creating indicators for measuring income and business

into the Network of Montanhao, that has also been adopted by Diadema communities.

7. Future Perspectives

The results enrolled above reassures how university can contribute with solidarity economy, bringing approaches, experiences and knowledge and integrating them in order to build a development model that includes communities at the basis of income of our society and creates real economic alternatives to improve their lives, something that our usual economic model never provides.

The university launched in February, 2014 an interdisciplinary Sustainability Center that enrolled professors, lecturers, administrative employees and students from all faculties, as well as communities and business companies. This initiative had its roots in the MUSP and is expected to enhance and enlarge the sustainability culture in our university, creating the container for discussing new social and scientific technologies to achieve a model of development that has sustainability as a guide principal. This center plans to create a Green Business Incubator and a Biomimicry Lab to develop new technologies based in Atlantic Rainforest and integrate former experiences and projects that were developed with communities like Montanhao, business companies and higher education.

References

[1]   ABRUCIO, F. (2002) Retratos Metropolitanos. A experiência do Grande ABC em perspectiva comparada. São Paulo: Fundação Konrad Adenauer.

[2]   ALVES, L. R. Culturas do trabalho. Comunicação para a cidadania. Santo André: Alpharrabio, 1999.

[3]   ANDRUSZ, G. D., Harloe, M., & Szelényi, I. (1996). Cities after Socialism. Urban and regional change and conflict in post-socialist societies. Cambridge, Massachusetts: Blackwell Publishers Inc.

[4]   ANTUNES, R. (2003). Adeus ao Trabalho? (9th.ed.). São Paulo: Cortez; Campinas: Editora da Universidade Estadual de Campinas.

[5]   BRASIL. Ministério do Trabalho. Atlas da Economia Solidária no Brasil 2005. Brasília: MTE, SENAES, 2006.

[6]   CHASE, G.W. and P. Rowland. 2004. The Ponderosa Project: Infusing Sustainability in the Curriculum, in P.F. Barlett and G.W. Chase. (eds), Sustainability on Campus: Stories and Strategies for Change. London: The MIT Press, 91–105.

[7]   CLOVER, D.E. 2002. Toward Transformative Learning: Ecological Perspective for Adult Edu- cation, in E. O´Sullivan, A. Morrel and M.A.O'Connor (eds), Expanding the Boundaries of Transformative Learning: Essays on Theories and Praxis. New York: Pelgrave, 159–72.

[8]   COUTINHO, L. (2003) O desafio urbano-regional na construção de um projeto de nação. Regiões e cidades, cidades nas regiões. São Paulo: Editora Unesp, Anpur, p.37-56.

[9] DANIEL, C. A. (2003). Ação política e diversidade de atores no universo social urbano. Regiões e cidades, cidades nas regiões. São Paulo: Editora Unesp – Anpur.

[10] EDELSTEIN, M.R. 2004. Sustaining Sustainability: Lessons from Ramapo College, in P.F. Barlett and G.W. Chase (eds) Sustainability on Campus: Stories and Strategies for Change. Cambridge: The MIT Press, 271–92.

[11] M'GONIGLE, M. and J. Starke. 2006. Planet U: Sustaining the World, Reinventing the University. Canada: New Society Publishers.

[12] MATARAZZO-NEUBERGER, W. M. 1994. Guildas, Organização e Estrutura da Comunidade: Análise da Avifauna da Represa Billings, São Paulo. PhD Thesis. São Paulo: Universidade de São Paulo.

[13] MITTERMEIER, R. A., N. Myers and C. G. Mittermeier. 1999. Hotspots – Earth's Biologically Richest and Most Endangered Terrestrial Ecoregions. Mexico City: CEMEX.

[14] ORR, D. W., 2004. Earth in mind: on education, environment, and the human prospect. Washington, DC: Island Press.

[15] SCHARMER, O., 2010. Seven acupunture points for shifting capitalism to create a regenerative ecosystem economy. Oxford Leadership Journal 1 (3): 1 -21.

[16] STERLING, S. 2001. Sustainable Education: Re-visioning Learning and Change. Totnes: Green Books Ltd.

[17] 2004. An Analysis of the Development of Sustainability Education Internationally: Evolution, Interpretation and Transformative Potential, in J. Blewitt and C. Cullingford (eds), The Sustainability Curriculum: The Challenge for Higher Education. London: Earthscan, 43–62.

[18] WALS, A.E.J. and B. Jickling. 2002. Sustainability in Higher Education: From Doublethink and Newspeak to Critical and Meaningful Learning. International Journal of Sustainability in Higher Education, 3 (3): 221–32.

[19] WALS, A.E.J and P.B. Corcoran. 2004. The Promise of Sustainability in Higher Education: A Synthesis, in P.B. Corcoran and A.E.J. Wals (eds), Higher Education and the Challenge of Sustainability: Problematics, Promise and Practice. Dordrecht: Kluver Academic Publishers, 223–25.

[20] SANTOS, M. (1982). Ensaios sobre a urbanização latino-americana. São Paulo: Hucitec.

[21] SODRÉ, M. (2002). Antropológica do Espelho. Uma teoria da comunicação linear e em rede. Petrópolis: Vozes.

13

Design of an Experimental Rig for Testing Multipurpose Blades

Bala Gambo Jahun[1], Fati Adamu Astapawa[2], Balogun Shuaibu Alani[3]

[1]Department of Agricultural and Bioresource Engineering Abubakar Tafawa Balewa University, Bauchi, Nigeria

[2]Department of Agricultural and Environmental Engineering, Modibbo Adama University of Technology, Yola, Nigeria

[3]Department of Mechanical Engineering, Federal Polytechnic, Bauchi, Nigeria

Email address:

bgjahun@yahoo.com (B. G. Jahun)

Abstract: The design of experimental rig for testing multipurpose blades was done to come up with a good blade that would be suitable for the pulverization of oil palm fronds and reduce toque, noise and vibration and fuel consumption during field operations. A good consideration on the strength of soil and fronds was done during the design. Mild steel was chosen for the frame design. From the design, work done to till the soil was found to be 34785.5Nm, diameter of the shaft was calculated to be 30mm, power transmitted by the belt and pulley is 104.4KN which falls between 20-150Kw and belt type D was selected. Belt length was 1,071.3mm which is within the standard length of 3127mm and belt D3127-IS: 2494 was selected. The electric motor to drive the blades carrier considering the power transmitted a 3Hp was selected and a gear motor to drive the experimental rig 4Hp was selected. It is expected that it would guide the manufacturers and stake holders in the oil palm industry to adopt better tractor mounted Mulcher for discarding oil palm fronds waste.

Keywords: Experimental Rig, Blades, Crop Residues, Fronds, Mulching

1. Introduction

In recent years, there has been an increasing concerned in having zero emission and with recent implementation of environmental policies by the government and increasing awareness of the benefits and importance of soil organic matter in sustaining crop production, agro-industrial wastes and crop residues are being returned to, or left in the field as soil ameliorants and sources of nutrients [1, 2]. The world's total annual production of OPW is estimated at about 184 million tonnes with about 5% annual increment [3]. The palm oil industry generates its wastes mainly in the form of lignocellulosic materials from the plantation and the palm oil milling processes. The extraction of 1 ton of crude palm oil (CPO) requires about 5 tonnes of fresh fruit bunches (FFB) which generates about 1.15 tonnes of EFB and 2.45 tonnes of palm oil mill effluents as residues [4]. The processing of an FFB (weighing 20-30 kg) generates about 20% of CPO, about 25% nuts (comprising about 5% kernels, 13% fibre and 7% shell) and about 23% EFB [5, 6]. These figures keep rising yearly as the demand for palm oil increases, and only about 10% of the generated OPW is utilized with the remaining 90% creating environmental burdens as their current disposal methods are unsafe. However, approaches of this kind carry with them various well known limitations to the soil and the infrastructure in the farm estate. It has become imperative to design suitable blades which would be used in a Tractor Mounted Mulcher by testing the blades using Multiuse blades Experimental Rig.

Organic mulching is one effective and established way to conserve soil and water. Utilization of oil palm residues such as pruned oil palm fronds (OPF) and empty fruit bunches (EFB) as a mulching material is a common conservation practice in oil palm plantations especially on non-terraced hill slopes [7, 8]. The popularity of using oil palm residues as mulching material is that oil palm produce large amount of biomass that have to be reused to avoid large amount of waste. Malaysia's palm oil industry produced 43 million tons of biomass [9].

The purpose of a tillage tool is to manipulate (change, move, or form) a soil as required to achieve a desired soil condition. There are three important initial factors to be considered in tillage tool design. This includes; initial soil condition, tool shape, and manner of tool movement as stated by [10]. They said that as a result of these three initial independent input factors, another two output factors would be generated, namely; the forces required for manipulating the soil and finally obtained the soil condition. All these five factors mentioned are directly connected to a tillage design. Among the initial three input factors, the designer has only complete control on the tool shape. The manner of the tool movement involves orientation of the tool, its path through the soil, and its speed along the path. For tools that travel in a straight line (i.e., not rotary or oscillating tools), the path is usually identified by merely specifying the depth and width of cut. Orientation of a tool with a particular shape may significantly affect both the soil manipulation and the forces. Most of the time, the linkage system used to position a tool and affects both depth of cut and orientation of the tool. When sufficient power is available, speed is the easiest design factor to vary. Increasing the speed generally increases draft but also affects soil movement and breakup.

Draft required during tillage is a function of soil properties, working depth, tool geometry, travel speed, and width of the implement. It is an important parameter for measuring and evaluating implement performance for energy requirements [11]. Several studies have been conducted to measure draft and energy requirements of tillage implements under various soil conditions [12]. Soil properties that contribute to tillage energy are moisture content, bulk density, cone index, soil texture, and soil strength.

One of the criteria used to assess the suitability of a tool for soil manipulation is the force required in dragging the tool through the soil [13-17]. The interaction between tillage tools and soil is of a primary concern to the design and use of these tools for soil manipulation [18]. The draft required for a given implement will also be affected by the soil conditions and the geometry of the tillage implements [15, 16, 19]. Hence, the soil-tool-tillage combination should be studied for a given location and tool geometry to optimize the tool performance and energy efficiency. A tillage tool, particularly the blade system, must pulverize the soil and mix it well with the oil palm residue to the desire degree and manipulate the soil sufficiently. The energy applied to the soil by the blades must be exploited efficiently in incorporating the crop residue into the soil. The power requirement per unit of soil tilled with the crop residue must be low. The capacity of the blade system must be high as reported by [20]. The soil parameters used to determine the performance of tillage blades are soil toss, soil volume disturbed, depth of penetration of the blades and soil condition.

2. Materials and Methods

A special experimental rig primarily consisting of a blade carrier suitable for mounting blades that operates vertically of different shapes was designed as shown in Figure 1. The experiment requires testing of four different blades and

picked the best among the blades borne in mind the torque, rate of pulverization, stress and strain on the blades, effective field capacity and fuel consumption. The overall size of the experimental rig was 2.4 x 1.2 x 3m with a weight of 1.5KN excluding the weight of the blades. The main frame is made of angle bar of 40 x 40mm with thickness of 2mm was used to support the load. The blades are to be mounted on the blade carrier of 590mm diameter having two holes were the blades are attached. The two holes were machined into round shapes. The blades would be fitted to the blade carrier with the aid of nuts and bolt and tightened firmly. An electric motor and a pulley were used to vary the speeds at 1000, 2000 and 3000 rpm. A dynamometer, a graduated ruler, stress strain gauges and fuel meter would be used to pull the data in the field.

In the working process, due to the advance movement of the equipment and the rotation movement of the rotor, the active parts penetrate into the soil and pulverize the oil palm crop residues with a particular shape of the blades in attachment. Under the action of centrifugal force, the soil slices are thrown over and mixes the residues with the soil at a depth of 30mm. As a result, the soil would be left behind the rotary rig in a loosened and mixed layer.

Proper calculation was done to avoid overload, and motor burn due to wrong estimation about load. The power transmission of the machine is by pulley and belt arrangement as shown in Figure 1. The speed of the blade carrier depends on:

i. The speed of the electric motor.
ii. The diameter of the pulleys.

The selected motor of 3hp delivers at speed of 1750 rpm which was connected to the pulleys. A diameter of the motor shaft pulley of 30mm was selected.

2.1. Parts of the Experimental Rig

The Experimental Rig is made up of the following units: the frame, electric motor, pulley, belt, chain and sprocket, driving wheel, disc plate, bevel gear, gear motor, bearing and coupler (Figure 1) and list of part is shown in Table 1 below.

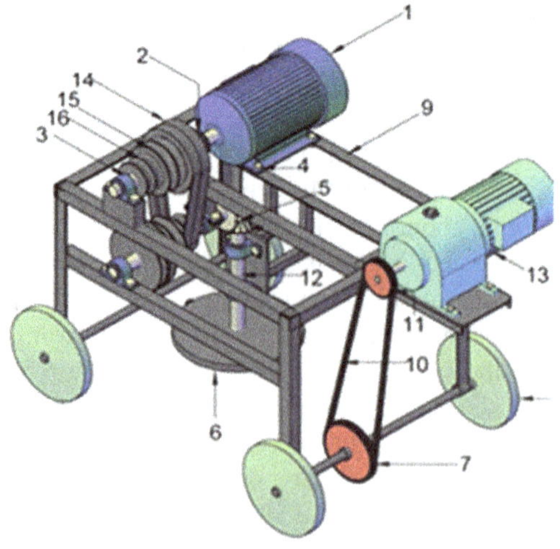

Figure 1. Diagram of the Experimental Rig

Table 1. *Parts of the Experimental Rig*

S/No	PART	QTY	SPECIFICATION
1	Electric Motor	1	3 Phase 3HP
2.	Coupling	1	D30x6mm, 60x6mm
3	Bearing	2	30x50mm
4.	Bolt/Nut	1	M12x50mm
5.	Bevel Gear	1	ID60x60, D30x60mm
6.	Disc Plate	1	OD570x19x50mm
7.	Gear	1	D200mmx50teeth
8.	Driving Wheel	4	D300x75mm
9.	Frame	1	50x50x5mm
10.	Chain Drive	1	Ø100mm
11.	Driving Sprocket	1	Ø100mmx30teeth
12.	Shaft	2	30x150mm 25x400mm
13.	Gear Motor	1	4HP, 100rpm
14	Belt	1	Ø300mm
15.	Large pulley	1	Ø210mm
16.	Small Pulley	1	Ø90mm

2.1.1. Electric Motor

Electric motor is the one of common device for rotating equipment and it is useful for smooth operation and makes our process faster and more efficient. 3Hp motor capacity and Speed of 1000, 2000 and 3000 rpm was considered for the load and speed.

2.1.2. Gear Motor

Gear motors are complete motive force systems consisting of an electric motor and a reduction gear train integrated into one easy-to-mount and -configure package. The reduction gear trains used in gear motors are designed to reduce the output speed while increasing the torque. The increase in torque is inversely proportional to the reduction in speed. Reduction gearing allows small electric motors to move large driven loads, although more slowly than larger electric motors. A 4Hp is used to drive the whole experimental Rig during operation.

2.2. Design Calculations

When analyzing a rotating system there are many calculations involved and a lot of assumptions made. This publication shows how the experimental rig was designed and methods considered. In designing a disc for machinery applications in soil and machinery dynamics, factors ought to be discussed as playing an important role like vibrations, thermal fatigue, analysis of blades etc. The starting point however, in rotating disc is always stresses due to inertia and these were primarily considered here.

2.2.1. Design of Disc

$$W_D = \frac{1}{2}\tau_s \times \pi \times Db \times \pi D \times z \quad (1)$$

Where;

W_D = Workdone to till the ground

τ_s = Shear strength of soil

D = diameter of disc

b = thickness of blade

z = factor of safety = 5

$$W_D = \frac{1}{2}\times 350\times 10^3 \times \pi \times 0.59\times 0.003\times \pi \times 0.59\times 5 = 34785.5 N_m$$

The power of the disc must be equal to the work to actualize tilling. Based on the equation above, the improved vibration response levels of discs rely on geometrically and physically accurate bladed disc models [22].

$$T = FR \quad (2)$$

Where;

T = Torque

F = force

R = speeds

$$1.13\times 10^3 \times \frac{0.59}{2} = 332.2 N_m$$

2.2.2. Determination of Disc Thickness

Increasing the thickness of the disc will make it stronger and less susceptible to bending or breaking,

$$X = \frac{2F}{D\sigma_s} \quad (3)$$

Where;

X = thickness of disc

σ = tensile strength of disc material

$$X = \frac{2\times 1.13\times 10^3}{0.59\times 55\times 10^3} = 0.1mm$$

3mm was chosen for flexural rigidity

2.2.3. Design of Shaft for Disc

A shaft is a rotating member, usually of circular cross section (either solid or hollow), transmitting power. It is supported by bearings and support gears, sprockets, wheels, rotors and is subjected to torsion and to transverse or axial loads, acting singly or in combination [21].

Using ASME CODE

$$d = \sqrt[3]{\frac{16}{\pi\tau_s}\sqrt{(k_b m_b)^2 + (k_t T)^2}} \quad (4)$$

Where;

d = diameter of shaft, mm

k_b = combined shock and fatigue factor applied to bending moment

k_t = combined shock and fatigue factor applied to torsional moment

T = Torque, N/m

m_b = bending moment

τ_s = torsional shear stress

$$d = \sqrt[3]{\frac{16}{\pi\times 205\times 10^6}\sqrt{(1.5\times 176.1)^2 + (1\times 332.2)^2}}$$

d = 22mm

30mm was chosen as the diameter of the shaft

2.2.4. Design of Shaft Based on Torsional Rigidity

Shafts are designed on the basis of torsional rigidity considerations. The total angle of twist θ in degrees is given by the Eq. (5). The permissible angle of twist or limiting value of twist for line shaft applications are 3° and 0.25° per m length of shaft respectively [23].

$$d = \sqrt[4]{\frac{584TL}{G\theta}} \qquad (5)$$

Where;

d = diameter of shaft, mm

T = applied torque, N/m

L = shaft length, mm

G = shear modulus of elasticity of the shaft material, N/m^2

θ = angle of twist (radians)

$$d = \sqrt[4]{\frac{584 \times 332.2 \times 0.53}{100 \times 10^9 \times 0.3}}$$

d = 43mm

45mm was chosen as the diameter of the shaft

2.2.5. Design of Shaft Based on Critical Speed

All rotating shaft, even in the absence of external load, deflect during rotation. The combined weight of a shaft and wheel can cause deflection that will create resonant vibration at certain speeds, known as Critical Speed. The Rayleigh-Ritz equation was used as in Eq. 6 which stated that maximum speed must not exceed 75% of critical speed [24]. Critical speed depends upon the magnitude or location of the load or load carried by the shaft, the length of the shaft, its diameter and the kind of bearing support.

$$N_c = \frac{30}{\pi} \sqrt{\frac{g}{\delta_{max}}} \qquad (6)$$

Where;

N_c = critical speed

δ_{max} = maximum deflection

For safety of operation let the speed equal to 75% of critical speed

$$d = \sqrt{\frac{100\pi}{0.7 \times 9.5 \times 78302.3}}$$

d = 25mm

30mm was chosen as the diameter of the shaft

2.2.6. Design of Bevel Gear

Torque application to a spiral bevel gear mesh induces tangential, radial, and separating loads on the gear teeth. For simplicity, these loads are assumed to act as point loads applied at the mid-point of the face width of the gear tooth. The radial and separating loads are dependent upon the direction of rotation and hand of spiral, in addition to

pressure angle, spiral angle and pitch angle. For transmitting motion between two perpendicular but coplanar shafts bevel gear were chosen [21].

I.   Number of teeth

$$\eta_T = \frac{48}{\sqrt{1 + (V.R)^2}} \qquad (7)$$

Where;

V.R = velocity ratio

η_T = 32

36 teeth was chosen

II. Determination of Weight of Gears

As typified by [25], the procedure consists of five basic steps: definition of gear geometry, calculation of applied loads, resolution of the loads into each structural member, sizing of required member cross-sectional areas, and calculation of component and total structural weight. These steps were considered in determination of the gears weight.

$$W_g = \rho Vg = 7850 \times \frac{\pi}{3} \times \left[R^2H - r^2h \right] g \qquad (8)$$

Where;

W_g = weight

ρ = density

V = velocity m/s

g = acceleration due to gravity

$$W_g = 7850 \times \frac{\pi}{3} \left[0.053^2 \times 0.038 - 0.039^2 \times 0.026 \right] \times 9.81$$

W_g = 6.4N

2.3. Bearing Selection

Each bearing should be selected according to the technical characteristics required by the specific application. Since the largest diameter obtained from the design is 45mm bearing number 209 was considered being the bore is 45mm.

2.3.1. Dynamic Radial Load

The choice of the appropriate type of bearing (radial, angular contact or thrust) is determined by the direction of the load which will act on the bearing. These bearings can accommodate a certain amount of axial load in addition to the radial load as its major advantage.

$$W = xVW_R + yW_A \qquad (9)$$

Where;

W_R = radial load, N

W_A = axial or thrust load, N

x = radial load factor

y = axial load factor

V = velocity factor

$$W = (0.56 \times 1 \times 1.13 \times 10^3) + 3 \times 134.6$$
$$W = 1.04 \text{KN}$$

2.3.2. Design of Basic Dynamic Radial Load Capacity

$$W_D = W \times K_s \qquad (10)$$

W_D = basic dynamic radial load capacity

K_s = service factor = 2 for mocerate shock load

$$W_D = (1.04 \times 2)$$
$$W_D = 2.08 \text{KN}$$

The dynamic and static basic capacities of bearing No. 209 are 25.5 and 18.3 respectively; therefore bearing number 209 is selected.

2.4. Selection of Belt and Pulley

Selecting the conveyor pulleys is started by specifying the belt tensions of the conveyor. Belt tensions come from the capacity calculations of the conveyor. Calculations can be made according to SFS-ISO5048.

$$\text{Power to be transmitted} = T\omega \qquad (11)$$

P = torque transmitted, N/m

T = tension on the tight and slack side of the belt, N

ω = velocity of the belt, m/s

$$P = 332.2 \times \frac{2\pi N}{60}$$

$$P = 332.2 \times \frac{2\pi 3000}{60}$$

$$P = 104.4 \text{KN}$$

From the design manual, the recommended belt for power range of (20-150) Kw is D. Therefore, belt type D was selected.

2.4.1. Determination of the Belt Length

To obtain good driving strength and good belt life, the belt pretension should be 1 to 8%, based on hardness and length of the belt [27]. For selected belt length, center distance, belt number has been determined by using the following Eq. 12 below.

$$L = \pi(r_1 + r_2) + 2X + \frac{(r_1 - r_2)^2}{X} \qquad (12)$$

L = total length of the belt pulley, mm

r_1 and r_2 = radii of the larger and smaller pulleys

x = distance between the centres of two pulleys, mm

$$L = \pi \times 150 + 600 + \frac{2500}{300} = 1,071.3 \text{mm}$$

For belt D the smallest standard length is 3127mm and therefore belt D3127- IS: 2494 was selected.

2.4.2. Maximum Tension in Belt

$$T_{max} = \sigma \times a \qquad (13)$$

T_{max} = maximum tension in belt, N

σ = maximum safe stress

a = width of the belt, mm

$$T_{max} = 2 \times 10^6 \times 504 \times 10^{-4}$$

$$T_{max} = 1030 \text{N}$$

2.4.3. Pulley Diameter and Selection

It is important that pulley is of sufficient diameter and the belt of sufficient section to transmit the required horsepower to the designed experimental rig. The minimum pulley diameters of a belt conveyor installation will be determined by the design and layout, stresses and splicing method of the belt. Based on torsional moment taking factor of safety as 5, the working stress 82MP the pulley diameter was calculated using Eq. 14 below.

$$d = \sqrt[3]{\frac{16T}{\pi\tau}} \qquad (14)$$

$$d = \sqrt[3]{\frac{16 \times 332.2}{\pi \times 82}}$$

d = 27mm

50mm was selected

For the small pulley, velocity ratio of 1: 2

$$\frac{N_1}{N_2} = \frac{1}{2} = \frac{D_2}{D_1} \qquad (15)$$

N_1 = shaft speed in rev/minute

N_2 = driven pulley speed in rev/minute

D_1 = diameter of the driving pulley, mm

D_2 = diameter of the driven pulley, mm

$D_1 = 2D_2 = 100$mm

2.5. Design of Middle Shaft

Weight of pulley and the volume of pulley are approximated by the following formula.

$$W = \frac{\pi}{3}\left[R^2H - r^2h\right] \qquad (16)$$

$$W = \frac{\pi}{3}\left[58^2 \times 230 - 33^2 \times 100\right]$$

$$W = 7 \times 10^{-4} \text{m}^3$$

$$\text{Weight} = \rho v g \qquad (17)$$

Where;

ρ = density

v = velocity, m/s

g = acceleration due to gravity

$$\text{Weight} = 7850 \times 7 \times 10^{-4} \times 9.81 = 54 \text{N}$$

2.5.1. Determination of Shaft Diameter Based on ASME CODE

The shaft diameter can be calculated in terms of external loads and material properties. However, the below equation is further standarised for steel shafting in terms of allowable design stress and load factors in ASME design code for shaft.

$$d = \sqrt[3]{\frac{16}{\pi\tau_s}}\sqrt{(k_b m_b)^2 + (k_t T)^2} \qquad (18)$$

Where;

d = diameter of shaft, mm

k_b = combined shock and fatigue factor applied to bending moment

k_t = combined shock and fatigue factor applied to torsional moment

T = Torque, N/m

m_b = bending moment

τ_s = torsional shear stress

$$d = \sqrt[3]{\frac{16}{\pi \times 205 \times 10^6}}\sqrt{(1.5 \times 126.5)^2 + (1 \times 332.2)^2}$$

$d = 21mm$

30mm was chosen

2.5.2. Determination of Shaft Diameter Based on Torsional Rigidity

$$d = \sqrt[4]{\frac{584TL}{G\theta}} \qquad (19)$$

Where;

d = diameter of shaft, mm

T = applied torque, N/m

L = shaft length, mm

G = shear modulus of elasticity of the shaft material, N/m^2

θ = angle of twist (radians)

$$d = \sqrt[4]{\frac{584 \times 332.2 \times 0.53}{100 \times 10^9 \times 0.3}}$$

$d = 43mm$

45mm was chosen

2.6. Selection of Electric Motor

Considering the power to be transmitted and to determine the appropriate electric motor for the experimental rig Eq. 20 was applied below.

$$P = (T_1 - T_2)V \qquad (20)$$

Where;

P = power transmitted

T_1 = tension on the tight side

T_2 = tension on the slack side

V = velocity of the belt in m/s

T_c = centrifugal tension acting tangentially

$T_c = mV^2$

For belt type D $m = 5.96N / m$

$$V = \frac{\pi DN}{60} \qquad (21)$$

D = diameter for the pulley shaft, mm

N = speed of pulley in rpm

$V = 2.6m / s$

$P = (989.7 - 267.5) \times 2.6$

$$P_{hp} = \frac{1878}{750} = 2.5Hp$$

3Hp was selected

Power required driving the Shaft

$$P = T\omega \qquad (22)$$

$$P = 1200 \times \frac{2 \times \pi N}{60}$$

$$P = \frac{1200 \times 2 \times \pi \times 50}{60}$$

$$P = 3142Nm / s$$

4Hp was chosen

2.7. Blades Shapes and Design

Four different blades were design (with different lifting angle of 90°, 120°, 60° and curved which serves as control) and develop with CAD methods like 3D modeling and analysis was used. Accordingly at the initial stage 3D Modeling was done on the basis of geometrical parameters normally which are similar to a commercially available blades using 3D CAD software. This model was analyzed through Solid Works for Finite Element Analysis particularly to investigate the main causes of wear and deformation as shown in Fig 2 below.

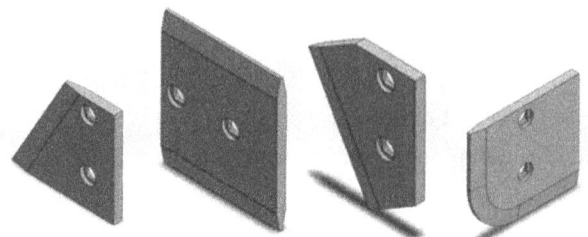

Figure 2. Different Blades with lifting angles of 60°, 90°, 120° and Curved.

3. Conclusion

The design and development of Multiuse laboratory blades testing experimental rig has been described. Appropriate design consideration and procedures have been taken into account to ensure the robustness of the machine. What finally counts is the performance of the product, not a good figure of some kind especially since there are a lot of failure conditions that are unpredictable by calculations. Nevertheless, accurate analyzes are necessary to compare different designs and to create a robust product that fulfills the customers´ demands.

The dispensation of oil palm fronds will be enhanced to achieve the cleaner environment, improve soil nutrient and control soil erosion on relatively large scale for domestic and industrial uses.

It will guide the manufacturers and stake holders to modify the components of the existing machine which entails promotion of technology transfer and adoption for the pulverization of oil palm residues.

It is expected that the patronage by peasant farmers and other users of the new machine so modified will reduce extremely the labour, drudgery and cost involved in steering the Tractor mounted Mulcher for making it a user friendly.

Agricultural sector in Malaysia and other parts of the world where palm trees are grown will be enhanced once again through the use of such modified machines aimed at adding values to the present Mulcher used by the farmers.

The economy of Malaysia will be enhanced since embracing of such implement will help in clean environment and rich soil that would increase oil palm production, easy accessibility during replanting period and eradicate breeding ground of the Rhinoceros Beetle, a serious pest for oil palm plantations.

Acknowledgements

This research was funded by the Howard Alatpertanian Sdn Bhd, Malaysia and wish to acknowledge the moral and financial support towards the research.

References

[1] Bundy, L., T. Andraski, and R. Wolkowski, *Nitrogen credits in soybean-corn crop sequences on three soils.* Agronomy journal, 1993. 85(5): p. 1061-1067.

[2] Rembon, F. and A. MacKenzie, *Soybean nitrogen contribution to corn and residual nitrate under conventional tillage and no-till.* Canadian journal of soil science, 1997. 77(4): p. 543-551.

[3] Shuit, S. H., et al., *Oil palm biomass as a sustainable energy source: a Malaysian case study.* Energy, 2009. 34(9): p. 1225-1235.

[4] Corley, R. and P. Tinker, *The palm oil.* World Agriculture Series, 2003.

[5] Corley, R. H. V. and P. Tinker, *The oil palm.* 2008: John Wiley & Sons.

[6] Yusoff, S., *Renewable energy from palm oil–innovation on effective utilization of waste.* Journal of cleaner production, 2006. 14(1): p. 87-93.

[7] Anderson, J. M., *Eco-friendly approaches to sustainable palm oil production.* J. Oil Palm Res., 2008: p. 127-142.

[8] Moradi, A., et al., *Effect of four soil and water conservation practices on soil physical processes in a non-terraced oil palm plantation.* Soil and Tillage Research, 2015. 145: p. 62-71.

[9] Chang, S. H., *An overview of empty fruit bunch from oil palm as feedstock for bio-oil production.* Biomass and Bioenergy, 2014. 62: p. 174-181.

[10] Abubakar, M. S. and Shittu, S. K, *Determination of Physical and Mechanical Properties of a soil related to the Design of Tillage Implement.* Proceedings of the 33rd National Conference and Annual General Meeting of the Nigerian Institution of Agricultural Engineers, 2012 33: p. 41-49.

[11] Chandon, K. and R. Kushwaha. *Soil forces on deep tillage tools.* in *The AIC 2002 Meeting CSAE/SCGR program Saskatoon, Saskatchewan, Canada July.* 2002.

[12] Grisso, R., M. Yasin, and M. Kocher, *Tillage implement forces operating in silty clay loam.* Transactions of the ASAE, 1996. 39(6): p. 1977-1982.

[13] McLaughlin, N., et al., *Energy inputs for conservation and conventional primary tillage implements in a clay loam soil.* Transactions of the ASABE, 2008. 51(4): p. 1153-1163.

[14] Naderloo, L., et al., *Tillage depth and forward speed effects on draft of three primary tillage implements in clay loam soil.* Journal of Food, Agriculture and Environment, 2009. 76(3): p. 382-385.

[15] Olatunji, O. and R. Davies, *Effect of weight and draught on the performance of disc plough on sandy-loam soil.* Research Journal of Applied Sciences, Engineering and Technology, 2009. 1(1): p. 22-26.

[16] Gill, W. R. and G. E. V. Berg, *Soil dynamics in tillage and traction.* 1967: Agricultural Research Service, US Department of Agriculture.

[17] Shen, J. and R. L. Kushwaha, *Soil-machine interactions: a finite element perspective.* 1998: Marcel Dekker Inc.

[18] Taniguchi, T., et al., *Draft and Soil Manipulation by Amoldboard Plow under Different Forward Speed and Body Attachments.* Transactions of the ASAE, 1999. 42(6): p. 1517.

[19] Yadav, B., I. Mani, and J. Panwar, *Design of tool carrier for tillage studies of disc in field conditions.* AGRICULTURAL MECHANIZATION IN ASIA AFRICA AND LATIN AMERICA, 2007. 38(2): p. 29.

[20] Khurmi, R. and J. Gupta, *Theory of machines.* 2005: S. Chand.

[21] Chan, Y. J., *Variability of blade vibration in mistuned bladed discs.* 2009, Department of Mechanical Engineering, Imperial College London.

[22] Naunheimer, H., et al., *Automotive transmissions: fundamentals, selection, design and application.* 2010: Springer Science & Business Media.

[23] Prabhakar, S., A. Sekhar, and A. Mohanty, *Transient lateral analysis of a slant-cracked rotor passing through its flexural critical speed.* Mechanism and machine theory, 2002. 37(9): p. 1007-1020.

[24] Kawalec, A., J. Wiktor, and D. Ceglarek, *Comparative analysis of tooth-root strength using ISO and AGMA standards in spur and helical gears with FEM-based verification.* Journal of Mechanical Design, 2006. 128(5): p. 1141-1158.

[25] Allen, S., Alfred, RH, and Herman, GL, *Machine Design.* Tata McGraw-Hill Publishing company Ltd. New Delhi, 2004 p. 101-127.

[26] Good, C. A., D. C. Viano, and J. L. Ronsky, *Biomechanics of volunteers subject to loading by a motorized shoulder belt tensioner.* Spine, 2008. 33(8): p. E225-E235.

14

Comparative Evaluation of Biodiesel Produced by the Improved Biodiesel Plant with a Centrifuge for Ester-Glycerol Separation and a Single Washing and Drying Unit

Nwogu Chukwunonso

Department of Mechanical Engineering, Michael Okpara University of Agriculture Umudike, Umuahia, Nigeria

Email address:

checknolly@yahoo.com

Abstract: This paper assessed the quality of biodiesel produced by the improved biodiesel plant. This was achieved by determining some physical properties of the biodiesel produced. The properties tested include: kinematic viscosity, specific gravity, flash point, fire point, cloud point and pour point, with their respective values determined as 5.260, 0.880, 124°C, 128°C, 4.5°C and -2°C. Comparison with ASTM D-6751, EN 14214 and Australian standards for biodiesel showed that the values obtained fall within acceptable limits. Furthermore, B5, B10, B20 and B30 blends of the biodiesel were prepared using direct blending method. The physical properties of the respective blends were also determined and the results showed that blending the biodiesel with petroleum based diesel improved its quality.

Keywords: Improved Biodiesel Plant, Physical Properties, Blends

1. Background

Diesel engine was developed in the 1890s by Rudolf Diesel out of a desire to improve upon inefficient, cumbersome and sometimes dangerous steam engines of the late 1800s. Since its development, the diesel engine has become the engine of choice for power, reliability, and high fuel economy, worldwide. Early experiments showed that pure vegetable oil could power early diesel engines for agriculture in remote areas of the world, where petroleum was not available at the time. The first public demonstration of vegetable oil based diesel fuel was at the 1900 World's Fair, when the French government commissioned the Otto Company to build a diesel engine to run on peanut oil. Since then, many other vegetable oils have been intensively studied as raw materials for biodiesel production. These include: soybean oil, palm oil, castor oil, *Parkia biglobbossa, Jatropha curcas*, sunflower oil, coconut oil, rapeseed oil, safflower oil, ground nut oil, Neem oil and cotton seed oil [1-6]. Shortly after Dr. Diesel's death in 1913 petroleum became widely available in a variety of forms, including the class of fuel we know today as "diesel fuel".

With petroleum being available and cheap, the diesel engine design was changed to match the properties of petroleum diesel fuel. The result was an engine which was fuel efficient and very powerful. Due to the widespread availability and low cost of petroleum diesel fuel, vegetable oil-based fuels gained little attention, except in times of high oil prices and shortages until World War II and the oil crises of the 1970's which saw brief interest in using vegetable oils to fuel diesel engines. Unfortunately, the newer diesel engine designs can no longer run on traditional vegetable oils, due to the much higher viscosity of vegetable oil compared to petroleum diesel fuel. The need to lower the viscosity of vegetable oils to a point where they could be burned properly in the diesel engine arose. Diesel fuel produced from vegetable oil is known as Biodiesel.

Methods used for reducing the viscosity of vegetable oils at the inception include pyrolysis and blending with solvents/emulsification of the fuel with water or alcohols, none of which provided a suitable solution. Hence, a Belgian inventor in 1937 proposed using transesterification to convert vegetable oils into fatty acid alkyl esters and use them as a diesel fuel replacement. The transesterification reaction is illustrated in Equation (1) below:

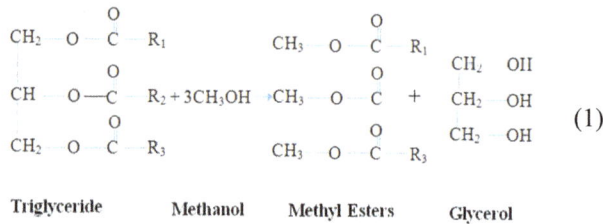

$$ (1) $$

Triglyceride   Methanol   Methyl Esters   Glycerol

Transesterification is the chemical process of converting an organic acid ester to another ester of the same acid [7, 8]. The process of transesterification converts vegetable oil into three smaller molecules which are much less viscous and easy to burn in a diesel engine. The transesterification reaction is the basis for the production of modern "biodiesel", which is the trade name for fatty acid methyl/ethyl esters produced by the reaction of vegetable oil with an alcohol and a chemical catalyst such as sodium hydroxide. Transesterification is the most attractive method of producing biodiesel, since this process decreases viscosity while maintaining the octane number and reduces polymerization during storage and combustion. The reaction flow chart for biodiesel production through transesterification is shown in Figure 1 below.

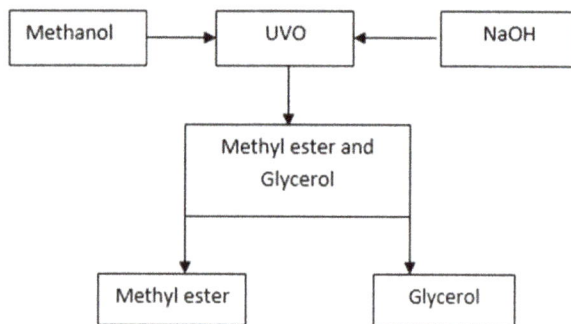

Figure 1. Biodiesel production flow chart.

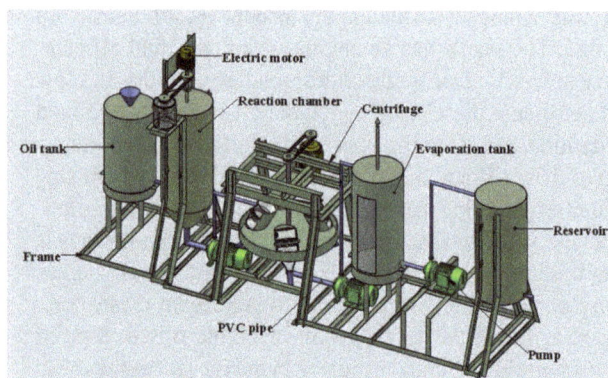

Figure 2. Isometric view of the improved plant [10].

Over the years, various plants have been developed for producing biodiesel from vegetable oil. Some of these plants include: Pacific biodiesel plant, Big Island biodiesel plant, Yoshida Kosan Biodiesel Plant, Keystone Biodiesel Plant and many others. Although each of these plants have unique design and features, they all have the same principle of operation and are all faced with the same challenge of prolonged methyl/ethyl ester-glycerol separation time and

relatively high energy consumption.

Based on this, an improved biodiesel Plant was developed which reduces the methyl/ethyl ester-glycerol separation time by introducing a centrifuge for the separation [9, 10]. The improved plant also consumes less energy since the neutralization washing and drying of the ester was designed to take place in a single vessel. Figure 2 below shows the isometric view of the improved plant with the unit labelled.

Apart from the high price of petroleum based diesel, other advantages of biodiesel include:

i. Biodiesel is a clean fuel, emitting over 40% less Carbon dioxide into the atmosphere compared to fossil fuels [11].
ii. It does not contain any Sulphur and as such does not emit any dangerous oxides into the atmosphere when it burns.
iii. It has high combustion efficiency and high lubricity.
iv. It reduces engine wear thereby increasing the life of the fuel injection equipment [12].
v. Biodiesel has a higher flash point of about 130°C compared to that of petroleum diesel which is about 64°C. Hence it is less liable to explode than petroleum diesel making them safer to handle.

Biodiesel can be used in pure form or blended with petroleum diesel at any level. Even a blend of 20% biodiesel and 80% petroleum diesel will significantly reduce carcinogenic emissions and gases that may contribute to global warming.

2. Methodology

2.1. Procedures for Quality Assessment of the Biodiesel Produced

Some physical properties of the biodiesel produced by the Improved Biodiesel Plant [10] were tested in accordance with ASTM D6751, EN14214 Australia standards. The parameters tested include;

i. Specific gravity
ii. Flash and Fire points
iii. Kinematic Viscosity
iv. Cloud and Pour points

2.1.1. Specific Gravity Test Procedure

Water is always used as a reference substance in the determination of specific gravity of liquids. Specific gravity is defined as the ratio of the density of a substance to the density of water. The density of the biodiesel produced was determined at ambient temperature (28°C). A density bottle of mass 50ml was weighed on the analytical balance and the initial weight of the bottle was noted. The sample was put in the density bottle, the spillage was cleaned and dried, and the bottle was weighed on the analytical balance. The process was repeated twice and the average value was determined as the result. Density of the sample was then computed using equation (2) below.

$$ D = \frac{M_2 - M_1}{V} \qquad (2) $$

Where D = Density in g/m³
M_2 = Mass of bottle and sample (g)
M_1 = Mass of bottle only (g)
V = Volume of the Liquid (m³)

2.1.2. Flash and Fire Points Test Procedure

Flash point of the biodiesel was determined using the setup apparatus comprising of an electric heater, beaker and thermometer. 80ml of the sample was introduced into a transparent Pyrex beaker placed on an electric heater. The beaker was fitted with the thermometer, clamped on a retort stand. Heat was applied gradually by turning the knob of the electric heater until the observed movement of the particles increased. A flame was constantly brought near the surface of the beaker until "a catch and disappearing" of flame on the surface of the hot liquid occurred. The temperature was noted as the flash point.

The biodiesel sample was further heated with continuous bringing of flame close to the vapor, the temperature at which the sample ignites without the flame disappearing from the surface of the sample was noted as the Fire Point.

2.1.3. Viscosity Test Procedure

The viscosity of the biodiesel was determined using a glass capillary kinematic viscometer. The viscometer was tightly clamped on a retort stand. 100g of the sample was collected into a Pyrex beaker and was gradually heated to a temperature above 40°C. The biodiesel sample was then transferred into the viscometer through the larger opening of the capillary tube and the fluid was allowed to cool until a temperature of 40°C was reached. Thereafter, suction was applied to the other end of the capillary tube to draw the fluid to the mark on the upper meniscus level of the capillary tube.

The fluid was allowed to run freely to the lower meniscus mark in the capillary tube. The efflux time for the fluid to flow from the upper meniscus mark to the lower meniscus mark was determined with the aid of a stopwatch. The test was triplicated for each sample and the kinematic viscosity was calculated using equation (3).

$$\text{Kinematic Viscosity} = kt \qquad (3)$$

Where k = Constant of the viscometer expressed in mm^2/s^2
t = Flow time in seconds of the liquid.

2.1.4. Cloud and Pour Points Test Procedure

The cloud and pour points of the biodiesel were analyzed by reducing their temperatures. The biodiesel sample was introduced into a transparent cylindrical glass and placed in a cooling bath containing crushed ice. The temperature of the biodiesel sample was monitored as it decreased, the temperature at which the clear sample loses its sharpness due to cloud formation was noted as the sample's cloud point.

Hence, the sample was further cooled with an attempt to pour the samples at every increase in temperature. The sample was inspected at every one-minute interval by tilting it horizontally. The temperature at which a sample could not pour freely for five seconds was noted as the pour point.

2.2. Biodiesel Standards

The properties of biodiesel should fall within certain acceptable limits in order to function properly in the conventional diesel engine without causing any form of harm to the engine. In accordance with ASTM D-6751, EN 14214 and Australian biofuels standard, biodiesel produced by any means/method should have the following specifications (Table 1).

Table 1. International Biodiesel Standards (Mohammed et al., 2013).

Properties	Units	ASTM D-6751	EN 14214	Australia
Viscosity, 40°C	mm²/sec	1.9-6.0	3.5-5.0	3.5-5.0
Density	gm/m³	n/a	0.860-0.900	0.860-0.900
Cetane number	-	47min	51min	51min
Flash point	°C	130min	120min	120min
Cloud point	°C	Report	Report	Report
Acid number	mgKOH/g	0.80max	0.5max	0.8max
Free glycerine	Wt.%	0.02max	0.02max	0.02max
Total glycerine	Wt.%	0.24max	0.25max	0.25max
Iodine number	-	-	120max	n/a
Oxidation stability	H	-	6min	n/a
Monoglyceride	Mass(%)	-	0.8 max	n/a
Diglyceride	Mass(%)	-	0.2max	n/a
Triglyceride	Mass(%)	-	0.2max	n/a
CFPP	°C	-	-	-4

3. Results and Discussion

The results of these tests as shown in Table 2 indicate that the specific gravity, flash point and kinematic viscosity of the biodiesel produced by the Improved Biodiesel Plant falls within acceptable limits as specified by ASTM D 6751, EN 1421 and Australian standards for biodiesel. The fire point, cloud point and pour point of the produced biodiesel are 128°C, 4.5°C and -2°C respectively. Although the ASTM D 6751, EN 1421 and Australian standards did not report any precise values for these properties, there is a notable reduction in their respective values of 261°C, 7.2°C and 2.5°C before the transesterification reaction occurred. It is obvious from Table 2 that the physical properties of the biodiesel produced by the improved plant are similar to those of the petroleum based diesel, compared to the properties of the raw vegetable oil. This is illustrated in Table 2 below:

Table 2. Physical properties of the Biodiesel in comparison with ASTM D 6751, EN 1421 and Australian standards.

PROPERTIES	Vegetable Oil	Biodiesel	ASTM D6751	EN 14214/Australia	Petroleum Diesel
Kinematic Viscosity @ 40°C	35.979	5.260	1.9 – 6.0	3.5 – 5.0	3.96
Specific Gravity @ 28°C	0.913	0.880	0.860 – 0.9	0.860 – 0.9	0.850
Flash Point (°C)	256	124	130min	120min	172
Fire Point (°C)	261	128	-	-	180
Cloud Point (°C)	7.2	4.5	-	-	-7
Pour Point (°C)	2.5	-2	-	-	-16

Table 3. Properties of biodiesel blends in comparison with ASTM D6751, EN 14214 and Australian standards.

PROPERTIES	B30	B20	B10	B5	ASTM D6751	EN 14214/Australia
Kinematic Viscosity @ 40°C	4.520	4.302	4.177	3.998	1.9 – 6.0	3.5 – 5.0
Specific Gravity @ 28°C	0.881	0.881	0.872	0.860	0.860 – 0.9	0.860 – 0.9
Flash Point (°C)	146	151	161	168	130min	120min
Fire Point (°C)	155	163	169	178	-	-
Cloud Point (°C)	2.5	0.2	-1	-5	Reported	Reported
Pour Point (°C)	-8	-10	-12	-14	Reported	Reported

These properties can further be improved through blending of the biodiesel with petroleum based diesel in different ratios. Biodiesel blends were prepared by direct blending method which involved mixing of 5ml, 10ml, 20ml and 30ml of the biodiesel produced with 95ml, 90ml, 80ml and 70ml of petroleum diesel respectively in a transparent bottle to produce B5, B10, B20 and B30 biodiesel blends. The physical properties of the various blends were also determined and the results are as shown in Table 3.

The values of the kinematic viscosity, specific gravity, flash point, fire point, cloud point and pour point obtained from the various blends are shown in Table 3 above. In the above table, B30 represents the blend with the highest quantity of biodiesel while B5 represents the blend with the least quantity of biodiesel. A little consideration of Table 2 and Table 3 reveals that although the properties of all the blends fall within acceptable limits, the blend B30 has properties closest to those of the unblended biodiesel (B100) compared with the other blends while the blend B5 has properties closest to those of the petroleum based diesel. This is simply because B30 contains the least quantity of petroleum diesel (70%) while B5 contains the highest quantity of petroleum diesel (95%). This implies that the blend B5 will be the most suitable blend for most diesel engines although its emissions will be the most harmful compared to other blends while the blend B30 may not be suitable for all engines but will be the safest among all the blends because of its very low emission profile.

The experimental results presented in Tables 2 and 3 do not show conformity with ASTM D 6751, EN 1421 and Australian standards for biodiesels alone, but also show conformity with results obtained by other authors from similar experiments performed on biodiesels produced by other plants/means. Some of such works include: Samuel *et al* [13] whose experiment gave Kinematic Viscosity, specific gravity and flash point values as 4.72, 0.879 and 139°C respectively and Alamu *et al* [1] with values of Kinematic Viscosity, specific gravity, flash point, cloud point and pour point as 4.839, 0.883, 167°C, 6°C and 2°C respectively.

4. Conclusion

The kinematic viscosity at 40°C, specific gravity at 28°C, flash point, fire point, cloud point and pour point of biodiesel produced by the improved biodiesel plant were determined experimentally as 5.260, 0.880, 124°C, 128°C, 4.5°C and -2°C respectively. These values were compared with the ASTM D6751, EN 14214 and Australian standards for biodiesels and found to fall within recommended range of values for the respective physical properties. The biodiesel

was further blended with petroleum based diesel in different ratios to obtain B5, B10, B20 and B30 blends respectively. Physical properties of the various blends were also determined and the result showed that B30 blend has properties closest to those of the unblended biodiesel (B100) compared with the other blends while the B5 blend has properties closest to those of the petroleum based diesel.

Hence, it can be concluded that the improved biodiesel plant is capable of producing biodiesel which can be used in conventional diesel engines without a need for engine modification.

References

[1]   Alamu O. J., Waheed M. A. and Jekayinfa S. O. "Alkali-catalysed Laboratory Production and Testing of Biodiesel Fuel from Nigerian Palm Kernel Oil". *Agricultural Engineering International*, 2007, 9: 1-11.

[2]   Aransiola E. F., Daramola M. O. and Ojumu T V. "Nigeria Jatrophacurcasoil seeds: prospects for biodiesel production in Nigeria". *International Journal of Renewable Energy Research.* 2(2), 2012, 317–325.

[3]   Akoh C. C., Chang S., Lee G. and Shaw J. "Enzymatic approach to biodiesel production". *J. Agric. Food Chem.* 55, 2007, 8995 – 9005.

[4]   Berchmans H. J. and Hirata S. *Biodiesel production from crude Jatrophacurcas L. seed oil with a high content of free fatty acids.* Bioresource Technology, 2008.

[5]   Robles-Medina A., Gonzalez-Moreno P. A., Esteban –Cerdán L and Molina-Grima E. "Biocatalysis: Towards evergreener biodiesel production". Biotechnol. Adv. 27, 2009, 398-408.

[6]   Hasibuan S., Ma'ruf A. and Sahirman. *Biodiesel from low grade used frying oil using esterification transesterification process.* MakaraSains 13(2), 2009, 105-110.

[7]   Knothe, G. "Analyzing biodiesel: Standards and other methods". *Journal of American Oil Chemists Society*, 83(10), 2006, 823-833.

[8]   Kemp, W. H. *Biodiesel: Basics and beyond: A comprehensive guide to production and use for the home and farm.* Tamworth, Ont.: Aztext Press, 2006.

[9]   Nwogu Chukwunonso. "Design and Development of an Industrial Centrifuge for Small and Medium Scaled Industries", *Innovative Systems Design and Engineering*, ISSN 2222-1727 Vol. 6, No. 10, 2015, 1-9.

[10] Nwogu C. N. and Obi A. I. "Development of an Improved Biodiesel Plant", *International Journal of Science and Research* (IJSR) ISSN (Online), 2015, 2319-7064.

[11] Okoro L., Sedoo V. Belaboh, Nwakama R. Edoye and Bella Y. Makama. "Synthesis, Calorific and Viscometric Study of Groundnut Oil Biodiesel and Blends", *Research Journal of Chemical Sciences* ISSN 2231-606, 2011.

[12] Emil A., Z. Yaakob, S. K. Kamarudin, M. Ismail and J. Salimon. "Characteristics and composition of jatropha curcas oil seed from Malaysia and its potential as biodiesel feedstock". *Eur. J. Sci. Res*, 29, 2009, 396-403.

[13] Samuel O. D., Waheed M. A., Bolaji B. O. and Dario O. U. "Production of Biodiesel from Nigerian Restaurant Waste cooking oil using Blender", *International Journal of Renewable Energy Research* Vol. 3, No. 4, 2013.

15

Microbial Fuel Cell for Electricity Generation and Waste Water Treatment

Marwan Mosad Ghanem, Omar Mohamed Al Wassal, Abdelrahman Ahmed Kotb, Mohamed Ayman El-Shahhat

STEM Egypt High School for Boys, 6th of October City, Egypt

Email address:

14153@stemegypt.edu.eg (M. M. Ghanem), 14159@stemegypt.edu.eg (O. M. Al Wassal), 14158@stemegypt.edu.eg (A. A. Kotb), 14105@stemegypt.edu.eg (M. A. El-Shahhat)

Abstract: Energy problem is a global issue that has a serious effect on many countries in the world. The demand for energy is currently growing far greater than the supply of the nationally generated energy. In order to overcome energy crisis and the output pollution of the generation, it is suggested to use the Microbial Fuel Cell (M.F.C). M.F.Cs are devices that use bacteria as a catalyst to oxidize organic and inorganic matters and generate electric current. With the modifications that are suggested by this study, there will be clean, available and suitable energy source. Besides having the property of being efficient, eco-friendly and cheaper than the other resources, it can utilize the sewage to generate electricity and produce clean water, which mean cleaner environment with a great supply of energy and clean drinkable water. Facing challenges that affect the development of the M.F.Cs is an important aspect to be studied, so our study suggests solutions nearly for all the challenges facing them like the types of the electrodes, output pollutants, the catalyst in the cathode chamber and the real application.

Keywords: Microbial Fuel Cell, Anode, Bacteria, Cathode, Agar, Biofilm, Catalyst

1. Introduction

Renewable energy field is the majority of the current age because of its importance in developing the humanity. The demand for energy nowadays is going far greater than the generated energy. There are many renewable and nonrenewable energy sources like the fossil fuels and wind energy. There is no effective way to keep us working with fossil fuels and reduce greenhouse emissions. The efficiency is an essential component of any plan to get us back on the track of balanced growth. So, new energy sources and technologies must be developed to achieve the required efficiency.

Also, their output pollution should be taken in mind as it may affect the environment. As a result, these new solutions should produce carbon dioxide at a lower rate. If we see the example that in man's life-essential activities, wastes are produced, a technology that can use these wastes and turn them into useful energy, is probably the most close-to-Nature form of energy production.

In order to decrease the pollution and find an eco-friendly energy source, it is suggested to use the microbial fuel cells (MFC). MFCs are devices that benefit from the natural metabolism of microbes to produce electrical power. There are many challenges facing the MFCs such as the output pollutants and how to reduce them, best electrodes to get the highest performance, suitable substrate to provide food for the bacteria (microbes), facing limiting factors of the MFCs, the cost of the catalyst and its performance, the exchange field, and the best design for all the project.

Unlike costly and difficult to be obtained energy alternatives, this device provides inexpensive energy source for any person at any time. As a result from these challenges, this research is proposed to be figured out to help solving them, that's why there can be found new, efficient and clean energy sources. Also, it is said to help finding clean environment with more energy sources and clean water from the technique of wastewater treatment by the M.F.C projects.

2. Related Works

Harnessing the metabolic activity of bacteria can provide energy for a variety of applications, once technical and cost obstacles are overcome. [10]. Microbial fuel cells (MFCs) can provide an answer to several of the problems which traditional wastewater treatment faces [14]. Butyrate is used as a substrate for the bacterial field. [11]. Power generated with acetate found to be higher when compared with other substrate [2, 11]. The waste water was used as a substrate to provide food for the bacteria and produce the highest output voltage [12]. (MFC) is one such renewable and sustainable technology that is considered to be one of the most efficient energy sources [15]. Ferricyanide (K3 (Fe (CN) 6) is frequently used as an electron acceptor in the MFCs due to its good performance and low over potential [8]. The main challenge in implementing MFC on a large scale is in maintaining low costs, minimizing hazards while maximizing power generation [16]. For cathodes, platinum (Pt), activated carbon (AC), graphite based cathodes and bio cathodes are used [3, 4]. For anode, carbon felt, graphite felt, carbon mesh and graphite fiber brush are frequently used due to their stability, high electric conductivity and large surface area [9, 8]. "Recently, some students from Harvard University experimented MFC device and they produced enough electricity to power a LED bulb for up to a year" [7]. The composition, concentration and type of the substrate affect the microbial community and power production [1, 13]. In most of the MFCs, acetate is commonly used as a substrate due to its inertness towards alternative microbial conversions (fermentations and methanogenesis) that lead to high efficiency and power output [13]. One of the limiting factors is cost of the electrode and membrane materials [9]. Temperature is an important limiting factor in the MFCs. [6].

If the challenges facing the MFCs could be solved, then a new efficient and clean energy source will be found to help in developing the humanity and getting new and magnificent world. Also, our group predicted the efficiency of the project to be 73:80% greater than the other attempts in the M.F.Cs field, which mean huge electrical production, with having the quality of being clean and green to the environment.

3. Methodology

3.1. Participants

We are a group of 4 stem students from STEM Egypt High school for boys in the 6th of October city. We are 4 males (100% males), M: 16 years old and we are completely Egyptian Students.

3.2. Research Design

Variables in our study vary according to the experiment that we want to do. All the variables are illustrated in every experiment below.

3.3. Measures

The multimeter was used to measure the output voltage, internal resistance, watts and the current. Meter to measure the length of the components of the prototype. Balance to know the mass of each material.

Materials

Table 1. Shows the materials used in building the prototype.

Material	Amount/ Size
Two glass-boxes	18*20*15cm
Short section of plastic pipe (polyethylene or PVC) for salt bridge	25cm
Agar	250 g
Salts (NaCl, KCl, KNO, NaOH, etc.).	125 g (For each one)
Graphite- Stainless steel-Carbon paper-copper- Aluminum)	2*10*15 cm
Mud (Bacteria)	-------
Bacterial substrate (Glucose- Sucrose- Acetate-Butyrate)	125 g (for each substrate)
Copper wire (plastic coated).	50 cm
Salt Water	2 liters
Wires with alligator clips	2 wires

3.4. Building the Prototype

Step1: Preparing the salt bridge from Agar: To make the salt bridge, agar was (whose percentage is 5 - 7%) with the potassium chloride (whose percentage is 93 - 95%) and water until they boiled, and let them get dry inside the pipe. This step is said to take about 45 minutes.

Step 2: (This step is said to take about 40 minutes). Conduct each component with the other to figure out the prototype as shown in figure 1 and figure 2:

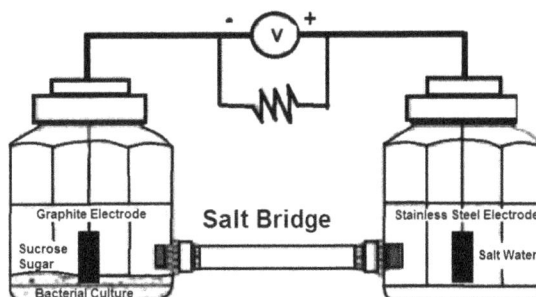

Figure 1. Shows the components of the prototype.

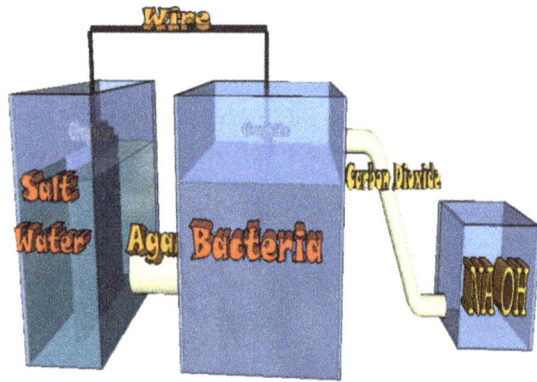

Figure 2. Shows the final shape of the prototype.

Experiment 1: The purpose from this experiment is to examine different anode types like (Graphite- Carbon paper) (Independent). It had been completed using mud as a source for the bacteria (controlled) and artificial wastewater as a source for the bacterial food (controlled). Also, a plat of Stainless steel will be used as a cathode electrode (controlled). After testing each electrode and collecting our data (in 40 minutes) the results were observed as shown in table 2:

Table 2. Shows the output voltage by different anode electrodes.

Electrode Type	Maximum power density (W/m^2) (Error ± 0.015)
Graphite	2.6
Carbon paper	0.8

From these results, it was observed that graphite plates can help producing the highest power density (Dependent), so it is recommended to use it as anode electrode to ensure the highest output voltage.

Experiment 2:

This experiment had been done to examine different substrates (Glucose – Sucrose – Acetate - Butyrate) (Independent) and find the difference in power density as shown in table 3. It had been done using Graphite as anode electrode and stainless steel as a cathode electrode (controlled). By collecting the data, it was found that sucrose can provide the highest power density, so it is advisable to use it as a substrate in the MFCs.

Table 3. Shows the output voltage by different substrates.

Substrate Type	I (current, mA) (Error ± 0.21)	P (mW/m^2) (Error ± 1.5)
Wastewater	4.85	26
Glucose	0.9	494
Sucrose	6.2	506
Acetate	1.27	23
Butyrate	0.46	305

Note: the reaction that takes place in the anode chamber can be explained by this formula:

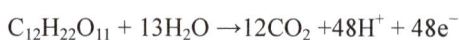

$$C_{12}H_{22}O_{11} + 13H_2O \rightarrow 12CO_2 + 48H^+ + 48e^-$$

Experiment 3:

The purpose of this experiment is to test different cathode electrodes to determine the best type. The electrodes that had been tested are the Aluminum, copper and stainless steel (They were considered as independent variables). The prototype will contain sucrose as a substrate and graphite plate as anode electrode (Controlled). Then, the output voltage had been measured with each electrode (Dependent). Note: Cathodic reaction:

$$12O_2 + 48H^+ + 48e^- \rightarrow 24H_2O$$

It was found that although the cathode does not take place in that reaction, but it has an effect on the output voltage. The data was collected as shown in table 4:

Table 4. Shows the output voltage by different cathode types.

Stainless steel	Copper	Aluminum
1.9 volt ±0.02	1.4 volt ±0.02	1.5 volt ±0.02

It had been observed that stainless steel can help producing the highest output voltage, so it is recommended to use it as a cathode plate in the MFCs.

Experiment 4:

By knowing the limiting factors of the MFCs, it was found that the temperature is an important factor in the project, so the suitable temperature for the project should be known to ensure the highest power density. In the prototype, graphite plate was used as an anode electrode, stainless steel as a cathode electrode and sucrose as a bacterial substrate (controlled). Then, the temperature of the prototype was changed inside the laboratory (Independent). It was observed that, the power density (dependent) was changed by changing the temperature according to table 5.

Table 5. Shows the output voltage and the power density at different temperatures.

T ± 1 (^{o}C)	OCV (mV) (Error ± 1.5)	Pmax (mW/m^2) (Error ± 0.15)
20	111	0.73
25	112	0.75
30	117	0.82
35	119	0.88
40	133	1.01
45	125	0.90

From these results, it is recommended to apply the project at temperature of 40°C.

Experiment 5:

In order to test how clean the prototype is, the prototype should obtain graphite plate as an anode electrode, stainless steel as a cathode electrode and sucrose as a bacterial substrate (controlled). Then, the temperature of the prototype was changed inside the laboratory to be 40°C. After that, the Sodium Hydroxide (NaOH) was exposed to the omitted Carbon Dioxide (CO$_2$) and it was observed that, they reacted with each other and produced Sodium Carbonate (NaCO$_3$) according to the following formula:

$$2NaOH + CO_2 \rightarrow Na_2CO_3 + H_2O$$

In such a way, the project had been definitely an eco-friendly project as it has nearly negligible output CO_2. Also, it produces Sodium Carbonate that can be used in different ways like the manufacture of glass, help in processing wood pulp to make paper, water softening, refining aluminum, laundry soaps and other household cleaning products, taxidermy. So, it will help in many industries.

Experiment 6: This experiment was done to measure how efficient is the prototype. It was completed by testing the output voltage of the prototype many times and testing an ordinary M.F.C at the same time and measure how the efficiency increased. As shown in Table 6, the prototype has a high efficiency. There was some negative results like test 4, but the other tests had positive values. We calculated the efficiency of our project and we found that the average equals 73%.

The anode chamber must be kept isolated from the outside environment. For long-term operation, electrodes should be constructed in a way that limits corrosion of copper wire due to contact with liquids. Power can be significantly increased by using a catalyst such as Nitrogen-Enriched Core-Shell instead of platinum for hydrogen adsorption that has lower cost than platinum.

Table 6. *Shows the output voltage of the M.F.C.*

	Ordinary M.F.C (V)	New Design (V)
Test 1	0.9	1.7
Test 2	1.08	1.87
Test 3	1	1.9
Test 4	1.65	1.15

4. Design

As the mentioned before, the design of the real project is a huge limiting factor facing the development of the M.F.Cs. But the mind said "No" and it was decided that there will be a magnificent design for the real application of the M.F.Cs. As shown in figure 3-b, the design consists of number of cells put beside each other and all of them are connected with three main pipes in the upper side. One to collect the output Carbon Dioxide and lead them to scrubbing chamber in which it will react with the Sodium Hydroxide to produce the Sodium Carbonate then, it will be taken to the factories to perform their work and produce useful materials. The other pipe is to collect semi-salty water to the treatment plants. The third one is to add salt water to the cathode chamber to perform its work. All the cells are connected to collect all their output energy. There is another one is to add organic matters or liquefied mud to the anode chamber to perform their work as a source for bacteria. In the down side, there is another pipe to collect the used organic matters or the liquefied mud, as shown in figure 3-c, to get them in their fields after using their huge energy, after that they will complete their round in the nature.

Any project have conditions to be applied and our project have some special conditions:

1- High temperature.

2- Availability of salt water and the organic matters or mud.

In order to provide the first condition and the suitable temperature all the day to make it 24-hours working system, the walls of the plant is said to be made from a material that can store heat energy all the day and release it at night.

This material is called "Thermstone". It is made from cement bricks, sand and silica. This brick characterizes by many of the benefits of thermal insulation which provides high compared with conventional concrete and clay bricks.

The roof of the plant is said to be made from a special type of glass to let sun rays enter the system as shown in figure 3-a. This glass is called "Laminated Glass" and it is hard to be broken because it's made of layers of safety glass bound together with a transparent adhesive.

To provide the system with a sufficient amount of heat energy, the system will be with a tracking system to make the roof movable according to the angle of the rays of the sun. Also, to get the highest possible benefit from this tracking system, there will be a number of solar cells working with the same tracking system to provide M.F.C system with the needed energy to lift the water and the liquefied mud to the cells. Furthermore, the solar cells system will provide additional energy out of the system.

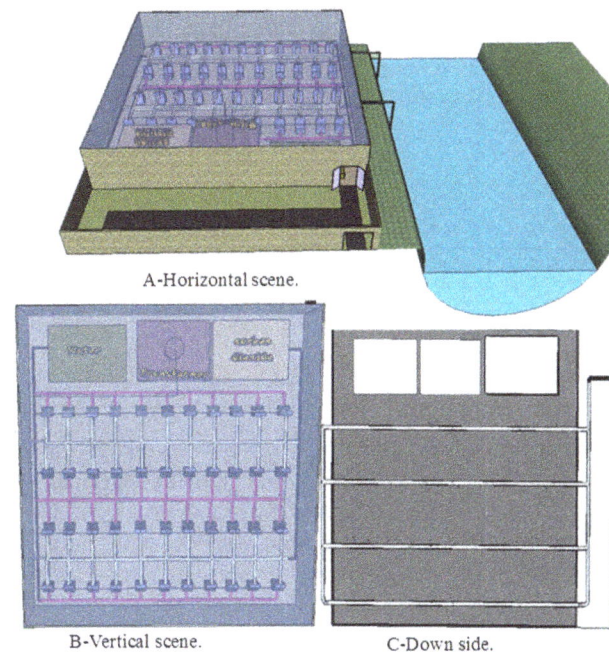

A-Horizontal scene.

B-Vertical scene.   C-Down side.

Figure 3. *Shows the design of the plant.*

5. Conclusion

With these modifications, our project will be a magnificent project for real application and green environment.

M.E.T is considered one of the best projects that helps in generating electricity because it depends on organic and agricultural wastes, such as Geobacter bacteria that is found in the mud, in generating electricity. The objective of the project is to achieve goals shown in figure 4.

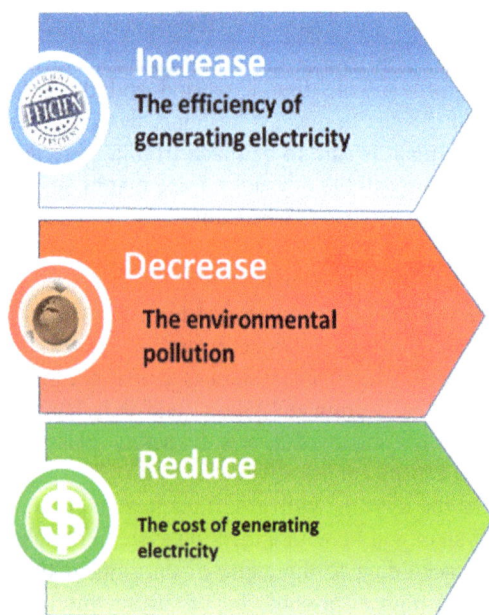

Figure 4. *Shows the goals of the project.*

There are many challenges that face our project. Platinum is one of the most difficult challenges facing MFC project because of its high cost as it may affect the whole project and impede its development. So, it is recommended to use alternative materials to be used as a catalyst to provide the project and make it more applicable in the real life. There was a suggested materials such as the Nitrogen-Enriched Core-Shell (N-Fe/Fe3C@C) and other materials as shown in table 7.

Table 7. *Shows different catalysts and their performance.*

	CR (%)
CC	26.9±1.1
CNT	24.1±3.0
Pt/C	39.8±2.6
N-Fe/Fe3C@C	43.6±0.8

That is why, we recommended using Nitrogen-Enriched Core-Shell as catalyst because of its low cost but there still be a problem which is the suitable laboratory to prepare it. If this project could be applied in the real life with these features, it is expected to help in producing clean energy with lower cost as shown in table 8.

Table 8. *Shows a comparison between the cost of the N-Fe/Fe3C@C and Platinum.*

Catalyst	Price (USD/g)
Platinum	$32.54
Nitrogen-Enriched Core-Shell	$20

Acknowledgement

This study is supported by STEM Egypt High School. The team aknowledges Eng: Ahmad Tawfik the useful suggestions in the study. Also, we would like to thank Dr. Saif Soliman, PhD in Chemistry, for providing the study by essential chemistry, Mrs. Israa Ali, master's degree in Biophysics, for her help in understanding the vital activities in the bacteria and Dr. Ahmed Akrab, PhD in chemistry, for his help in knowing the chemical reactions in the device.

References

[1] Cheng, S., and Logan, B. E. (2011). Increasing power generation for scaling up single-chamber air cathode microbial fuel cells. Bioresource Technology 102, 4468-4473.

[2] Chae, K.-J., Choi, M.-J., Lee, J.-W., Kim, K.-Y., and Kim, I. S. (2009). Effect of different substrates on the performance, bacterial diversity, and bacterial viability in microbial fuel cells. Bioresource Technology 100, 3518-3525.

[3] Chen, G.-W., Choi, S.-J., Lee, T.-H., Lee, G.-Y., Cha, kim, C.-W. (2008). Application of biocathode in microbial fuel cells: cell performance and microbial community. Applied Microbiology and Biotechnology 79, 379-388.

[4] Du, Z., Li, H., and Gu, T. (2007). A state of the art review on microbial fuel cells: A promising technology for wastewater treatment and bioenergy. Biotechnology Advances 25, 464-482.

[5] Environ. Sci. Technol, Electricity Generation Using an Air-Cathode Single Chamber Microbial Fuel Cell in the Presence and Absence of a Proton Exchange Membrane, 2004, 38, 4040-4046.

[6] Gonzalez del Campo, A., Lobato, J., Cañizares, P., Rodrigo, M., & Fernandez Morales, F. (2013). Short-term effects of temperature and COD in a microbial fuel cell. *Applied Energy, 101,* 213-217. doi: 10.1016/j.apenergy. 2012.02.064.

[7] Justa, Aditi. *Harvard students harness electric power from bacteria in soil*. Eco Friend, June 12, 2010. Web. Nov. 6, 2011.   http://www.ecofriend.com/entry/harvard-studentsharness-electric-power-from-bacteria-in-soil/.

[8] Logan, B. E., and Regan, J. M. (2006). Microbial Fuel Cells—Challenges and Applications. Environmental Science & Technology 40, 5172-5180.

[9] Logan, B. (2010). Scaling up microbial fuel cells and other bioelectrochemical systems. Applied Microbiology and Biotechnology 85, 1665-1671.

[10] Logan, B. E., et al., *Biological hydrogen production measured in batch anaerobic respirometers (vol 36, pg 2530, 2002).* Environmental Science & Technology, 2003. 37(5): p. 1055-1055.

[11] Liu, H., Cheng, S. A. and Logan, B. E. (2005a). Production of electricity from acetate or butyrate using a single-chamber microbial fuel cell. Environ. Sci. Technol., 39(2), 658–662.

[12] Moon, H., Chang, I. S. and Kim, B. H. (2006) Continuous electricity production from artificial wastewater using a mediator-less microbial fuel cell. Bioresource Technol., 97, 621–627.

[13] Pant D, V. B. G., Diels L, Vanbroekhoven K. (2010). A review of the substrates used in microbial.

[14] Rabaey, K. and Verstraete, W. (2005). Microbial fuel cells: novel biotechnology for energy generation. Trends Biotechnol., 23(6), 291–298.

[15] Sanford, Galen. *Make Electricity, Not Sludge*. Blue Tech Blog, June 15, 2010. Web. Nov.19, 2011. http://bluetechblog.com/2010/06/15/make-electricity-not-sludge/.

[16] Schwartz, K. (2007). Microbial fuel cells: Design elements and application of a novel renewable energy sources. Basic biotech. ells. Enzyme and Microbial Technology 47, 179-188.

[17] Tsuchiya, H. and Kobayashi, O. (2004). Mass production cost of PEM fuel cell by learning curve. Int. J. Hydrogen Energy, 29(10), 985–990.

[18] Wei, Y., Van Houten, R. T., Borger, A. R., Eikelboom, D. H. and Fan, Y. (2003). Minimization of excess sludge production for biological wastewater treatment. Wat. Res., 37(18), 4453–4467.

16

Analysis of solar heating system for an aquaponics food production system

Kevin R. Anderson[1], Maryam Shafahi[1], Arthur Artounian[1], Adam Chrisman[2]

[1]Mechanical Engineering Department, Solar Thermal Alternative Renewable Energy Lab, College Engineering, California State Polytechnic University, Pomona, CA, USA
[2]SUNEARTH Inc., Fontana, CA, USA

Email address:

kranderson1@csupomona.edu (K. R. Anderson), maryam.shafahi@email.ucr.edu (M. Shafahi), artounianarthur@gmail.com (A. Artounian), AChrisman@SunEarthInc.com (A. Chrisman)

Abstract: Aquaponics is a sustainable farming technology that combines aquaculture and hydroponics, growing fish and plants together in a symbiotic environment. Aquaponics sets an excellent example for an efficient multidisciplinary solution to the real world problems such as drought, polluted environment and food contamination. In this paper we present an aquaponics system heated by solar thermal energy in order to maintain the fish living environment at 21 °C. The paper presents an *f*-chart based analysis demonstrating the feasibility of the system. The results show for a collector area of 22 m^2 that an annual solar fraction of 94% is needed to support an 833 liter aquaponics system.

Keywords: Solar Thermal Energy, Aquaponics, Sustainable, Renewable, f-Chart

1. Introduction

The State of California in the USA is facing its most severe drought emergency in decades while more than 70% of its water consumption is attributed to agriculture. This state needs more low-demand and water-efficient agriculture systems to overcome its long-term water crisis. Aquaponics, an increasingly popular farming system, produces fish and crops with 10% of the amount of water used in traditional farming [1]. Additionally, crop production represents the largest source of groundwater nitrate in the majority of agricultural lands in California which has been raising public health concerns [2]. Utilizing aquaponics requires 90% less water and little fertilizers for the plant growth. It provides faster growth rate, crop maturity and yields, and better quality of the crops. Moreover, aquaponics is able to grow crops in places where ordinary horticulture and aquaculture is impossible due to poor or contaminated soil or water. Another beneficial factor is minimizing the growing area [3-5]. Aquaponics sets an excellent example for an efficient multidisciplinary solution to the real world problems such as drought, polluted environment and food contamination. It is a rescue plan involving specialists from agriculture, science, engineering, and business.

The aquaponics system described herein grows tilapia and gold fish with a combination of different crops such as romaine lettuce, celery, Swiss chard, ruby chard and lettuce The optimum water temperature for producing tilapia is between 70 to 75 °F [4]. In order to maintain the appropriate temperature for the fish, the fish tanks of the aquaponics system must be equipped with efficient heat exchangers. Additionally, the water quality should be monitored continuously to ensure the level of pH, temperature, Ammonia, Nitrite and Nitrite fulfills the aquaponics rather limited ecosystem's demands. The long-term goal of this research is to establish a local model which will predict the performance of the systems in terms of plant yield and fish growth accounting for the influence of weather, fish and plant type, as well as water and energy consumption. The two existing systems are outdoors utilizing gravel and rafting bed for plant growing. This research is sustainable since we can find a market for our truly organic crops and fish. The current test set-up is shown in Figure 1.The ultimate goal of this research is to build a solar thermal heating system for the aquaponics system. This paper presents the results of a solar thermal design analysis of such a system.

Figure 1. Aquaponics sustainable food production facility.

2. Materials and Methods

The present paper is a continuation of the work presented in [7,8]. The solar thermal heating system aquaponics test set-up at Cal Poly Pomona is shown schematically in Figure 2.

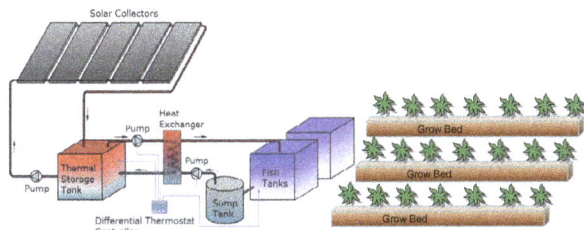

Figure 2. Solar thermal heating system schematic.

The solar thermal system to be integrated into the aquaponics system is used to offset the energy cost of powering auxiliary heaters during cold weather. This will allow aquaponics system to be used in regions where the weather fluctuates vastly over 24 hours. Figure 2 shows the fish living environment and aquaponics grow beds. The waste matter harvested from the fish tanks is diverted to the grow beds in order to fertilize the crops. Figure 3 shows the details of how the fish tanks are integrated to the grow beds using the aerator, bio-filter and clarifier. The fish are housed in the fish tanks, which is held at a desired temperature of 70 °F by the solar thermal energy system so that the fish can survive cold evening temperature variations.

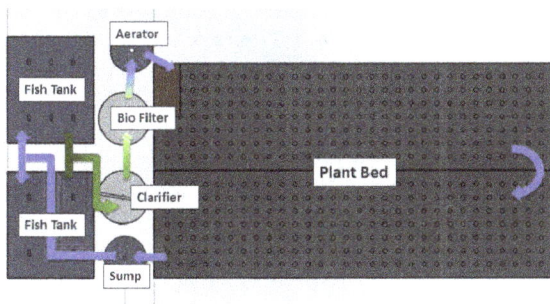

Figure 3. Aquaponics sustainable food production facility.

The system shown in Figure 2 consists of the following hardware:
- Solar Collectors: an array of 5 SUNEARTH Oasis-PP

solar thermal collectors with a gross area of 236 square feet tilted at 34 degrees from the horizontal
- Thermal Storage Tank: an insulated 220 gallon storage tank with an integrated drain-back unit into the closed-loop system
- Heat Exchanger: a counter-flow heat exchanger between the storage tank and the fish tanks isolate the solar thermal heating loop from the fish supply water
- Data logger: a four-channel thermocouple data logger is used to record the inlets and outlets of the heat exchanger
- Pump: the solar collector loop flow rate is regulated by a timer, which functions only during daylight hours
- Differential Thermostatic Controller: the pumps directing the flow from the storage tank into the heat exchanger are activated by a multiple-relay differential thermostatic controller according to the following algorithm; if a 10 °F delta between the thermal storage tank and fish tanks is detected or when the fish tank temperature drops below 68 °F

Figure 4 shows a typical temperature variation in the aquaponics fish tanks versus time.

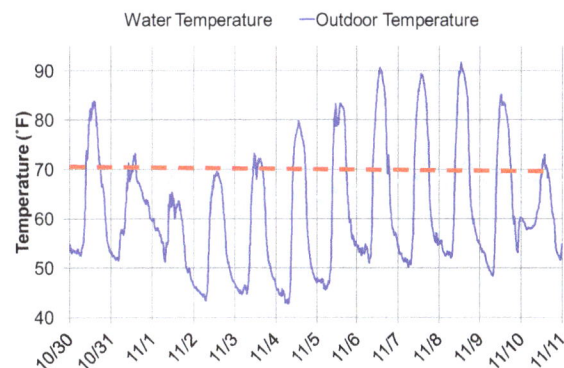

Figure 4. Temperature variations in fish tanks.

The data shown in Figure 4 was recorded over two weeks. This data is for the aquaponics system utilizing 100% auxiliary pool heaters to meet the load i.e. zero solar thermal corresponds to 600 W per month (or 7.2 kW annually) required by the pool heaters, which is taken from measurement on the current system configuration.

Clearly, using pool heaters is not a green/renewable/sustainable solution for large scale aquaponics infrastructures. Thus, herein a design a solar thermal heating system to offset the cost of auxiliary pool heaters is presented. In the remaining sections of this paper we discuss the proposed design and analysis of the solar thermal based heating system for the aquaponics fish tanks.

3. Solar Thermal Analysis

The *f*-chart analysis method of [8] was used to perform a feasibility on the solar thermal heating of the system shown in Figure 2. The *f*-chart method is a correlation of the results of many hundreds of thermal performance simulations of solar heating systems. The resulting simulations give *f*, the fraction of the monthly heating load (for space heating and hot water) supplied by solar energy as a function of two dimensionless parameters, X (collector loss) and Y (collector gain). X is related to the ratio of collector losses to heating loads, and Y is related to the ratio of absorbed solar radiation to the heating loads. The basic equations of the f-chart method per [9] are summarized below, beginning with the expressions for the collector loss, X and the collector gain, Y

$$X = F_R U_c \frac{F_R'}{F_R}\left(T_{ref} - \overline{T}_a\right)\Delta\tau\frac{A_c}{L} \quad (1)$$

$$Y = F_R(\tau\alpha)_n \frac{F_R'}{F_R}\frac{(\overline{\tau\alpha})}{(\tau\alpha)_n}\overline{H_T}N\frac{A_c}{L} \quad (2)$$

The f-chart equations for the fraction f of the monthly space and water heating loads supplied by solar energy for liquid based systems (such as that one considered herein) is given by

$$f = 1.029Y - 0.065X - 0.245Y^2 + 0.0018X^2 + 0.0215Y^3 \quad (3)$$

Figure 5 shows a f-chart for liquid systems. As shown in Figure 5, solar thermal design engineers can utilize the f-chart by computing the X, Y values and then determining the corresponding f value.

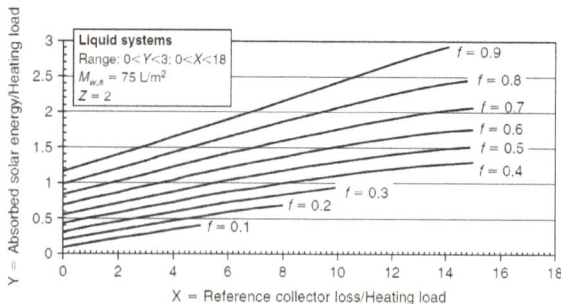

Figure 5. The f-chart for liquid based solar heating systems [10].

The fraction F of the annual heating load supplied by solar energy is the sum of the monthly solar energy contributions divided by the annual load and is given as follows:

$$F = \frac{\sum fL}{\sum L} \quad (4)$$

where the following nomenclature is used in Eqns. (1) through Eqn. (4)

A_c = area of solar collector (m^2 or ft^2)

f = fraction of the monthly heating load carried by solar energy (%)

F_R' = collector heat exchanger efficiency factor (%)

F_R = collector heat removal factor (%)

F = fraction of the annual heating load supplied by solar energy (%)

U_c = collector loss coefficient (W/m^2-K or BTU/hr-ft^2-°F)

$\Delta\tau$ = total number of seconds (SI) or hours (IP) per month

$\overline{T}_a$ = monthly average ambient temperature (°C or °F)

T_{ref} = empirically derived reference temperatuer (100 °C or 212°F)

L = monthly total heating load for hot water (GJ or MMBTU)

$\overline{H_T}$ = monthly averaged daily insolation incident on collector surface per unit area (MJ/m^2 or BTU/ft^2)

N = number of days in the month

$(\overline{\tau\alpha})$ = monthly average transmittance-absorptance product (%)

$(\tau\alpha)_n$ = normal transmittance-absorptance product (%)

4. Solar Thermal Results

The solar thermal design simulation software tool PolySun [11] was used to generate the results presented herein. Figure 6 shows a layout of the system as analyzed in PolySun.

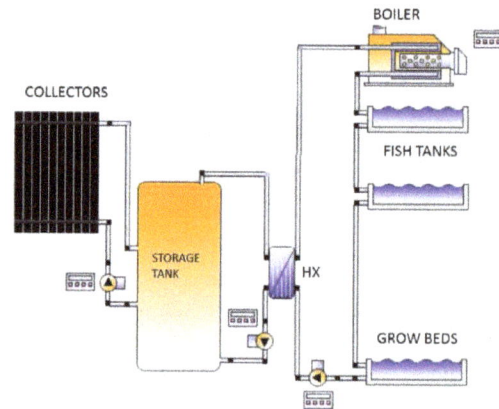

Figure 6. Schematic layout of solar heating system for the aquaponics food production system.

As shown in Figure 6, the system analyzed is composed of the following: collectors:
5 unglazed 48 ft^2 of area each, total collector area = 236.25 ft^2 at a tilt angle of 34°. There are two fish tanks 220 gal each, two commercial scale grow beds with a capacity of 250 plants, two filtration devices; clarifier/settling tanks, and a mineralizer /bio-filter. As shown in Figure 6, a boiler is used to interface the water in the fish tanks to the water in the heat exchanger. This provides the required isolation of the contaminated fish tank water from the storage tank water. The boiler is specified at 1.8 kBTU/hr with an efficiency of 35 %. The heat exchanger which interfaces the storage tank to the fish tanks is a shell/plate style rated at a thermal resistance of R = 2×10^{-4} K/W. The main pump in the solar thermal loop has a flow rate of 19 gpm with a pressure drop of 28 psi and power consumption of 4997 kBTU. The pump in the fishtank is 16 gpm, 1.94 psi pressure drop and a power draw of 84 kBTU. Finally, the storage tank drainback system is rated at 150 gal.

Figure 7 shows the solar energy input to the system. Figure 7 plots solar energy into the system (kBTUs) vs. month of the year.

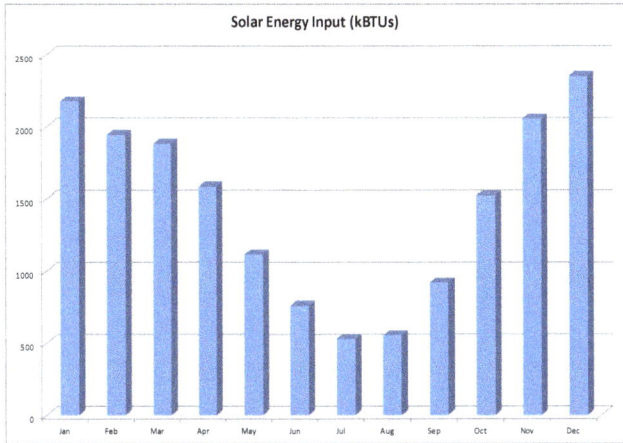

Figure 7. *Solar energy input to the system (values in kBTUs).*

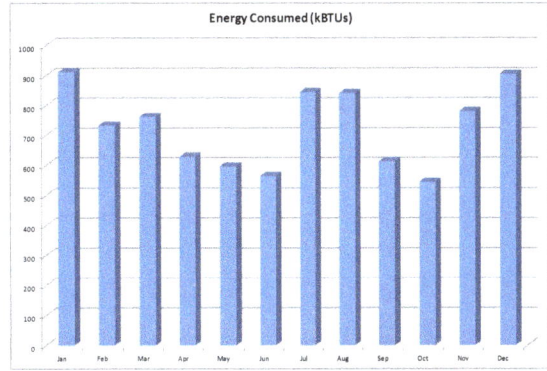

Figure 8. *Energy consumed by the system (values in kBTUs).*

Figure 8 plots the energy consumed by the system (kBTUs) vs. month of the year. Table 1 shows the f-chart calculations for this solar heating system for an aquaponics food production application of thermal energy harnessing and storage.

Table 1. *f-chart Analysis summary.*

Month	Avg. H_T (MJ/m²)	Avg. T_a (°C)	L (GJ)	P_L	P_s	Solar fraction f	fL (GJ)
Jan	10.1	13	1.72	28.9	2.3	0.95	1.63
Feb	13.1	13	1.89	23.9	2.4	0.84	1.59
Mar	17.3	16	2.28	21.3	2.9	0.88	2.01
Apr	21.8	17	2.55	18.1	3.2	0.90	2.30
May	23.1	19	2.35	19.7	3.8	0.97	2.28
Jun	23.7	22	2.10	20.7	4.2	1.00	2.10
Jul	25.7	24	1.80	24.2	5.5	1.00	1.80
Aug	23.5	25	1.60	26.8	5.7	1.00	1.60
Sep	19.0	25	1.65	25.4	4.3	1.00	1.65
Oct	15.0	20	1.75	26.4	3.3	1.00	1.75
Nov	11.4	16	1.80	26.1	2.4	0.88	1.58
Dec	9.3	12	1.78	28.4	2.0	0.86	1.53
Total			23.27				21.82

From Table 1 it is found that the fraction F of the annual heating load supplied by solar energy is

$$F=\sum fL/\sum L=21.82/23.27=94\%$$

Figure 9 shows the fraction of solar energy to the system f per month

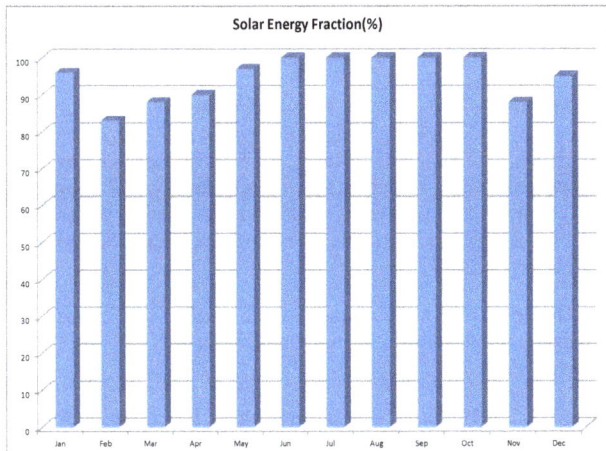

Figure 9. *Solar energy fraction carried by the system, f (%).*

The heat exchanger penalty factor analysis was carried out using the following expression per [10]

$$F_{HX}=\frac{F'_R}{F_R}=\left[1+\left(\frac{F_RU_c}{(\dot{m}c_p)_c}\right)\left(\frac{(\dot{m}c_p)_c}{(\dot{m}c_p)_{min}\varepsilon}-1\right)\right]^{-1} \quad (5)$$

where
F_{HX} = heat exchanger penalty factor

$(\dot{m}c_p)_c$ = thermal capacitance of the collector

$(\dot{m}c_p)_{min}$ = minimum thermal capacitance

ε = heat exchanger effectiveness.

Application of Eqn. (5) to the current system using a Stainless-steel plate/shell heat exchanger rated at 790 provided the data of Table 2.

Table 2. Heat exchanger penalty factor analysis

ε	Penalty F_{hx}	NTU	A_{HX} (m^2)	Energy Collection Q_u (GJ/yr)
0.21	0.963	0.25	0.12	0.10
0.30	0.974	0.40	0.19	0.15
0.40	0.980	0.58	0.28	0.23
0.50	0.984	0.82	0.40	0.32
0.60	0.987	1.13	0.55	0.44
0.70	0.989	1.55	0.76	0.61
0.80	0.990	2.22	1.08	0.87
0.90	0.991	3.42	1.67	1.35
0.99	0.992	7.83	3.83	3.09
1.00	0.992	--	--	--

where NTU = number of transfer units for the heat exchanger, A_{HX} = heat exchanger area (m^2), and Q_u = the useful energy collected (GJ/yr). The results of ε-NTU of Table 2 are seen to be in agreement with standard references in Heat Exchanger design [12]. The average temperature of the fish tanks is shown in Figure 10.

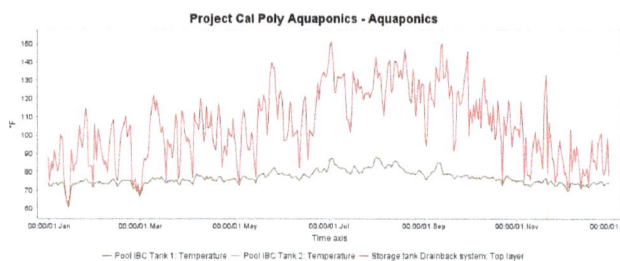

Figure 10. Aquaponics fish tank temperature transient.

Figure 10 illustrates that the solar thermal system as designed herein will hold the fish tanks in the aquaponics system at an average of 70 °F as desired.

5. Conclusions

This paper has presented the analysis of a solar thermal heating system proposed for an aquaponics sustainable food production system. The f-chart analysis method based on [9] suggests that 94% of annual energy is carried by the solar thermal collectors / heat exchanger / storage tank system. This option of using solar thermal is a viable green energy solution as opposed to the current baseline design on using pool heaters to heat the aquaponics fish tanks, which requires on average 7.2 kW annual to maintain the fish tanks at the proper temperature. This paper has also presented a heat exchanger design simulation showing that an effectiveness of 75% provides NTU=2. This value is in agreement with standard results for heat exchanger design a found in [12]. Future work involves the procurement and installation of the solar thermal heating system analyzed herein. From whence on-site data for thermal performance can be gathered and disseminated.

References

[1] Goodman, Community and Economic Development, Master's thesis, Massachusetts Institute of Technology, 2011

[2] Kristin N. Dzurella, Josué Medellín-Azuara, Vivian B. Jensen, Aaron M. King, Nicole De La Mora, Anna Fryjoff-Hung, Todd S. Rosenstock, Thomas Harter, Richard Howitt, Allan D. Hollander, Jeannie Darby, Katrina Jessoe, Jay Lund, G. Stuart Pettygrove, Nitrogen Source Reduction to Protect Groundwater Quality, Center for Watershed Sciences, University of California, Davis, July 2012

[3] Rakocy, J., Masser, M., Losordo, T., Recirculating Aquaculture Tank Production Systems: Aquaponics- Integrating Fish and Plant Culture, SRAC Publication No. 454, 2006

[4] Tyson, R. V., Reconciling pH for Ammonia Biofilteration in a Cucumber/Tilapia Aquaponics System Using a Perlite Medium, PhD thesis, University of Florida, 2007

[5] Blidariu F., Grozea A., Increasing the Economical Efficiency and Sustainability of Indoor Fish Farming by Means of Aquaponics - Review, Animal Science and Biotechnologies, 2011, 44 (2)

[6] "Application of Solar Power in Sustainable Food Production Systems" by Matt Shekels, Daniel Woolston, Dr. Maryam Shafahi, Dr. Kevin Anderson, Mechanical Engineering, California State Polytechnic University at Pomona poster presented at SOLAR 2014, San Francisco, CA, July 6-10, 2014

[7] "Application of Solar Power on Sustainable Food Production System" by Dr. Maryam Shafahi, Dr. Kevin R. Anderson, Jeff Moore, Darius Shu, Hadasa Reyes, Erik Mora, Roslina Hussin, Department of Mechanical Engineering, California Polytechnic State University, Pomona. Southern California Conferences for Undergraduate Research (SCCUR), Saturday, November 22, 2014.

[8] J.A. Duffie and W. A. Beckman, Solar Engineering of Thermal Processes, John Wiley and Sons, New York, NY, USA, 2nd edition, 1991.

[9] Klein, S.A., 1976. A design procedure for solar heating systems. Ph.D. Thesis, Chemical Engineering, University of Wisconsin, Madison.

[10] Kalogirou, S, 2009. Solar Energy Processes and Systems, 1st Ed., Elsevier Publications, London, UK.

[11] PolySun Simulation software http://www.velasolaris.com/

[12] Heat Transfer, Incropera and Dewitt, 2nd Ed. McGraw-Hill, 1991, NY, NY.

17

Cost optimization of hybrid stand-alone power system for cooled store in Kirkuk

Sameer Saadoon Al-Juboori[1, *], Ali Hlal Mutlag[1], Ehsan Fadhil Abbas Al-Showany[2]

[1]Electronic and Control Engineering Dept., Kirkuk Technical College, Kirkuk, Iraq
[2]Refrigerating and Conditioning Engineering Dept., Kirkuk Technical College, Kirkuk, Iraq

Email address:

sameersaadoon@yahoo.com (S. S. Al-Juboori), ali33hl@yahoo.com (A. H. Mutlag), ehsanfadhil@ymail.com (E. F. A. Al-Showany)

Abstract: However, the design, control, and optimization of the hybrid systems are usually very complex tasks; the stand-alone hybrid solar–diesel power generation system is recognized generally more suitable than systems that only have one energy source for supply of electricity to off-grid applications. A proposed PV system has been designed and optimized using HOMER software computer model to supply a potato cooled store in Kirkuk city in Iraq. The result obtained from the optimization gives the cost of energy (COE) is 0.639 US$/kWh with 2axis trucking system and 0.692 US$/kWh with no trucking system. Energy cost is 0.796 US$/kWh when the load is supplied by the diesel generator alone.

Keywords: Homer, Stand Alone, Hybrid, Kirkuk, Off-Grid, Trucking

1. Introduction

Alternative energy resources such as solar and wind have attracted energy sectors to generate power on a large scale. A drawbacks common to wind and solar options, is their unpredictable nature and dependence on the weather and climatic changes, and the variations of solar and wind energy may not match with the time distribution of demand [1]. One of the major worldwide concerns of the utilities is to reduce the emissions of traditional power plants by using renewable energy and to reduce the high cost of supplying electricity for remote areas. Hybrid power systems can provide a good solution for such problems because they integrate renewable energy along with the traditional power plants. Renewable energy is defined as the energy generated from natural resources such as sunlight, wind, rain, and geothermal heat, which are renewable. Hybrid power systems usually integrate renewable energy sources with fossil fuel based generators to provide electrical power. They are generally independent of large electric grids which are used to feed loads in remote areas. Hybrid systems offer better performance, flexibility of planning and environmental benefits comparing to the diesel generator based stand-alone system. Hybrid systems also give the opportunity for expanding the generating capacity in order to cope with the increasing demand in the future. Remote areas represent a big challenge to electric power utilities.

Hybrid power systems provide an excellent solution to this problem as one can use the natural sources available in the area e.g. the wind and/or solar energy and thereby combine multiple sources of energy to generate electricity [2-4]. The optimal design of hybrid renewable power systems is usually defined by economic criteria. But there are also technical and environmental criteria to be taken into an account to improve decision-making. In this paper a discussion on different criteria will introduce the non-economical perspectives in addition to the economic criteria [5,6]. Besides of the shortage supply, the combining power generation with fossil fuels has also harmed environment through the emissions of greenhouse gases (GHG) and other pollutants. Renewable energy can play an essential role in mitigating the ongoing shortage supply and achieving the ultimate goal of replacing fossil fuels with emission free power generation [7, 8].

2. Homer Algorithm Package

HO HOMER is a computer model that simplifies the task of evaluating design options for both off-grid and grid-connected power systems for remote, stand-alone, and distributed-generation (DG) applications.

HO HOMER's optimization and sensitivity analysis algorithms allow one to evaluate the economic and technical feasibility of a large number of technology options and to account for variation in technology costs and energy resource availability for both conventional and renewable-energy technologies [4]. HOMER models a power system's physical behavior and its life-cycle cost, which is the total cost of installing and operating the system over its life span. It allows the modeler to compare many different design options based on their technical and economic merits. It also assists in understanding and quantifying the effects of uncertainty or changes in the inputs. [9]

3. Optimal Size of the Proposed System Using HOMER

Potato is one of the most important food crops in Iraq. The objective of the study in [10] is to establish (19×11×6) m cooled store to save 300 tons of potato crop in Kirkuk city in Iraq and to identify cooling load necessary to keep the crop fresh. The daily estimated consumption of potato in this city is 15 tons.

The aim of this paper is to design a hybrid power system to supply the cooled store.

The system consists of; PV modules, diesel generator, batteries, charge controller, inverter, and the necessary wiring and safety devices. The system feasibility analysis was performed using the HOMER software.

4. The Hybrid System Model

In order to design stand-alone renewable hybrid power systems, there are four main aspects to be considered:
- the demand/load characterization,
- the potential of renewable and conventional energy generation,
- the restrictions of the system, and
- the optimization criteria.

The optimization criteria considered are mainly economic aspects: Net Present Cost (NPC) and Cost of Energy (COE) typically. Also technical variables and environmental factors define the configuration of the system and consequently its performance and viability. Various aspects must be taken into account when working with stand-alone hybrid systems for generation of electricity. Reliability and cost are two of these aspects; it is possible to confirm that hybrid stand-alone electricity generation systems are usually more reliable and less costly than systems that rely on a single source of energy [11-14]. It has been proven that hybrid renewable electrical systems in off grid applications are economically viable, especially in remote locations [15-19]. In addition, climate can make one type of hybrid system more profitable than another type. For example, photovoltaic hybrid systems (Photovoltaic–Diesel–Battery) are ideal in areas with warm climates [20].

4.1. Load Profile

The load profile of the cooled store in Kirkuk city is shown in Figure 1. The total daily average load is 667 kWatt-hours [10].

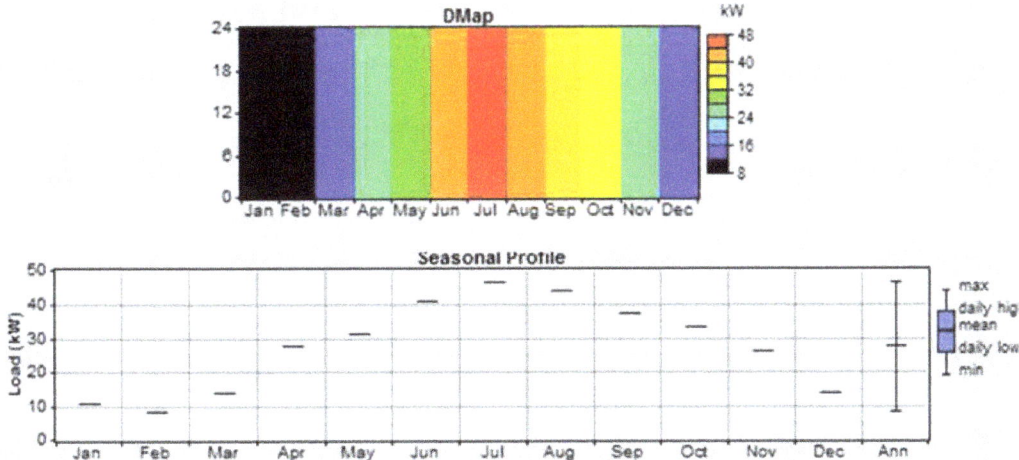

Figure 1. The Load Profile.

4.2. System Equipment Configuration

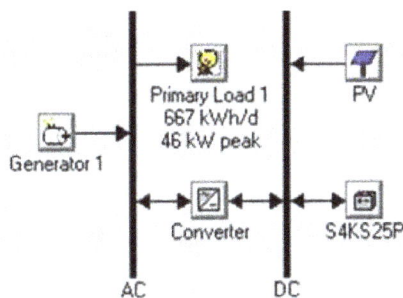

Figure 2. The equipments considered in the optimization design.

Figure 2 shows the considered equipments in the optimization. They're photovoltaic solar cells, converter, battery bank and loading system.

4.3. Solar Data

Solar inputs data for HOMER are taken as monthly averaged daily insolation incident on a horizontal surface (kWh/m^2/day) from NASA's Surface Meteorology, NASA gives average values over a 22 year period[21].The solar insolation is taken for 35° 28Nlatitude and44° 23Elongitude of the proposed site in Kirkuk city in Iraq. Figure 3 shows the solar resource profile over one year.

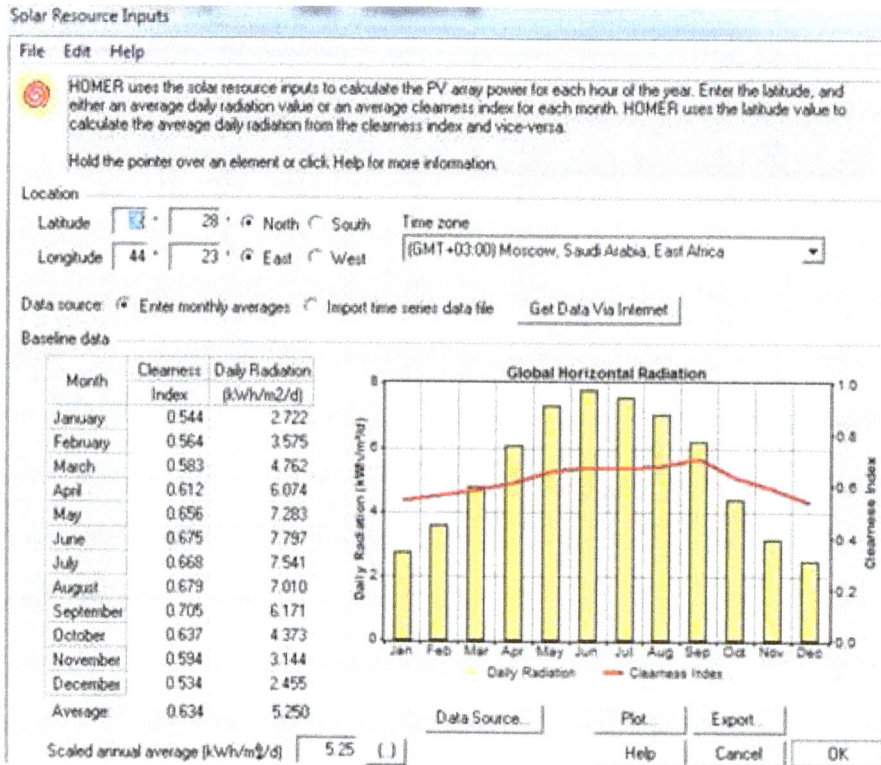

Figure 3. *Solar Resources Profile.*

4.4. PV Array Data

The PV array capital and replacement costs were specified with 16000 US$ and 15000 US$, respectively. Maintenance cost was considered for the panels around 1000 US$/yr. A derating factor of 80% and 20 years lifetime was considered as shown in Figure 4.

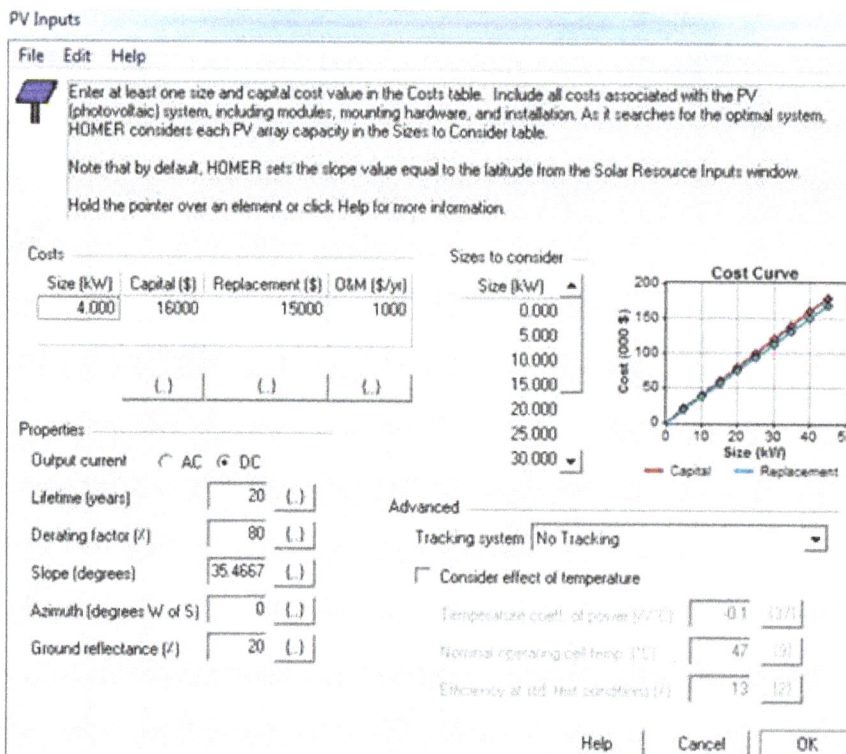

Figure 4. *PV array data.*

4.5. Battery Storage

The battery chosen is the Surrette4ks25p series. It has a nominal voltage of 4V and nominal capacity of 1900Ah (2.4

kWh). Each string consists of 3 batteries in series to get 12V DC. Batteries specifications and data were shown in Figure 5.

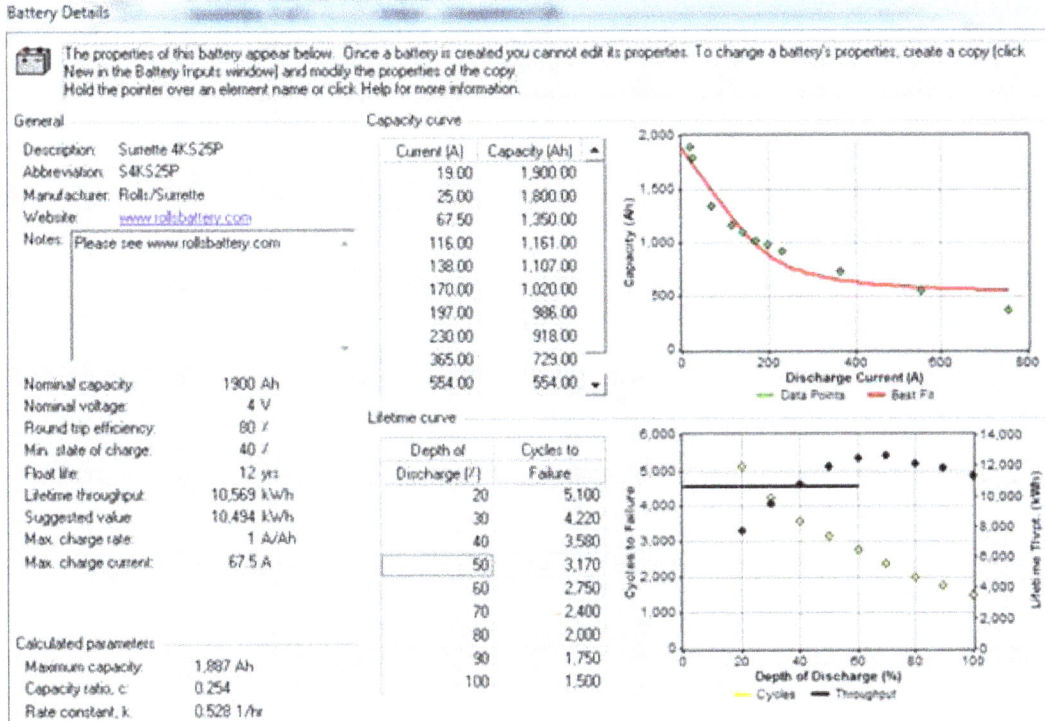

Figure 5. Batteries specifications and data.

4.6. Converter

The inverter and the rectifier efficiencies were assumed to be 90% and 85% respectively for all the considered sizes

considered. The considered sizes varied from 0 kW to 50kW. The converter inputs are shown in Figure 6.

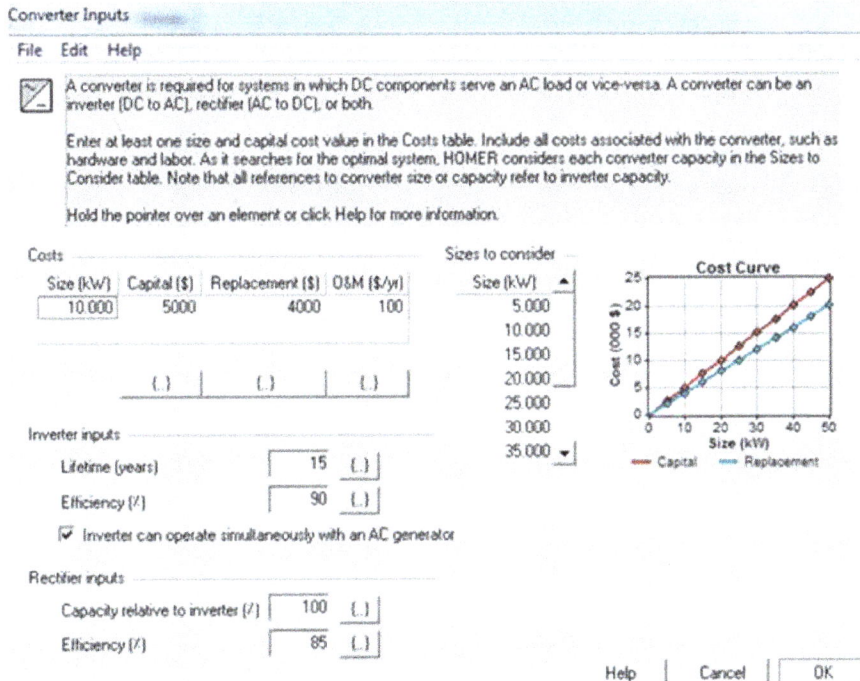

Figure 6. The converter input data.

5. Hybrid System Controller

Using homer software which was gives three study cases which were implemented considering trucking system type effects.

5.1 Case One: No Tracking System

The simulation overall results in case of no tracking system is shown in Table 1.The optimum total net present cost NPC and the cost of energy unit COE are 2,154,920$ and 0.692 $/kWh respectively. Categories can be shown by system components, cost types and in details. Figure 7 shows the optimal simulation results by components.

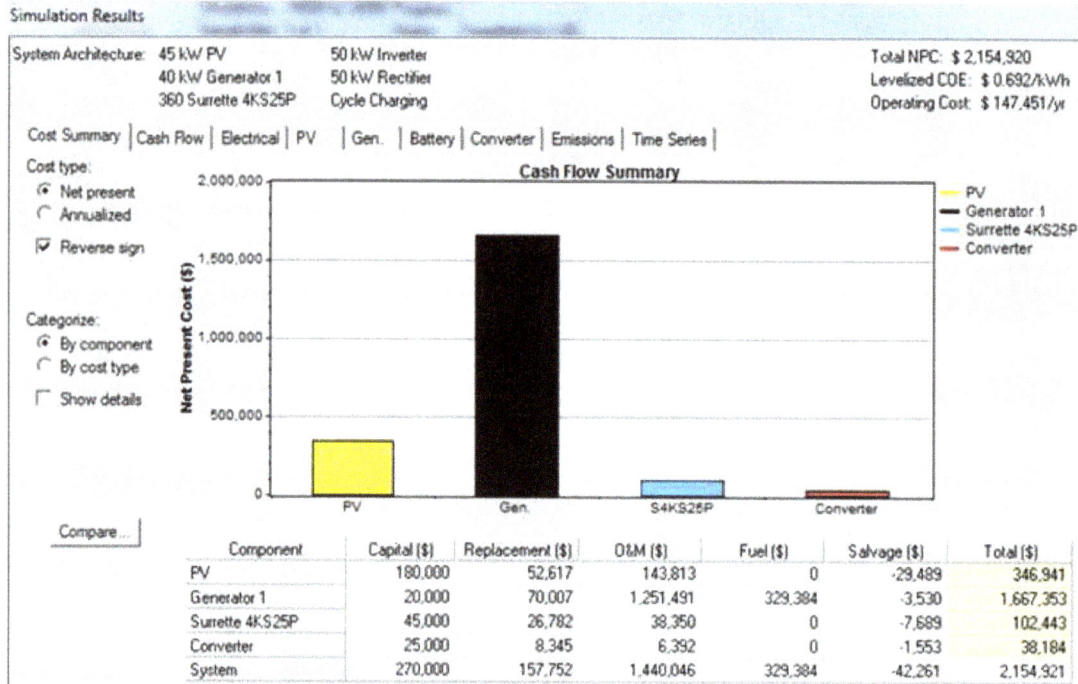

Figure 7. The optimal simulation results by components.

Table 1. The simulation overall results in case of no tracking system.

PV	Gen. [kW]	Batteries	Conv. [kW]	Initial Capital	Operating Cost[$/yr]	Total NPC	COE [$/kWh]	Ren. Frac.	Diesel [L]	Gen. [hrs]
45	40	360	50	270.00	147.451	2154920	0.692	0.20	64416	4895
45	40	360	50	267.5	149.051	2172876	9.688	0.20	64530	4973
40	40	360	50	250.0	151.052	2180954	0.701	0.17	67060	5083
40	40	360	50	247.5	152.385	2195497	0.705	0.17	67201	5148
35	40	360	50	230.0	155.047	2212023	0.711	0.13	69800	5288
35	40	360	50	227.5	155.434	2214466	0.711	0.13	69805	5311
45	45	126	50	243.25	154.957	2224347	0.715	0.18	66912	4751
30	40	360	50	207.5	157.919	2226232	0.715	0.10	72005	5455
45	45	360	50	272.5	153.109	2229749	0.716	0.18	66397	4559
40	45	126	50	223.25	157.219	2233033	0.717	0.15	69461	4863
30	40	360	50	210.00	158.261	2233105	0.717	0.10	72124	5464
45	45	117	50	242.125	156.299	2240146	0.720	0.18	67164	4807
40	45	117	50	222.125	158.257	2245177	0.721	0.15	69585	4909
35	45	126	50	203.25	159.882	2247078	0.722	0.11	72033	4992

The production percentage from PV array and diesel generator is 28% and 72% respectively. Figure 8 shows the production details. Daily generator output is shown in Figure 9.

System Architecture: 45 kW PV   50 kW Inverter
                       40 kW Generator 1   50 kW Rectifier
                       360 Surrette 4KS25P   Cycle Charging

Total NPC: $ 2,154,920
Levelized COE: $ 0.692/kWh
Operating Cost: $ 147,451/yr

Cost Summary | Cash Flow | Electrical | PV | Gen. | Battery | Converter | Emissions | Time Series |

Production	kWh/yr	%
PV array	77,696	28
Generator 1	195,021	72
Total	272,717	100

Consumption	kWh/yr	%
AC primary load	243,454	100
Total	243,454	100

Quantity	kWh/yr	%
Excess electricity	0.00202	0.0000
Unmet electric load	18.7	0.0077
Capacity shortage	61.1	0.0251

Quantity	Value
Renewable fraction	0.199
Max. renew. penetration	534 %

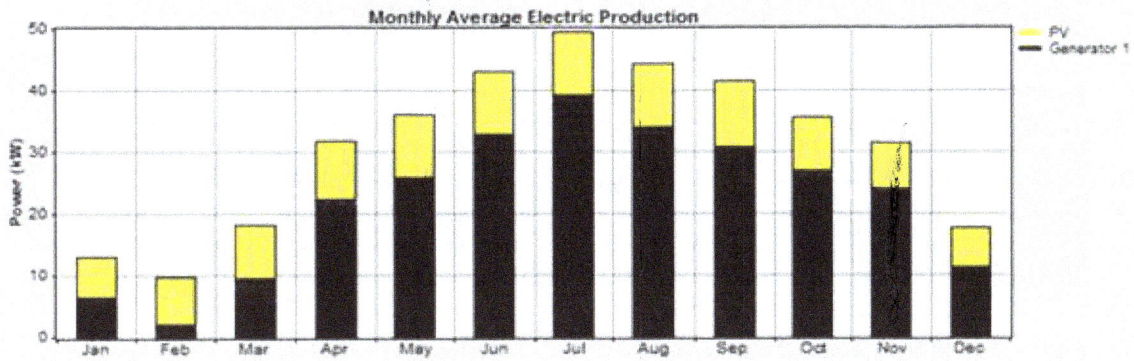

Figure 8. Case study1 production details.

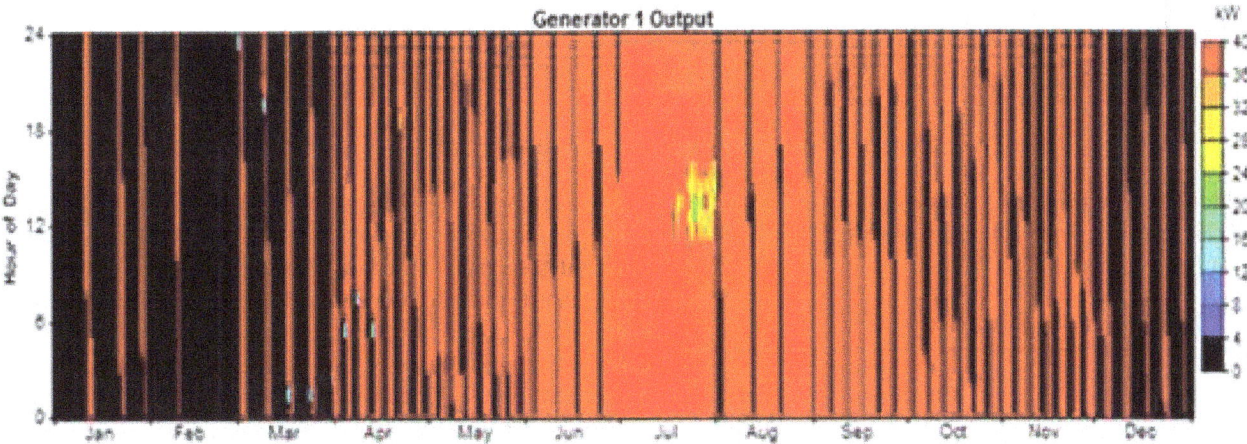

Figure 9. daily generator output.

Daily convertor output and batteries state of charge are shown in Figure 10.

System Architecture: 45 kW PV   50 kW Inverter   Total NPC: $ 2,154,920
                      40 kW Generator 1   50 kW Rectifier   Levelized COE: $ 0.692/kWh
                      360 Surrette 4KS25P Cycle Charging   Operating Cost: $ 147,451/yr

Cost Summary | Cash Flow | Electrical | PV | Gen. | Battery Converter | Emissions | Time Series |

Quantity	Inverter	Rectifier	Units
Capacity	50.0	50.0	kW
Mean output	9.6	3.4	kW
Minimum output	0.0	0.0	kW
Maximum output	43.8	27.0	kW
Capacity factor	19.1	6.8	%

Quantity	Inverter	Rectifier	Units
Hours of operation	5,843	2,913	hrs/yr
Energy in	92,977	35,256	kWh/yr
Energy out	83,681	29,968	kWh/yr
Losses	9,296	5,288	kWh/yr

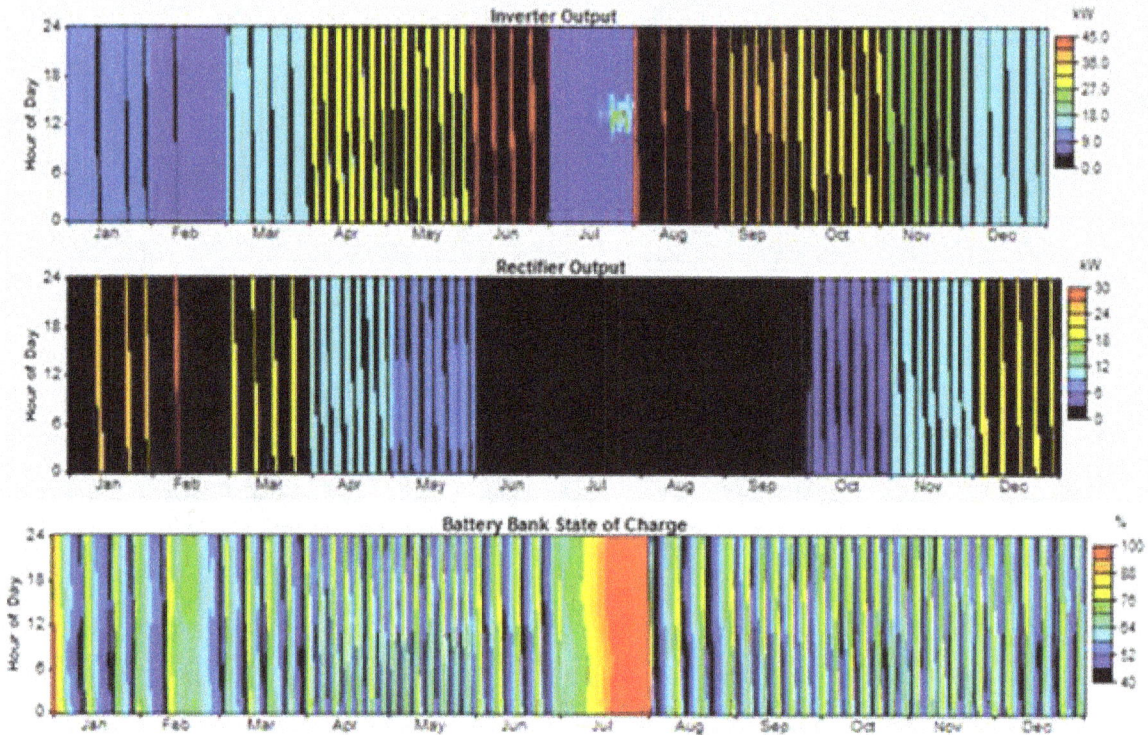

Figure 10. Convertor daily output and batteries state of charge.

To understand how friendly environment is our system in case study 1, the important pollutants were calculated. The emissions in kg/yr are shown in Table 2.

Table 2. The most important pollutants in kg/yr.

	Emission [kg/yr]
Carbon dioxide	169630
Carbon monoxide	419
Unburned hydrocarbons	46.4
Particulate matter	31.6
Sulfur dioxide	341
Nitrogen oxides	3736

5.2 Case Two: Two Axis Tracking System

Sun tracking is one of the methods which can boost the total collected energy from sun by 10-100%. Sun tracking systems move the solar panel based on hourly and seasonal movement of the sun in order to absorb the highest possible amount of energy [22]. Table 3 shows the simulation overall results in case of two axis tracking system effects. The optimum total net present cost NPC and the cost of energy unit COE are 1,9888,411$ and 0.639 $/kWh respectively. Using two axis tracking system increased PV percentage electric production from 28% to 35% as shown in Figure 11. This result will reduce CO_2 emission.

Table 3. The simulation overall results in case of two axis tracking system.

PV	Gen. [kW]	Batteries	Conv. [kW]	Initial Capital	Operating Cost[$/yr]	Total NPC	COE [$/kWh]	Ren. Frac.	Diesel [L]	Gen. [hrs]
40	40	126	50	220750	138278	1988411	0.639	0.26	60221	4748
	50	117	50	64625	188778	2477842	0.796	0.00	92760	5660

System Architecture: 40 kW PV   50 kW Inverter   Total NPC: $1,988,411
40 kW Generator 1   50 kW Rectifier   Levelized COE: $0.639/kWh
126 Surrette 4KS25P   Cycle Charging   Operating Cost: $138,278/yr

Cost Summary | Cash Flow Electrical | PV | Gen. | Battery | Converter | Emissions | Time Series |

Production	kWh/yr	%
PV array	95,371	35
Generator 1	180,117	65
Total	275,487	100

Consumption	kWh/yr	%
AC primary load	243,422	100
Total	243,422	100

Quantity	kWh/yr	%
Excess electricity	269	0.0978
Unmet electric load	50.6	0.0208
Capacity shortage	183	0.0750

Quantity	Value
Renewable fraction	0.260
Max. renew. penetration	510 /

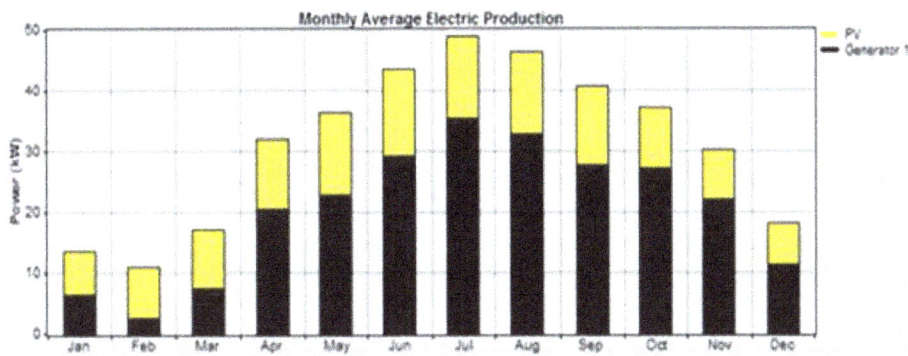

Figure 11. Case study2 production details.

5.3. Case Study Three: Power supplied by Diesel Generator

The optimum total net present cost NPC and the cost of energy unit COE are 2,477,843$ and 0.796 $/kWh respectively. Figure 12 and 13 show the monthly electric generation and the optimal simulation results by components respectively.

System Architecture: 50 kW Generator 1   50 kW Rectifier   Total NPC: $2,477,842
117 Surrette 4KS25P   Cycle Charging   Levelized COE: $0.796/kWh
50 kW Inverter   Operating Cost: $188,778/yr

Cost Summary | Cash Flow Electrical | Gen. | Battery | Converter | Emissions | Time Series |

Production	kWh/yr	%
Generator 1	280,481	100
Total	280,481	100

Consumption	kWh/yr	%
AC primary load	243,473	100
Total	243,473	100

Quantity	kWh/yr	%
Excess electricity	0.00290	0.00
Unmet electric load	0.00	0.00
Capacity shortage	0.00	0.00

Quantity	Value
Renewable fraction	0.00
Max. renew. penetration	0.00 /

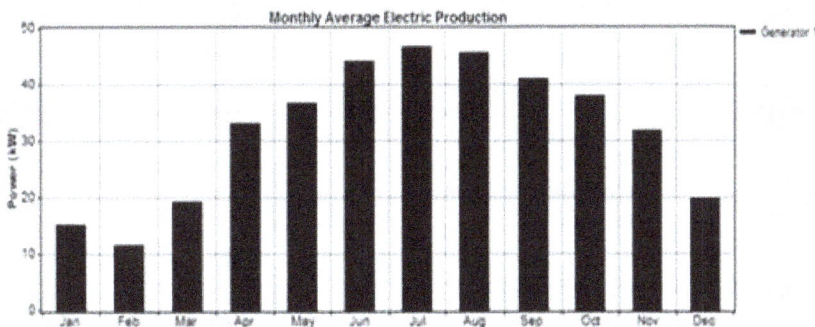

Figure 12. the monthly electric generation

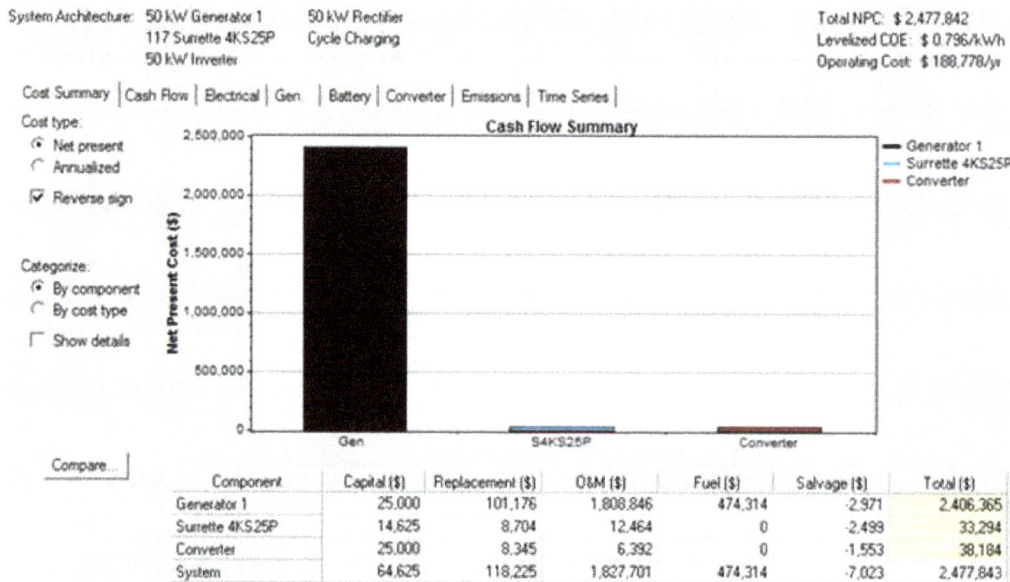

System Architecture: 50 kW Generator 1   50 kW Rectifier   Total NPC: $ 2,477,842
                     117 Surrette 4KS25P   Cycle Charging   Levelized COE: $ 0.796/kWh
                     50 kW Inverter   Operating Cost: $ 188,778/yr

Figure 13. Case study 3 optimal simulation results by components.

Total net present cost, cost of energy unit, CO2 emission and percentage of electric production for all studied cases are summarized in Table 4.

Table 4. Total net present cost, energy unit cost and CO2 emission for all studied Cases.

Case	Description	Power System Diagram	Total Net Present Cost[$]	Cost of Energy COE [$/kWh]	CO₂ Emission [kg/y]	Production Percentage PV%	Generator %
1	Power supply: Hybrid. Tracking sun: No trucking system.		2,154,920	0.692	169,630	28	72
2	Power Supply: Hybrid. Tracking sun: Two axis trucking system.		1,988,411	0.639	158,580	35	65
3	Power Supply: Diesel Generator.		2,477,842	0.796	244,268	0.0	100

6. Conclusions

The stand-alone hybrid solar power generation system is recognized as a viable alternative to conventional fuel-based remote area power supplies. It is generally more suitable than systems that only have one source of energy for supply of electricity to off-grid applications. All the optimization systems are ranked according to net present cost. All other economic outputs are calculated for the purpose of powering the store and finding the best net present cost.

Results shows that the initial capital cost depends on the size of the PV panel, the number of the batteries used and the size of the converter.

Sun tracking is one of the methods which can boost the total collected energy from sun.

In Table 4, case studies 1&2 show that PV electric production increased by 7% and CO2 emissions decreased by 6% when using 2 axis sun tracking system.

References

[1] Hongxing Yang, Optimal sizing method for stand-alone hybrid solar–wind system with LPSP technology by using genetic algorithm, Solar Energy 82 (2008) 354–367.

[2] Essam A. Al-Ammar, Nazar H. Malik and Mohammad Usman, Application of Using Hybrid Renewable Energy in Saudi Arabia, ETASR - Engineering, Technology & Applied Science Research Vol. 1, _o. 4, 2011,pp: 84-89.

[3] Y. Bhikabhai, Hybrid Power Systems And Their Potential In The Pacific Islands, SOPAC Miscellaneous, Report No. 406, 2005.

[4] E. Mohamed, "Hybrid Renewable Energy Systems for the Supply of Services in Rural Settlements of Mediterranean Partner Countries HYRESS project", 4th European Conference PV-Hybrid and Mini-Grid 2008.

[5] M. Martínez-Díaz1and R. Villafáfila-Robles, Study of optimization design criteria for stand-alone hybrid renewable power systems, Renewable Enrgy and Power Quality Journal RE & PQJ, International Conference on Renewable Energies and Power Quality ICREQ13, Bilbao, Spain, 20th to 22th March, 2013.

[6] Wei Zhou, Current status of research on optimum sizing of stand-alone hybrid solar–wind power generation systems, Applied Energy 87 (2010) 380–389.

[7] Mohamed EI Badawe, OPTIMAL SIZING, MODELING, AND DESIGN OF A SUPERVISORY CONTROLLER OF A STAND-ALONE HYBRID ENERGY SYSTEM, M.Sc. thesis, Faculty of Engineering & Applied Science, St. John's, Newfoundland, Canada, 2012.

[8] Md. Mizanur Rahman, Integration of centralized photovoltaic (PV) system into a rural electric feeder at Laxmipur in Bangladesh: Challenges and viability, The Third International Renewable Energy Congress, December 20-22, 2011, Hammamet, Tunisia.pp:373-379.

[9] Majid Alabdul Salam, Optimal sizing of photovoltaic systems using HOMER for Sohar, Oman, International Journal of Renewable Energy Research, Vol.3, No.2, 2013. pp: 301-307.

[10] Al- Jomaily A.K. and Abbas Ehsan, Theoretical study for selecting solar cooling system in storage used to saving Potato in Kirkuk city, Wasit Journal for Science & Medicine, Vo.6 Issue No.2 , pp190-210, 2013.

[11] Mir-Akbar Hessamin, Hugh Campbell, Christopher Sanguinetti, "A feasibility study of hybrid wind power systems for remote communities", Energy Policy, Volum 39, Issue 2, 2011, pp: 877-886.

[12] Jose L. Bernal-Agustı n, Rodolfo Dufo-Lo pez, Simulation and optimization of stand-alone hybrid renewable energy systems, Renewable and Sustainable Energy Reviews 13 (2009) 2111–2118.

[13] Muselli M, Notton G, Louche A. Design of hybrid-photovoltaic power generator, with optimization of energy management. Solar Energy 1999;65(3): 143–57.

[14] Bagen, Billinton R. Evaluation of different operating strategies in small standalone power systems. IEEE Trans Energy Convers 2005;20(3):654–60.

[15] Sameer Al-Juboori, Design Hybrid Micropower System In Mistah Village Using Homer Model, International Journal of Advanced Research in engineering & Technology(IGARET), Volume 4, Issue 5, July – August 2013, pp. 218-230.

[16] Sameer Al-Juboori, Ali H. Mutlag and Rana H. Abduljabbar, Evaluation of Stand Alone Remote Area Hybrid Power system, International Journal of Electrical and Electronics Engineering (IJEEE), Vol. 3, Issue 1, Jan 2014, 61-68.

[17] Elhadidy MA, Shaahid SM. Parametric study of hybrid (wind + solar + diesel) power generating systems. Renew Energy 2000;21(2):129–39.

[18] Gutie´ rrez Vera J. Options for rural electrification in Mexico. IEEE Trans Energy Convers 1992;7(3):426–30.

[19] Wichert B. PV–Diesel hybrid energy systems for remote area power generation — a review of current practice and future developments.Renew Sustain Energy Rev 1997;1(3):209–28.

[20] Shaahid SM, Elhadidy MA. Opportunities for utilization of stand-alone hybrid (photovoltaic + diesel + battery) power systems in hot climates. Renew Energy 2003;28(11):1741–53.

[21] NASA, "Surface meteorology and Solar Energy." [Online]. Available: http://eosweb.larc.nasa.gov/sse/

[22] Hairul Nissah Zainudin, Comparison Study of Maximum Power Point Tracker Techniques for PV Systems, Proceedings of the 14th International Middle East Power Systems Conference (MEPCON'10), Cairo University, Egypt, December 19-21, 2010, Paper ID 278, PP: 750-755.

18

Solar Radiation Estimation from the Measurement of Sunshine Hours over Southern Coastal Region, Bangladesh

Shuvankar Podder[1], Md. Minarul Islam[2]

[1]Department of Electrical and Electronic Engineering, Bangladesh University ofEngineering and Technology, Dhaka, Bangladesh
[2]Department of Electrical and Electronic Engineering,Shahjalal University of Science and Technology, Sylhet, Bangladesh

Email address:

podder.shuvankar@gmail.com (S. Podder), minarbuet@gmail.com (Md. M. Islam)

Abstract: In this study,the global solar radiation over the southern coastal region of Bangladesh is estimated from the duration of relative sunshine hours. Five models are considered to estimate the solar irradiance. These models are modified form of classical Angstrom – Prescott regression equation. A quadratic logarithmic model, relating the relative solar radiation and the relative sunshine hours is proposed for southern coastal region of Bangladesh. NASA Surface Meteorology and Solar Energy (SSE)have record of solar radiation data all over the world, measured from satellite. As Bangladesh Meteorological Department or any other organization has no record of measured solar radiation data for the considered locations, the estimated solar irradiance from the proposed regression model is compared with the data recorded by NASA SSE. Also t – statistics is applied to the estimated results to determine whether or not they are statistically significant at a particular confidence level.

Keywords: Solar Radiation, Sunshine Hours, Coastal Region, Nonlinear Relation, Hybrid

1. Introduction

According to renewable energy policy 2009, the Government of Bangladesh is committed to facilitate both private and public sector investments in renewable energy projects to substitute indigenous non- renewable energy supplies and scale up contributions existing renewable energy based electricity production. The policy envisions that 5% of the total energy production will have to be achieved by 2015 and 10% by 2020 using renewable resources. There is a good scope for solar, wind, biomass and mini-hydro power generation in Bangladesh. Among these solar and wind possess most potential for electricity generation [1].

The reliability of electric power encourages hybridization of two or more renewable energy systems because of its intermittent nature.Solar-wind hybrid system is an universal one. Bangladesh Power Development Board (BPDB) has launched 7.5 MW off-grid solar-wind hybrid systems in Hatiya Island, Noakhali. BPDB has planned to install 1 MW off-grid solar-diesel based hybrid plant in Kutubdia Island and a 500 kW photovoltaic plant at Sandwip. 8 MW grid-connected and 2 kW off-grid photovoltaic plants are ongoing projects at Rangamati and Noakhali, respectively.BPDB has

also lined up installation of MW range wind power stationatCox's Bazar [1]

Adequate assessment of renewable resource data are essentials for planning and designing renewable energy based power systems. At present, solar radiation data are available from (1) Renewable Energy Research Centre (RERC), Dhaka University; it has recorded long-term hourly solar irradiance of Dhaka city with Eppley Precision Pyrometer. (2)Bangladesh Meteorological Department (BMD); it has 35 sunshine recording stations [2] situated generally in towns and cities. BMD has no record of solar radiation on the abovementioned solar projects areas. Solar radiation reaching the earth's surface depends upon climatic conditions. Thus a mathematical model can be developed relating climatic factors with solar radiation. A number of studies [3-4] have computed solar radiation from observation of cloud cover. Other studies [5-8] have estimated solar radiation from sunshine hours. BMD has record of daily bright sunshine hours at the abovementioned places.

Some studies havecorrelated sunshine hours and solar radiation over some major cities in Bangladesh [9-10]. But no attempt has yet been made to estimate solar radiation of the places in Bangladesh where the prospect of solar-wind hybrid system has promising potential. In this paper, five models relating solar radiation and relative sunshine hours have been analyzed andsolar radiation is predicted at Rangamati, Sandwip, Noakhali, Kutubdia and Cox's Bazar. Resultsare compared with the data reported by NASA Surface Meteorology and Solar Energy (SSE) [11] on that places.

2. Experimental Data

Solar radiation arriving on the horizontal earth surface and duration of bright sunshine hours are two main experimental data in this study.Data of sunshine hours for the years 1983 to 2013 have been collected from BMD.As solar radiation data is not available at BMD, they have been collected from NASA SSE. In this paper, the target locations for analyzing solar radiation are Rangamati, Sandwip, Noakhali, Kutubdia and Cox's Bazar. The geographical parameters of those five locations are shown in Table 1. The relative sunshine hours of five places are shown in Table 2.

Table 1. Geographical parameters.

Stations	Latitude	Longitude	Elevation
Rangamati	22.63	92.2	14
Sandwip	22.48	91.48	7
Noakhali	22.70	91.10	12
Kutubdia	21.82	91.86	50
Cox's Bazar	21.58	92.02	3

Table 2. Relative sunshine hours.

Month	Ranga-mati	Sand-wip	Noak-hali	Kutub-dia	Cox's Bazar
January	0.6635	0.6826	0.6225	0.7362	0.7998
February	0.6906	0.6758	0.6626	0.7370	0.7880
March	0.6361	0.6490	0.6324	0.6962	0.7255
April	0.6050	0.6181	0.5929	0.6379	0.6814
May	0.4679	0.4789	0.4804	0.5270	0.5358
June	0.3194	0.3513	0.2821	0.3156	0.2929
July	0.2815	0.3270	0.2657	0.2886	0.2710
August	0.3609	0.3739	0.3486	0.3514	0.3355
September	0.4227	0.4239	0.3935	0.4711	0.4665
October	0.5545	0.5725	0.5584	0.6031	0.6292
November	0.6527	0.6890	0.6598	0.7252	0.7560
December	0.6677	0.6780	0.6478	0.7556	0.7620

3. Methodology

According to World Meteorological Organization(WMO) the sunshine duration is defined as the period during which the direct solar irradiance exceed a threshold value of 120 W/m²-day or 2.88 KWh/m²-day [12]. Solar radiation of a certain period is proportional to sunshine duration. Different models describing solar radiation and sunshine hours are paraphrased here.

3.1. Angstrom Model

The relation between solar radiation and sunshine duration was first proposed by Angstrom in 1924.The original Angstrom equation is given by [13]

$$\frac{\overline{H}}{\overline{H_c}} = a + b\frac{\overline{n}}{\overline{N}} \quad (1)$$

Where $\overline{H}$ = monthly average daily global radiation (Wh/m²/day), $\overline{H_c}$ = monthly average clear sky daily global radiation for the location, $\overline{n}$ = monthly average daily maximum bright sunshine duration in hours, $\overline{N}$ = actual sunshine duration in a day in hours, and a, b are empirical coefficients. These coefficients are location specific.

A basic difficulty in this model is to determine $\overline{H_c}$, clear sky radiation. To avoid this difficulty a modified model was presented by Prescott [14] in 1940.

3.2. Angstrom-Prescott Model

Popularly known Angstrom-Prescott model is given by

$$\frac{\overline{H}}{\overline{H_c}} = a + b\frac{\overline{n}}{\overline{N}} \quad (2)$$

where,$\overline{H}, \overline{n}, \overline{N}, a, b$ are same as equation (1) and$\overline{H_o}$= monthly average daily extraterrestrial radiation at the specific location. The ratio of solar radiation at the surface of the Earth (H) to extraterrestrial radiation (H_0), that is, H/H_0, is called the clearness Index and the ratio n/N is referred to as the cloudless index.

Monthly average daily extraterrestrial radiation is calculated from following equation:

$$H_0 = \frac{24}{\pi}G_{sc}\left(1 + cos\frac{360n}{365}\right)\left(cos\varphi cos\delta sin\omega_s + \frac{\pi\omega_s}{180^o}sin\varphi sin\delta\right) \quad (3)$$

where,
G_{sc} = the solar constant = 1.367 kW/m²
n = the day of a year (a number between 1 to 365, starting from 1st January)
φ = the latitude in degree
δ = the solar declination in degree
ω_s =the sunset hour angle in degree.
The solar declination is calculated according to the following equation:

$$\delta = 23.45^o \sin\left(360^o \frac{284+n}{365}\right) \quad (4)$$

The sunset hour angle is calculated using the following equation:

$$cos\omega_s = -tan\varphi tan\delta \quad (5)$$

The average H_0 for the month is calculated as follows:

$$H_{0,avg} = \frac{\sum_{n=1}^{N}H_0}{N} \quad (6)$$

where,
$H_{0,avg}$ = the average extraterrestrial horizontal radiation for the month in kWh/m²/day
N =the number of days in the month
The maximum possible sunshine duration N in hours for a horizontal surface is given by:

$$N = \frac{2}{15}\omega_s \quad (7)$$

3.3. Akinoglu and Ecevit Model

Akinoglu BG et el [15] constructed a quadratic relation between H/Ho and n/N from modified Angstrom model. According to Akinoglu BG et el:

$$\frac{\overline{H}}{\overline{H_o}} = a + b\frac{\overline{n}}{\overline{N}} + c\left(\frac{\overline{n}}{\overline{N}}\right)^2 \qquad (8)$$

3.4. Newland Model

Newland et el [16] separated global solar irradiance into its components for the southern coastal region Macau, China. He showed that a non linear relation between (n/N) and (H/Ho) gives better prediction of global irradiance. His proposed relation is

$$\frac{\overline{H}}{\overline{H_o}} = a + b\frac{\overline{n}}{\overline{N}} + c\log\left(\frac{\overline{n}}{\overline{N}}\right) \qquad (9)$$

3.5. Ampratwum and Dorvlo Model

Ampratwum et el [17] studied five stations in Oman and proposed a logarithmic relationship between (n/N) and (H/Ho). His proposed model is

$$\frac{\overline{H}}{\overline{H_o}} = a + b\log\left(\frac{\overline{n}}{\overline{N}}\right) \qquad (10)$$

3.6. Proposed Model

In this paper, another non-linear model is proposed. The proposed model relating (H/Ho) and (n/N) is

$$\frac{\overline{H}}{\overline{H_o}} = a + b\log\left(\frac{\overline{n}}{\overline{N}}\right) + c\log\left(\frac{\overline{n}}{\overline{N}}\right)^2 \qquad (11)$$

The coefficients a,b,c of different models are calculated by least square regression. Five models considered for estimating global solar irradiance on five southern coastal region of Bangladesh aretabulated in Table 3. MATLAB simulationis used in determining regression coefficient.

Table 3. *Considered models.*

Models	Regression EquationS
Angstrom-Prescott	$\frac{\overline{H}}{\overline{H_c}} = a + b\frac{\overline{n}}{\overline{N}}$
Akinoglu and Ecevit	$\frac{\overline{H}}{\overline{H_o}} = a + b\frac{\overline{n}}{\overline{N}} + c\left(\frac{\overline{n}}{\overline{N}}\right)^2$
Ampratwum and Dorvlo	$\frac{\overline{H}}{\overline{H_o}} = a + b\log\left(\frac{\overline{n}}{\overline{N}}\right)$
Newland	$\frac{\overline{H}}{\overline{H_o}} = a + b\frac{\overline{n}}{\overline{N}} + c\log\left(\frac{\overline{n}}{\overline{N}}\right)$
Proposed Model	$\frac{\overline{H}}{\overline{H_o}} = a + b\log\left(\frac{\overline{n}}{\overline{N}}\right) + c\log\left(\frac{\overline{n}}{\overline{N}}\right)^2$

4. Model Performance

Stone [18] concluded that the t-statistics test might be taken as a statistical indicator for the evaluation and comparison of solar models. The smaller the value of t, the better is the model's performance. If the calculated value of t-

stat is less than a critical value t_c, then it can be concluded that estimation is significant to (n-1) degree of freedom at the (1- α) confidence level. Stone recommend that t-statistics may be used in conjunction with Mean Bias Error (MBE), Root Mean Square Error (RMSE) and Mean Absolute Percentage Error (MAPE) to access the relative model performance.The mostly used statistical indicator MBE, RMSE and MAPE are defined as

$$MBE = \left(\sum_{i=1}^{n=12}(H_{ic} - H_{im})\right)/n \qquad (12)$$

$$MAPE = \left(\sum_{i=1}^{n=12}|H_{ic} - H_{im}|\right)/n \qquad (13)$$

$$RMSE = \left\{\left(\sum_{i=1}^{n=12}(H_{ic} - H_{im})^2\right)/n\right\}^{1/2} \qquad (14)$$

$$r = \frac{\sum_{i=1}^{n=12}(H_{ic}-\overline{H}_c)(H_{im}-\overline{H}_m)}{\sqrt{\left(\sum_{i=1}^{n=12}(H_{ic}-\overline{H}_c)^2\right)\left(\sum_{i=1}^{12}(H_{im}-\overline{H}_m)^2\right)}} \qquad (15)$$

where,H_{ic} , and H_{im} are the estimated and measured monthly average global solar radiation for the ith month. The average of the deviations $E (= H_{ic} - H_{im})$ is MBE and gives information about the long- term performance of the correlations. MAPE is a measure of the goodness of each correlation, while RMSE measures the short-term prediction quality of the correlations [19].

5. Result and Discussion

As shown in Table 1, the five study regions are geographically close to one another andthey are mainly southern coastal belt of Bangladesh. Therefore, a general relation between solar radiation and sunshine hours can be developed for these places.

The regression coefficients of five models for the considered locations are shown in Table 4.The physical significance of the regression coefficients `a' and `b' is that `a' is a measure of the overall atmospheric transmission for total cloud conditions (n/N=0), and is a function of the type and the thickness of the cloud cover, while `b' and `c' are the rate of increase of (H/Ho) with (n/N). The sum (a+b) denotes the overall atmospheric transmission under clear sky conditions.

Statistical evaluations of five models are summarized inTable 5. It is seen that the regional correlation has minimum error in all models. MBE, RMSE and MAPE are lowest in "*Akinoglu and Ecevit*" model for all locations. Also value of r is highest in "*Akinoglu and Ecevit*"-model for all locations indicate that this model best fit the sunshine hour data with solar radiation. The value of t-stat lies far below the critical value t_c(at α=0.01) indicating correlation models performance is statistically significant at 99% level of significance.

The proposed model in this paper shows statistically good performance. The value of t-stat for all locations in this model is lower than linear and logarithmic models. This indicates that proposed model is better in estimating solar radiation than that proposed by Angstrom and Ampratwum et   el.

Table 4. *Regression coefficients in different models.*

	Model	a	b	c	$a+b+c$
	Angstrom-Prescott	0.1733	0.6619		0.8352
	Akinoglu & Ecevit	0.4166	-0.4181	1.0930	1.0915
Rangamati	Ampratwum &Dorvlo	0.7299	0.3041		1.034
	Newland	-0.6133	1.5878	-0.4370	0.5375
	Proposed	0.9107	0.8560	0.3508	2.1175
	Angstrom-Prescott	0.1305	0.6898		0.8203
	Akinoglu & Ecevit	0.3240	-0.1174	0.7807	0.9873
Sandwip	Ampratwum &Dorvlo	0.7249	0.3401		1.0650
	Newland	-0.4791	1.3937	-0.3517	0.5629
	Proposed	0.8689	0.8155	0.3309	2.0153
	Angstrom-Prescott	0.1757	0.6457		0.8214
	Akinoglu & Ecevit	0.3952	-0.3909	1.0983	1.1026
Noakhali	Ampratwum &Dorvlo	0.7083	0.2815		0.9898
	Newland	-0.6030	1.5796	-0.4186	0.5580
	Proposed	0.9061	0.8490	0.3372	2.0923
	Angstrom-Prescott	0.1423	0.6750		0.8173
	Akinoglu & Ecevit	0.2578	0.1845	0.4647	0.9070
Kutubdia	Ampratwum &Dorvlo	0.7293	0.3291		1.0584
	Newland	-0.2779	1.1545	-0.2390	0.6376
	Proposed	0.8472	0.7483	0.2835	1.8790
	Angstrom-Prescott	0.1730	0.5868		0.7598
	Akinoglu &Ecevit	0.2792	0.1299	0.4245	0.8336
Cox's Bazar	Ampratwum & Dorvlo	0.6889	0.2861		0.9750
	Newland	-0.1898	0.9957	-0.2048	0.6011
	Proposed	0.7819	0.6459	0.2413	1.6691

The estimated annual radiations on considered locations along with measured radiation are shown in Table 6.

A general relation relating relative sunshine hours (n/N) and relative solar radiation (H/Ho) has been developed according to proposed model for these southern coastal regions. To determine the coefficients of general relation, least square regression has been used combining all the data of five locations. The proposed general correlation equation for southern coastal region is

$$\frac{\overline{H}}{H_o} = 0.8111 + 0.6301 log\left(\frac{\overline{n}}{N}\right) + 0.2157 log\left(\frac{\overline{n}}{N}\right)^2 \quad (16)$$

To validate the proposed general correlation, the estimated solar radiation for considered five locations along with measured radiation has been shown in figure 1 to 5. From those figures it is evident that the predicted radiations according to equation (16) are sufficiently close to that measured by NASA SSE.

Table 5. *Statistics of five models.*

	Model	r	MBE	RMSE	MAPE	t-stat	t_c
	Angstrom-Prescott	0.9595	0.0108	0.1855	0.0334	0.1933	
	Akinoglu and Ecevit	0.9770	0.0061	0.1243	0.0019	0.1642	
Rangamati	Ampratwum and Dorvlo	0.9226	0.0163	0.2611	0.0208	0.2076	3.106
	Newland	0.9751	0.0065	0.1322	0.0221	0.1639	
	Proposed	0.9729	0.0067	0.1399	0.0234	0.1587	
	Angstrom-Prescott	0.9676	0.0069	0.1527	0.0283	0.1502	
	Akinoglu and Ecevit	0.9692	0.0036	0.1351	0.0233	0.0875	
Sandwip	Ampratwum and Dorvlo	0.9534	0.0110	0.1959	0.0377	0.1858	3.106
	Newland	0.9686	0.0042	0.1383	0.0242	0.1000	
	Proposed	0.9678	0.0047	0.1416	0.0251	0.1102	
	Angstrom-Prescott	0.9516	0.0158	0.2151	0.0016	0.2443	
	Akinoglu and Ecevit	0.9676	0.0098	0.1575	0.0012	0.2068	
Noakhali	Ampratwum and Dorvlo	0.9155	0.0209	0.2863	0.0026	0.2425	3.106
	Newland	0.9676	0.0114	0.1617	0.0014	0.2354	
	Proposed	0.9670	0.0119	0.1668	0.0289	0.2378	
	Angstrom-Prescott	0.9763	-0.0035	0.1591	0.0288	0.0726	
	Akinoglu and Ecevit	0.9859	-0.007	0.1359	0.0228	0.1732	
Kutubdia	Ampratwum and Dorvlo	0.9458	0.0047	0.2469	0.0461	0.0630	3.106
	Newland	0.9869	-0.007	0.1366	0.0235	0.1768	
	Proposed	0.9854	-0.006	0.1387	0.0250	0.1578	
Cox's Bazar	Angstrom-Prescott	0.9771	0.0015	0.1534	0.0288	0.0330	3.106

Model	r	MBE	RMSE	MAPE	t-stat	t_c
Akinoglu and Ecevit	0.9849	-0.004	0.1256	0.0196	0.0965	
Ampratwum and Dorvlo	0.9443	0.0098	0.2542	0.0511	0.1276	
Newland	0.9841	-0.002	0.1276	0.0207	0.0578	
Proposed	0.9827	-0.002	0.1321	0.0229	0.0418	

Table 6. Comparison of estimated and measured radiation

Stations	Models	Annual Estimated Radiation	Annual Measured Radiation
	Angstrom-Prescott	4.7266	
	Akinoglu and Ecevit	4.7220	
Rangamati	Ampratwum and Dorvlo	4.7321	4.72
	Newland	4.7224	
	Proposed	4.7225	
	Angstrom-Prescott	4.5642	
	Akinoglu and Ecevit	4.5619	
Sandwip	Ampratwum and Dorvlo	4.5693	4.56
	Newland	4.5625	
	Proposed	4.5630	
	Angstrom-Prescott	4.5741	
	Akinoglu and Ecevit	4.5681	
Noakhali	Ampratwum and Dorvlo	4.5792	4.56
	Newland	4.5698	
	Proposed	4.5703	
	Angstrom-Prescott	4.7707	
	Akinoglu and Ecevit	4.7669	
Kutubdia	Ampratwum and Dorvlo	4.7789	4.77
	Newland	4.7669	
	Proposed	4.7677	
	Angstrom-Prescott	4.6915	
	Akinoglu and Ecevit	4.6863	
Cox's Bazar	Ampratwum and Dorvlo	4.6998	4.69
	Newland	4.6878	
	Proposed	4.6883	

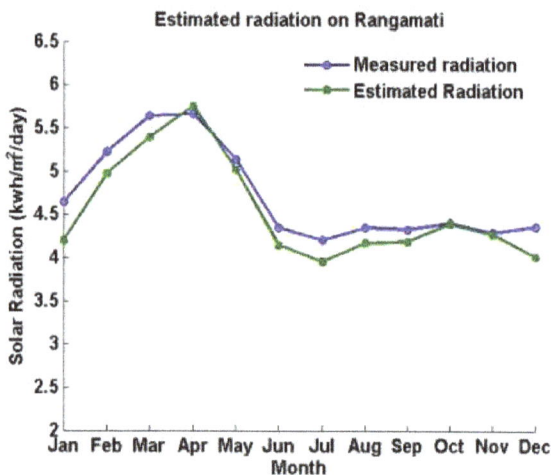

Figure 1. Estimated and measured radiation on Rangamati.

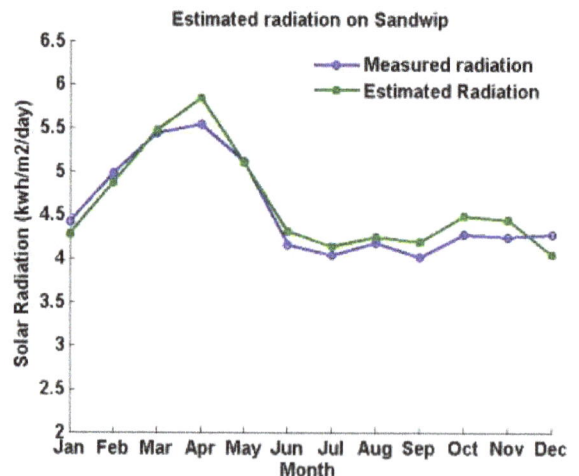

Figure 2. Estimated and measured radiation on Sandwip.

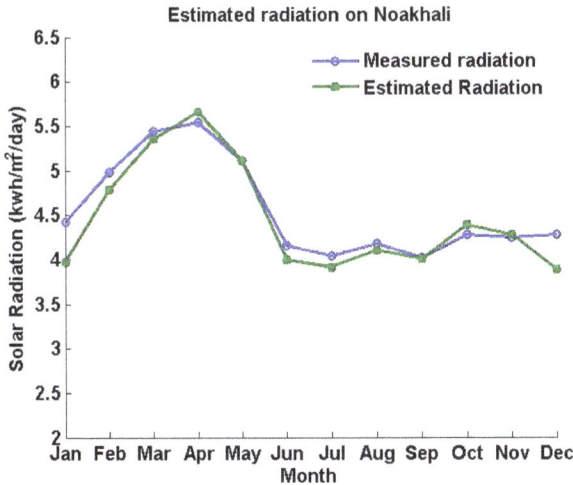

Figure 3. Estimated and measured radiation on Noakhali.

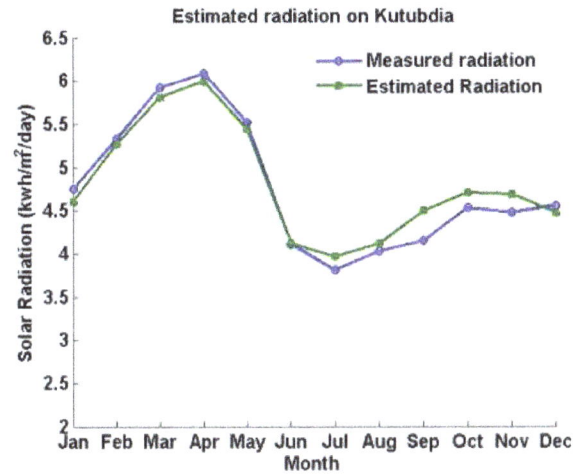

Figure 4. Estimated and measured radiation on Kutubdia.

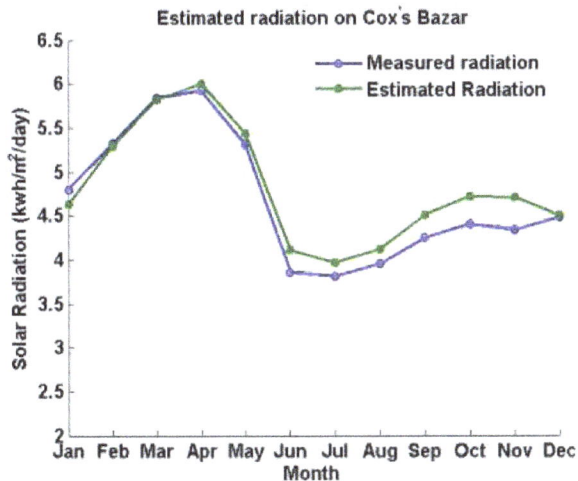

Figure 5. Estimated and measured radiation on Cox's Bazar.

The accuracy of the estimated radiation according to equation (16)has also been determined by statistical means. The MBE, RMSE, MAPE and t-stat for estimated radiation according to proposed equation has been shown in Table 7. It

is found fromTable 7 that the values of t-stat are far below from t_c at 99% confidence level. This indicates that the general correlation is statistically significant.

Table 7. *Statistics of proposed equation.*

Stations	r	MBE	RMSE	MAPE	t-stat	t_c $(\alpha=0.01)$
Rangamati	0.972	-0.106	0.171	0.030	2.638	3.106
Sandwip	0.961	0.062	0.164	0.032	1.358	3.106
Noakhali	0.960	-0.102	0.199	0.023	1.991	3.106
Kutubdia	0.984	0.033	0.155	0.028	0.719	3.106
Cox's Bazar	0.979	0.126	0.198	0.037	2.752	3.106

In this paper, correlation between relative sunshine hour and solar radiation has been developed for southern coastal region of Bangladesh. Using this correlation, global solar radiation on any southern coastal region of Bangladesh can be estimated from relative sunshine hours. To determine the correctness of proposed relation, solar radiation has been estimated on another southern coastal region Patenga, Chittagong. Patenga is situated at 22.7^0 latitude and 91.8^0 longitudes.

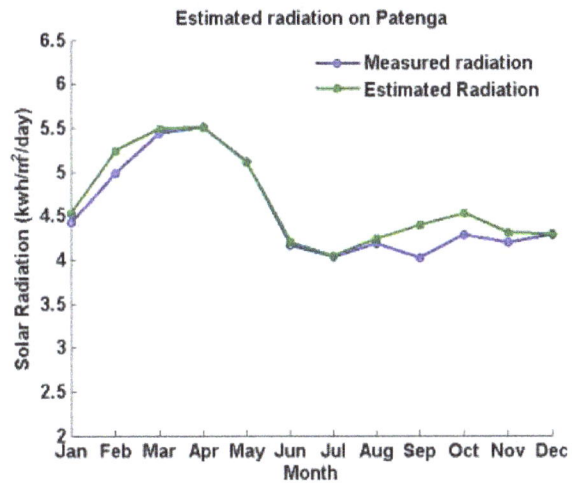

Figure 6. Estimated and measured radiation on Patenga.

Table 8. *Radiation on Patenga.*

Month	n/N	H_0	$H_{measured}$	$H_{estimated}$
January	0.7246	7.1764	4.4207	4.5251
February	0.7345	8.2196	4.9811	5.2379
March	0.6491	9.4823	5.4428	5.4914
April	0.5696	10.4854	5.5048	5.5032
May	0.4809	10.9736	5.1137	5.1079
June	0.3311	11.0921	4.1595	4.1955
July	0.3054	10.9962	4.0356	4.0380
August	0.3724	10.6154	4.1825	4.2380
September	0.4549	9.7876	4.0227	4.3914
October	0.5751	8.5646	4.2823	4.5271
November	0.6648	7.2901	4.2493	4.3000
December	0.7215	6.8279	4.2811	4.2910
Average			4.5563	4.6539

Table 9. Statistics of estimated radiation on Patenga.

Station	r	MBE	RMSE	MAPE	t-stat	$t_c(\alpha=0.01)$
Patenga	0.9749	0.1024	0.1558	0.0238	2.8905	3.106

The estimated solar radiation on Patenga according to equation (16) along with measured solar radiation has been shown in figure 6. The relative sunshine hours, measured solar radiation and estimated solar radiation over the year on Patengahas been shown in Table 8. The statistical parameters MBE, RMSE, MAPE and t-stat for estimated solar radiation according to proposed correlation on Patengahas been shown in Table 9. It is found that estimated global solar radiation is statistically satisfied.

6. Conclusion

In this analysis, five models relating global solar radiation and relative sunshine hours have been considered for predicting the global solar radiation pattern over the southern coastal region of Bangladesh. The level of performance of five models has been studied by statistical measures. The t-statistics have been applied to test the significance of applicability of these models.

A nonlinear logarithmic model has been proposed for estimating the global solar radiation from sunshine hour data. Statistical tests show that proposed model gives fairly good result and can be applied to southern coastal areas of Bangladesh. Few articles correlate the global solar radiation with sunshine ours over Bangladesh. But developing a nonlinear model for estimating solar radiation over southern coastal region of Bangladesh is quiet new. This work emphasis on this region considering the potential of generating electricity from hybrid solar-wind based renewable energy system. The accuracy of prediction can be further developed by considering the fog density, cloud cover and atmospheric scattering effect.

References

[1]   Development of Renewable Energy Technologies by BPDB, accessed at 15 January 2015, :available http://www.bpdb.gov.bd/bpdb/index.php?option=com_co%20ntent&view=article&id=26&Itemid=24

[2]   Solar and Wind Energy Resource Assessment (SWERA) project report, pp. 13-14, accessed at 10 January 2015,available http://www.dlr.de/tt/Portaldata/41/Resources/dokumente/institut/system/publications/SWERA_10km_solar_finalreport_by_DLR.pdf

[3]   S. Rangarajan, M. S. Swaminathan and A. Mani, Computation of solar radiation from observations of cloud cover.Solar Energy, Vol. 32, No. 4, pp. 553-556, (1984)

[4]   A. Nyberg, Determination of global radiation with the aid of observations of cloudiness. Acta Agric.Scand, Vol. 27, pp. 297-300.

[5]   M. R. Rieltveld, A new method for estimating the regression coefficients in the formula relating solar radiation to sunshine. Agr.Meteorol. vol - 19, pp243-252(1978).

[6]   K. K. Gopinathan, A general formula for computing the coefficient of the correlation connecting global solar radiation to sunshine duration. Solar Energy, Vol. 41. No. 6, pp. 499-502. (1988)

[7]   P. R. Benson, M. V. Paris, J. E. Serry and C. G. Justus, Estimation of daily and monthly direct, diffuse and global solar radiation from sunshine duration measurement , SolarEnergy, Vol. 32, No.4, pp:523-535(1984).

[8]   H. Ogelman, A. Ecevit and E. Tasdemiroglu, A new method for estimating solar radiation from bright sunshine data. Solar Energy, Vol. 33, No.6, pp. 619-625, (1984)

[9]   H R Ghosh, S M Ullah, S K Khadem, N C Bhowmik and M Hussain, Measurement and estimation of sunshine duration for Bangladesh. Renewable Energy Research Center, University of Dhaka, Bangladesh

[10]   Debazit Datta, Bimal Kumar Datta, Empirical model for the estimation of global solar radiation in Dhaka, Bangladesh. International Journal of Research in Engineering and Technology, Volume: 02 Issue: 11, pp. 649-653.

[11]   NASA Surface Meteorology and Solar Energy; accessed at 15 January 2015, fromhttps://eosweb.larc.nasa.gov/cgi-bin/sse/sse.cgi

[12]   Guide to Meteorological Instruments and Method of Observation, accessed at 15 January 2015, availablehttp://www.wmo.int/pages/prog/gcos/documents/gruanmanuals/CIMO/CIMO_Guide-7th_Edition-2008.pdf

[13]   A. Angstrom, Solar and terrestrial radiation, Quarterly Journal of the Royal Meteorological Society, vol. 50, no. 210, pp. 121-126, 1924

[14]   J. A. Prescott, "Evaporation from water surface in relation to solar radiation," Transactions of The Royal Society of South Australia, vol. 40, pp.114–118 (1940).

[15]   Akinoglu BG, Ecevit A. Construction of a quadratic model using modified Angstrom coefficients to estimate global solar radiation. Solar Energy 1990, Vol. 45, pp. 85-92.

[16]   Newland FJ. A study of solar radiation models for the coastal region of South China. Solar Energy 1989, Vol. 43(4), pp. 227-235.

[17]   David B. Ampratwum, Atsu S.S. Dorvlo, Estimation of solar radiation from the number of sunshine hours, Applied Energy, Vol. 63 (1999), pp. 161-167

[18]   Stone RJ, Improved statistical procedure for the evaluation of solar radiation models. Solar Energy 1993, Vol.51, pp. 289 – 291

[19]   B.G. Akinoblu andA. Ecevit, A further comparison and discussion of sunshine-based models to estimate global solar radiation. Energy, Vol. 15, No.10, pp. 865-872, 1990

[20]   Srivastava SK, Sinoh OP, Pandy GN, Estimation of global solar radiation in Uttar Pradesh (India) and comparison of some existing correlations. Solar Energy (1993), Vol. 51, pp. 27 – 29.

19

Comparison of District Heating Systems Used in China and Denmark

Lipeng Zhang[1, 2, *], Oddgeir Gudmundsson[2], Hongwei Li[1], Svend Svendsen[1]

[1]Civil Engineering Department, Technical University of Denmark, Anker Engelunds Vej Building 118, Kgs.Lyngby, Denmark
[2]Danfoss A/S, District Energy Division, Application Center, Nordborgvej 81, Nordbrg, Denmark

Email address:
lipz@byg.dtu.dk (Lipeng Zhang)

Abstract: China has one of the largest district heating (DH) markets in the world with total district heat sales in 2011 amounting to 2,810,220 TJ. Nevertheless, it still has great potential for further expanding its DH supply, due to rapid urbanization and the demand to improve the quality of life. However, the current DH system in China is in great need of system improvements, technology renovation, and optimization of operations and management. As one of the world's leading countries in terms of DH supply, Denmark has state-of-the-art DH technologies and rich experience in the design and operation of DH systems. Experiences learned from the Danish DH system are useful for improving the current Chinese DH system. This article provides an overview of the technological differences between the two countries, focusing on: a) heat generation, b) the DH distribution network, c) DH network control, and d) the end consumer. The paper looks at the obvious differences between these two countries in terms of DH supply and concludes that there is significant, achievable potential for improvement regarding both energy efficiency and user comfort in the Chinese DH system, through technological advancement and implementing the operational know-how of more modern DH systems.

Keyword: District Heating, Energy Efficiency, Technical Measure, China, Denmark

1. Introduction

Denmark is one of the most energy-efficient countries in the world. A wide range of pro-active, energy-saving measures have decreased energy consumption and increased the use of renewable energy and technological development. Since the 1980s, Denmark's energy consumption has consequently remained steady, while the economy has continued to grow. The widespread use of district heating (DH) and combined heat and power (CHP) has made a major contribution to Denmark's drive towards efficiency and energy self-sufficiency (Dyrelund, 2012). The country's DH system combines space heating (SH) and domestic hot water (DHW) and runs continuously throughout the year. Denmark develops diverse heat generation technologies, powered by renewables and otherwise wasted energy (Lund et al., 2010)(Alberg et al., 2010)(Mathiesen et al., 2012)(Münster et al., 2012), as well as gradually reducing fossil fuel. Furthermore, well-oriented and supportive policies issued by the Danish government have resulted in the technical success. Commercial companies carried out the research and development of DH-relevant products and solutions, along with universities, consultancies, as well as trade associations—all made substantial contributions to the revolution of DH technology.

The Chinese DH system had developed based on standard Soviet-era technology, which provided only heat, not DHW. There is considerable potential for improving the Chinese DH system and reducing greenhouse gas (GHG) emission. Coal, as the dominant heat source fuel, has resulted in a series of environmental, health, and economic challenges (U.S.Environmental Protection Agency, 2008). Furthermore, this kind of single heat source also heavily highlights issues of supply security, since energy consumption keeps increasing along with rapid urbanization and industrialization. The huge growth of the DH sector has made China the fastest growing DH market in the world. However, heat generation, distribution energy efficiency, heat demands, and fulfillment of user comfort requirements are not comparable with some European DH systems, such as Denmark. In China, heating energy consumption for $1m^2$ is almost 2 times that of developed countries in the same latitude (Liu et al., 2011)(Xu et al., 2009). Currently, China's heat reform is still in process, with the aim of improving building energy efficiency, updating the overall DH system, as well as establishing new heat metering and billing mechanisms based on actual consumption. Meanwhile, Danish DH experience will be a good resource from which China can learn.

This article looks at the obvious technical differences by comparing the main elements of DH systems between Denmark and China. It aims to identify the potential within the Chinese DH system, along with opportunities for integration of Danish DH technologies. It is important to note that these technical measures, that are essential to the Danish system, are appropriate and feasible for China at a practical level.

1.1. Historical Perspective and Future Prospects

1.1.1. Denmark

Since the first waste incineration and CHP plant-based DH system was built in Denmark in 1903 (DBDH,2013), the Danish DH supply has gone through moderate development over the past 100 years. In 1973, the worldwide oil crisis tremendously affected the Danish economy, due to nearly 100% importation of foreign oil. The Electricity Supply Act of 1976 implemented the policy that all new power capacity after 1976 had to be CHP and the Heat Supply Act of 1979 ensured the least cost integration of power, heat, gas, and waste sectors in Denmark (Gerlach, 1991). The development of CHP on both a large scale (city-wide) and small scale (communities and institutions) and the associated DH have been booming since the 1980s (Mortensen, 1992). Such measures significantly increased energy supply efficiency and enhanced energy supply security, which has helped Denmark become energy independent since 1997 (Christensen, 2008). In 2012, the Danish government set forth an ambitious energy target: by the year 2035, the electricity and heat supply will be covered 100% by renewable energy and, by the year 2050, all the energy supply in Denmark should be 100% from renewable sources(Danish Energy Agency, 2013). DH once again became one of the key measurements and the share of total DH supply will increase from 60% to 70%, with the rest of the heating demand met by heat pumps (Lund and Mathiesen, 2009).

The future trend of the Danish DH is expected to be towards 4th generation DH (4GDH) (Lund et al., 2014), which is defined by smart thermal grids utilizing low quality energy like renewables, with optimized combinations of heat sources to supply appropriate lower temperatures to low-energy demand buildings through a high-efficiency DH network (Li and Svendsen, 2012).

1.1.2. China

During China's first five-year plan period (1953-1958), the first batch of thermal power plants were constructed, aided by the Soviet Union. In 1958, Beijing established China's first thermal power plant to supply heat to a few public buildings; this was the starting point of China's urban central heating. Afterwards, central heating utilizing CHP as the heat

generation came almost to a standstill for quite a long period, due to unexpected errors related to heat capacity. In the 1970s, the number of CHP plants began to increase again. However, these plants typically belonged to factories and enterprises, mainly meeting their own heat demands (Xu, 2010). During the early 1980s and into the late 1990s, CHP plants grew rapidly. Since the 1980s, CHP units started to supply heat for public, residential and commercial buildings. In 1986, the state council of China released the No. 22 document (Xu, 2000), which set the general direction for the development of CHP. Moreover, the central government increased funding and policy support. In this way, CHP was promoted. After the late 1990s, more and more heat-only boilers (HOBs) were built, gradually equaling CHP as the heat generation units, later even surpassing CHP. Although CHP should be preferred, due to better primary energy usage, there can be certain conditions that favor HOBs when it comes to DH, especially in the starting phase. HOBs played a transition role; after CHPs were built to supply the base load, they can be used efficiently for peak load. In 2007, China's total hot water DH sales amounted to 1,586,410 TJ, central HOBs contributed 1,047,750 TJ and accounted for 66%, and CHP represented 33% at 522,880 TJ (Xu, 2010).

The development of DH in China has gone hand in hand with rapid urban expansion and economic growth over the past ten years. In 2008, out of China's 655 cities, approximately 329 were equipped with DH facilities (Baeumler et al., 2012). The district heated floor space has expanded rapidly from 2.16 billion square meters in 2004 to 4.74 billion square meters in 2011 (China National Bureau of Statistics, 2012). At the same time, CHP more than doubled in capacity between 2001 and 2005, rising from 32 GW to 70 GW (IEA, 2007).

In China's 12th five-year plan report, improving energy efficiency is specifically mentioned as an important issue (Thomson, 2014). The DH sector has received further focus by policies that, among other things, actively promote urban clean energy retrofitting, strengthen building energy-efficiency retrofitting, develop CHP and DH, and eliminate a number of small coal-fired boilers, along with the phasing out of decentralized heating coal stoves in rural areas by encouraging the utilization of renewable energy. A prior policy of eliminating scattered coal boilers by consolidating them into large central heating systems with high energy efficiency and pollution control will continue (Lo and Wang, 2013)(Price et al., 2011).

1.2. Climate and Heating Periods

1.2.1. Denmark

Denmark has a temperate marine climate with mild winters and cool summers. The coldest month is January with average daytime temperatures of 2°C and nighttime temperatures of 2.9°C During the winter, strong wind can quickly change the outside temperatures (Global Talent, 2013). The theoretical heating period is from October to the following April. Nevertheless, the fact is that an internal building heating system, connected to a DH system, can be turned on or off according to heat consumers' comfort; actually the heat users can decide for themselves how long the heating period is. In addition, during the non-heating period, the DH supplies water for DHW preparation only.

1.2.2. China

China stretches over a large area with various winter climates classified from warm to severe cold. Figure 1 shows the climate zones map of China (Gao et al., 2014) and Table1 gives the population information and the proportion of residential and commercial buildings in the different climate zones. Cold and severe cold zones cover about 70% of

national territory and account for 43% of the total residential and commercial buildings in the country (Baeumler et al., 2012). All 13 provinces and cities belong to the cold and severe cold climate zones (Ministry of Construction of China & General Administration of Quality Supervision Inspection and Quarantine of the P. R. China, 2012), which are geographically located north of the Qinling Mountain Range and the Huaihe River. The cold and severe cold zones are defined as having at least 90 days of average outdoor temperature at or below 5°C (the Ministry of Construction of China & State Bureau of Technical Supervision, 1993).

Table 1. Distribution of population, residential and commercial buildings in different climate zones of China[21]

Climate zones	Inhabitants (million)	Residential and commercial buildings ratio
Severe-Cold and Cold zones	550	43%
Hot-Summer and Cold-Winter zone	500	42%
Hot-Summer and Warm-Winter zone	160	12%
Temperate zone	90	3%

Figure 1. Climate zones map of China

Generally, in China, the heating season is specified from October to the following March. Despite being shorter or longer for some areas, the average heating period is around 150 days. Once the heating season is over, the DH supply is turned off, meaning that DHW preparation by heat from DH is uncommon. The common DHW solutions in China are: a) small and decentralized DHW systems, such as HOBs that generate DHW and supply a building block, e.g. gas-fired boilers produce hot water for a residential community or b) individual water heaters that produce hot water in each household, such as solar, electric, or gas water heaters.

2. Methods

In this paper, the technical comparison elaborates four main DH elements: heat production, DH distribution network, DH network control, as well as the end consumer. In addition, the specific DH technologies, successfully

applied in Denmark, are analyzed, having potential for development in Chinese DH systems. However, wholesale adoption of the exact technologies would not be wise, as different national situations must be considered; otherwise the advantages of the technologies would be compromised potentially leading to failure.

2.1. Heat Generation and Fuel Sources

When discussing heat generation technology for DH, it is notable that the technology is independent of the heat source and many different fuels can supply the system, such as renewables, waste to energy, and fossil fuels. The only requirement is that the temperatures of the heat sources are sufficiently high to heat the buildings. This capability both increases the security of supply and allows for optimization of the cost of heat generation, a remarkable advantage with which individual heating solutions cannot compete. Historically, DH has developed in relation to CHP. There is a clear benefit to having a CHP plant supplying the heat to the DH network. If the fuel source for the DH is fossil-based, it has also been shown that the CHP creates the lowest carbon footprint of all fossil-fuel burning plants (Orchard, 2009).

2.1.1. Denmark

In the Danish DH system, heat is supplied from either CHP plants or heating plants. 665 CHP plants include 16 centralized CHP plants in large cities, whereas most decentralized CHP plants are in small cities or for private supply for enterprises or institutions. CHP supplies 77% of the heat, with the remaining 23% (Odgaard, 2013) being supplied by various other heat-only devices, such as biomass boilers, geothermal heat plants, solar plants, and waste heat from industry. A total of 230 DH plants can be found in Denmark (Dansk Fjernvarme, 2013). In 2011, the energy supply composition for DH source was composed of: recycled heat including indirect use of renewables 69.8%, direct renewables 19.3%, and others 10.9% (EURO HEAT & POWER, 2014).

Figure 2 shows the development of energy source for DH between the years 2001 and 2010 in Denmark (Werner and Frederiksen, 2013). A clear trend of a gradual decrease in the use of fossil fuels is evident in the DH sector; in contrast, the increased usage of renewable energy, such as biomass, waste energy, geothermal and solar energy, provided environmentally friendly heat to the DH network. The inspiration from these examples is that during the progressive establishment of a smart energy consumption pattern, the orientation directives of the Danish government played an important role. They masterfully use the economic levers of taxes and subsidies to set the national energy development direction.

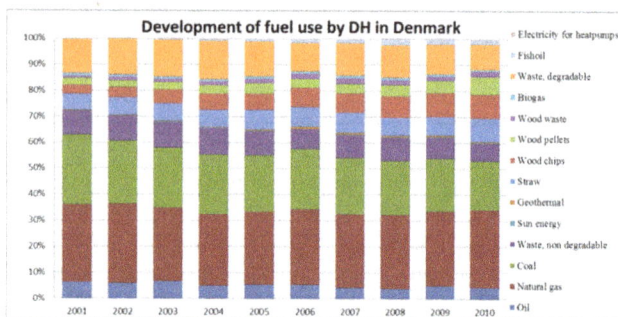

Figure 2. Developement of energy source for DH in Denmark 2001-2010

Biomass accounted for approximately 70% of renewable-energy consumption in 2010, mostly in the form of straw, wood chips, and pellets, while biogas accounted for less (Bertelsen and Tafdrup, 2014). The "Danish Biomass Action Plan of 1993" had forced power plants to use biomass to generate power and heat, a political decision that reoriented biomass energy consumption, resulting in a four-fold increase from 1980 to 2005 (Jørgensen, University of

Copenhagen). The Avedøre 2 CHP plant is known as the world's largest and most efficient biomass-fuelled CHP plant (Ottosen and Gullev, 2004). Two units in this plant with a total capacity of 810 MW of electricity and 900 MW of heat, run on a wide variety of biomass fuels, as well as less coal, oil, and natural gas. In 2027, the plant is expected to run 100% on biomass (Wikipedia, 2014).

Denmark is at the forefront of the development of large-scale solar DH systems in Europe. Of the top 10 large-scale solar heating plants in Europe, nine are located in Denmark. The number 1 plant, Marstal Fjernvame solar DH system, was established in 1996 (SDH, 2013). This is, so far, the largest solar heating plant in the world, with 33,300m^2 of ground-mounted flat panel collectors, with a thermal capacity of 23,300 KW. There are two energy systems (new and existing) that, together, have an annual production of 56,000MWh of heat (Extranet, n.d.). Water thermal energy storage systems can provide seasonal and diurnal storage for the energy systems. According to the "solar thermal strategy" of the Danish Energy Agency, in 2030, 10% of Danish DH load will come from solar thermal and in 2050, nearly 40% of the DH load is estimated to come from solar heat generation (Runager and Nielsen, 2009).

There is great potential for developing geothermal DH in Denmark due to the presence of assessed geothermal resources in large parts of the country. In 2012, DH extracted and used about 300TJ of geothermal heat. Table 2 shows three representative geothermal plants in Denmark to demonstrate the development status of deep geothermal. In addition, shallow geothermal will likely expand in the coming years, especially in areas with no DH or natural gas supply. Furthermore, current ground source systems cover more horizontal collectors, as well as a small proportion of borehole heat exchangers, when considering groundwater protection and drinking water quality (Mahler et al., 2013).

Table 2. Three representative geothermal plants utilized deep geothermal energy in Denmark

	Year	Location	Heat capacity	Flow volume	Temp.	Depth	Saline amount
1	1984	Thisted,Denmark	7MW	200m³/h	44°C	1.24 km	15%
2	2005	Copenhagen,Denmark	14MW	235m³/h	73°C	2.6 km	19%
3	2013	Søderborg,Denmark	12MW	350m³/h	48°C	1.2 km	15%

Over many years of policy-making in Denmark, waste has experienced a role reversal from being a health problem in the 1960s to a resource since 2000. Waste incineration is the method for recovering energy from waste. Danish waste incineration plants are connected to the energy grid, providing DH and electricity to the Danish market, while, at the same time, decreasing the volume of waste by up to 70% (Andersen and Mortensen, Copenhagen Cleantech Cluster). Municipal solid waste (MSW) and household waste are an important source of heat for the DH sector. In Denmark, all MSW is incinerated, and household waste is not allowed to go to landfills. Typically, Danish incineration plants generate approximately 2 MWh of heat and 2/3 MWh of electricity from every ton of waste incinerated, that implies that the

operation of waste incineration plants produces nearly 80% heat and 20% electricity (Vestforbrænding, 2013). Therefore, waste incineration plants are well suited as a heat source for DH. Moreover, with a high priority on efficient energy usage, waste heat from industry has also become an important heat source for the DH sector. For example, in the town of Fredericia, the DH network distributes waste heat from local chemical plants to 55,000 households (Eldrup, 2013).

Denmark has achieved a highly efficient energy system based on CHP, which is already widespread and successful in this country, due to consistent prioritization over the past few years. On the other hand, Denmark never stops pursuing renewable heat sources for DH, as well as developing thermal storage technology. In this way, Denmark already has the foundation stones for delivering high efficiency for its DH systems and, in addition, it will continue to consolidate and optimize over the coming years.

2.1.2. China

There are three main heating production modes in China: 1) CHP plants and DH plants, 2) HOBs, and 3) small, scattered HOBs or individual stoves for single buildings or individual households. Since DH has evident economic benefits in highly populated areas, the first two modes are common in cities, where the fuels are coal, natural gas, or oil. The third mode is the heating solution generally used in suburban and rural areas by burning coal, oil, or crop waste.

Figure 3 shows the proportional change in the trend of heat sources in northern China from 1996 to 2008. The heated area in northern China nearly quadrupled from 2.4 billion m^2 in 1996 to 8.8 billion m^2 in 2008, the proportion of individual coal stoves drastically decreased from 50% in 1996 to less than 10% in 2008, and the percent of natural gas gradually rose to 5% of total heating areas in 2008 (China National Bureau of Statistics). The share of the heating supply coming from CHP accounts for one third of the heat supplied in the DH sector, while the remaining heat comes from HOBs, mostly fueled by coal.

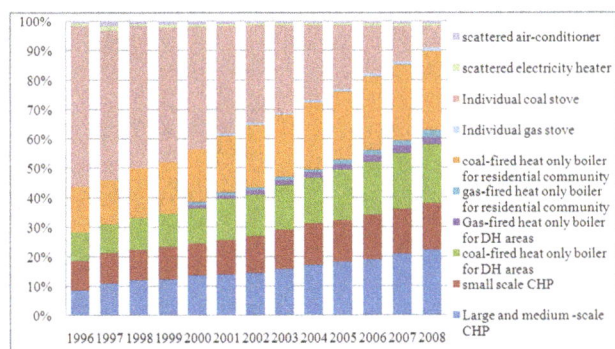

Figure 3. Proportional change in the trend of heat sources in northern China from 1996 to 2008.

For CHP plants, large-scale, high-capacity installations are actively encouraged, in order to realize the goal of saving energy and reducing emissions. This implies that large CHP plants will more easily obtain construction approvals and financial support than smaller ones. Currently, large extraction-condensing steam turbines, namely 200MW, 300MW, and 600MW (Tsinghua University building energy research center, 2011), are the leading type in China. This kind of CHP system can generally offer a 2.0-10.0 heat-to-power ratio and 60%-80% overall efficiency.

Coal is the dominant fuel source in the Chinese heating sector; this situation will continue in the coming years. Burning non-clean coal influences the efficiency of the boilers and causes excessive consumption of coal, as well as environmental pollution. According to the data from China's State Statistics Bureau in 2008, the national heating sector consumed 145.4 million tons of raw coal—about 91% of the total energy supply of the sector, in addition to 5% petroleum products and 4% natural and other gases, of which around one third is used in low efficiency HOBs (China National Bureau of Statistics).

Coal-fired HOBs are reported to have an efficiency of 60-65% (WADE, 2010), which can be considered quite low compared to the efficiency levels experienced in Western Europe. Currently, China's main cities have planned to restrict new heating plants to gas-fired technology. This fuel conversion is a long-term solution to deal with the consistent pollution issues faced today and to improve the efficiency of boilers. For Beijing city, the long-term plan is that gas heating will cover 51% of Beijing's heating areas in 2015. At the same time, heating areas of Beijing will expand from 680 million m^2 in 2010 to 850 million m^2 in 2015 (Beijing Heating group, 2011). Beijing has pledged to shut down most coal-fired boilers in central city areas before 2016, as part of its efforts to reduce fine particle pollution, especially during the heating season. This will result in a nearly 5-million-ton reduction in coal use, compared to 26.35 million tons in 2010 ("Beijing shuts coal-fired boilers for clearn air," 2013). However, according to Li et al., (2009), this fuel conversion policy would lead to a significant increase in overall costs if building energy efficiency is not simultaneously taken into consideration.

Since coal is the main fuel for CHP plants and HOBs, this brings up a series of challenges for health, the environment, and the economy. On the other hand, the situation will continue in the coming years. At the same time, urbanization and industrialization are speeding up along with the economic growth, such that China faces the great challenge of energy supply security. According to *2011 Annual Report on China Building Energy Efficiency* (Tsinghua University building energy research center, 2011), utilizing the surplus heat from industrial processes as the heat source in DH sector, otherwise discharged into the environment (Ajah et al., 2007), would enable China to realize energy goals and meet the challenges. Table 3 lists available surplus heat from the industrial processes around cities (Tsinghua University building energy research center, 2011).

Table 3. Available surplus heat from industrial processes around cities of China

Code	Available low quality energy	Temp. level	Extractable heat amount
1	Gas emissions from coal and gas combustion	50-180°C	10%-20% of fuel total calories
2	Heat emission from the condenser of power plant	20-40°C	70%-200% of generated electrical energy
3	Surplus heat from industrial production, e.g. industrial furnaces, steel plants, non-ferrous metals plants, chemical plants	30-200°C	30%-80% of consumed energy in the plant
4	Heavy after-sewage water treatment	20°C	Recycled water per ton can release heat of around 12kwh if temperature lowered to 10°C

Fang et al., (2013) introduce a demonstration project and present the huge potential to utilize surplus heat from industrial processes in China's DH sector. Li et al., (2011) introduce a new method for improving energy efficiency and the capacity of the DH system.

2.2. DH Distribution Network

Distribution cost is a critical factor for the profitability of a DH system (Gebremedhin, 2012). Furthermore, heat loss is a major issue for the distribution pipelines. According to Werner and Frederiksen, (2013), annual relative heat loss is influenced by four factors: total heat transmission coefficient from the insulation heat resistance, average pipe diameter, distribution temperature level, and the linear heat density.

2.2.1. Linear Heat Density

In (Persson, 2010), the definition of linear heat density is: Qs (GJ), heat sold annually in a DH system, divided by the trench length of the piping system L (m), which is symbolized by equation (1), with the unit GJ/m. This ratio indicates the level of DH distribution system utilization, and is a good indicator of the ratio of revenue to distribution cost.

$$\text{Linear heat density} = Qs/L \quad (GJ/m) \qquad (1)$$

As is well known, most of the cost of a DH system lies in the distribution pipe work. Regions with high linear heat density can allocate more infrastructure costs to DH pipeline, thereby maintaining the competitiveness of DH. In Denmark, 80% of the DH companies face an average heat density within the interval of 1.2 – 5 GJ/m/year (Finn Bruus and Halldor Kristjansson, 2004), while, according to (Baeumler et al., 2012), the average heat load density in China is about 38.88 GJ/m/year. Table 4 shows the annual average linear heat density in Denmark and China based on equation (1) and the data from Euroheat & Power (EURO HEAT & POWER, 2014). China has higher linear heat density than Denmark because densely populated cities with high-rise buildings are always in the DH supplied areas.

Table 4. Linear heat density in Denmark and China in 2007 and 2011

Year	Country	Total DH sales (TJ)	Trench length of DH pipeline system (km)	Linear heat density (GJ/m)
2007	Denmark	94,271	27,851	3.38
	China	2,250,150	102,986	21.85
2011	Denmark	101,940	30,288	3.37
	China	2,810,220	147,338	19.07

2.2.2. Denmark

For a large-scale Danish DH network, the complete pipeline system generally consists of the transmission system, distribution system, and municipal network, with different companies being in charge of each part-a good example is Copenhagen's DH system.

The Danish DH systems have evolved over time by utilizing innovative methods to reduce distribution heat loss. One of the main contributing strategies has been to operate the distribution network with relatively low supply temperatures and as high a differential temperature between forward and return pipes as possible; this both insures good cooling of the supply and minimizes mass flows in the system, which results in pump savings later on. Currently, the operating temperatures of the DH system are typically 70/40°C during the heating season and 65/25°C during the non-heating period. Moreover, Denmark is in the transition from 3rd generation DH to 4th generation DH, the DH system will be working under 50/20°C in the future, instead of the former 70/40°C (Brand and Svendsen, 2013). An international research center, 4DH, has carried out relevant research work (4DH, 2014.). Further investigation is seeking even lower supply temperatures. Low network temperature increases the quality match between heating demand and supply (Li and Svendsen, 2012), minimizes heat loss in the distribution network, improves the network's economic feasibility, and enables easier adaption to renewable energy (Dalla Rosa et al., 2013)(Dalla Rosa et al., 2011).

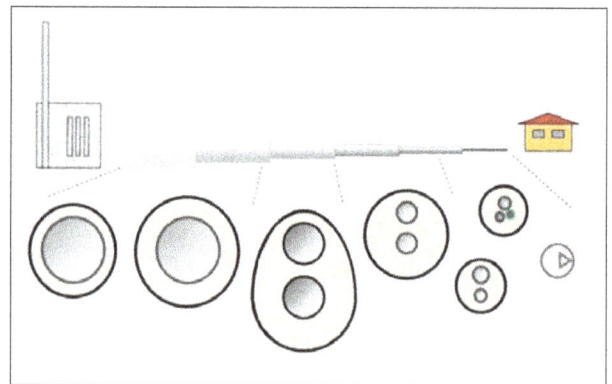

Figure 4. Innovation of DH distribution pipes in Denmark

Another notable contribution is the invention of the concept of pre-insulating steel pipes and covering the insulation with a water-resistant casing. Moreover, the service pipes are designed with optimized geometry (Figure 4), in order to reduce the relatively high heat loss of small pipes and consolidate the competitiveness of DH in low heat density areas, where single-family houses or new energy-efficient buildings are common. Transmission pipes from local CHP/DH plants generally have larger dimensions and use signal piping made of pre-insulated steel. The distribution pipeline from local heating substations likewise uses mainly pre-insulated steel pipe work; if the pipe size is small enough, the twin-pipe structure will be used. For municipal networks, with final distribution based on all plastic pipe work and insulation, twin-pipe or even triple-pipe is used. The triple-pipe is described as two forward lines and one return line, generally combined with a booster pump in the house, not only in order to achieve smaller heat loss than traditional service pipes, but also to provide better hot water comfort. Twin-pipe means that two pipes are located within a common circular insulation with an outer casing. Twin-pipe is energy-efficient because the return pipe is arranged close to the temperature field generated by the supply pipe. In that case, heat resistance due to coinciding temperature fields becomes greater, resulting in lower heat loss from the return pipe (Werner and Frederiksen, 2013). Heat loss savings of 37% and investment cost reduction of 12% can be achieved by using twin pipes instead of two single pipes (Finn Bruus and Halldor Kristjansson, 2004). In addition, to achieve large differential temperatures and hydraulic balance in a DH system, it is necessary to have high-efficiency heat exchangers and control valves installed in the network.

2.2.3. China

China greatly extended the DH pipelines along with the expansion of heating areas, rapid economic development, and accelerating urbanization. It could be said that large dimensions and high temperature levels are the characteristics of the Chinese DH network. Single pipe is the most common structure. One of the reasons for this is that high heat density in urban areas needs a larger dimension of pipes. In addition, more and more DH transmission systems utilize directly buried pre-insulated pipes, instead of the former concrete trenches where the insulation foam for pipelines was applied on site. Typically, the service pipe is steel and is covered by insulation foam with rigid polyurethane. The outer protection is commonly a glass steel tube or high-density polyethylene tube. A DH temperature range of 115~130°C for supply and 50~80°C for return is typical (Ministry of Housing and Urban-Rural Development of China, 2010).

Heat loss is a big issue influencing the efficiency of a DH network. According to (Yan et al., 2011), around 30% of total supply heat is lost in Chinese DH systems due to hydraulic imbalance and water leakage. Around 30% of total supply heat is lost in Chinese DH system due to hydraulic imbalance and water leakage. Tsinghua University has conducted research to clarify energy loss items by taking a typical Beijing residential building as an example. In order to meet the annual heat demands, 0.30 GJ per square meter is required, whereas heat generation has to produce 0.45 GJ to ensure 18 °C statutory indoor temperature in the heating season (Tsinghua University building energy research center, 2011). That means 33% of total produced heat is lost when heat is delivered from heat generation to end users.

Among the factors, hydraulic imbalance accounts for quite a significant amount of total energy loss, and excess heat supply could be one result. This reflects the general lack of automatic control measures in the Chinese DH system. For the hydraulic balance of the DH network, it is important to have accurate flow control to substations, buildings, and end users, so that the heat demand can be better matched with exact energy consumption. Since in China the common case is one in which a large substation supplies heat to a group of high-rise or multi-story buildings, which contain a number of apartments in a building. For this kind of large and complex

DH system, it is inevitable to have hydraulic imbalance, if there are no control or adjustment devices in the individual branches: proximal end users get more flow than needed, and the distal receive less than required. Moreover, the apartments in a single building are located in different orientations and positions; therefore, the indoor temperature differs from room to room. Under this condition, when the distal end user's basic thermal comfort is met, the proximal user's environment might be overheated. Furthermore, there are no adjusting devices at the ends of the internal building's heating system. Naturally, proximal heat users are likely to open a window to bring the indoor temperature down, as needed, according to their comfort. Additionally, if the supplied heat cannot be adjusted according to the varying weather, the heat could be excessively supplied during the entire heating season. Hydraulic balance in a DH system can be achieved by using the differential pressure of the system to insure an adequate flow through the branches; usually, valves and pumps are the basic components for solving the hydraulic imbalance issue. Boysen and Thorsen, (2007), analyze how to establish hydraulic balance in a DH system. Weather compensation controllers can adjust the produced heat according to weather changes to meet the heat demands.

2.3. DH Substations-Network Control Methods

Since an increasing number of DH experts recommend applying the indirect-connection in modern DH systems (Thorsen and Gudmundsson, 2012), substations play a key role in securing energy-efficiency and no-risk operation. Substations provide hydraulic separation between heat generation and heat consumers, thereby avoiding the contamination of internal building heating systems.

2.3.1. Denmark

When comparing substation technology between Denmark and China, one finds that a large community-level substation is the most common case in China and one substation generally supplies heat to 50,000-200,000 m^2 floor area (Xu et al., 2009). In Denmark, a substation may be a customer substation, which is installed in each building, referred to as a building-level substation. The sub-station may even be in each flat, apartment, or single-family house, known as a flat station. At the same time, a typical Denmark substation will supply both space heating and DHW. Generally, the closer the control equipment is to the heat consumer the better the network control that is achieved. Moving the control components towards the heat consumers has been a continuous trend in Danish DH systems and has played a crucial role in the increased efficiency and economic performance of the Danish DH industry.

The importance of good control in order to achieve high energy efficiency cannot be stressed enough. As DH is a hydraulic system, the draw off by one consumer will inevitably have consequences for the other consumers, therefore, the closer the control is to the consumer the less affected they become. The optimum control is achieved when all consumers have their own substation (Thorsen, 2010).

Moreover, the additional benefits of utilizing small substations are that they can be pre-manufactured and insulated, and achieve great space savings. Compared to large substations, building-level substations improve energy efficiency and allow for the application of more advanced solutions.

Table 5. Large substation and building level station based on a real case in China

Heating zone	Heating area (m²)	Scenario 1: Large substation		Scenario 2: Building-level substation		
		Heat capacity	Unit	Heat capacity	Unit	Sum
1#	48390	2900kw	1	200kw	5	
				250kw	2	
				300kw	1	11 units, total 3100kw
				400kw	2	
				500kw	1	
2#	48381	2900kw	1	100kw	1	
				150kw	1	
				200kw	8	14 units, total 3100kw
				250kw	2	
				200kw	5	
3#	48547	2900kw	1	200kw	6	
				250kw	5	14 units, total 3350kw
				300kw	3	
4#	47020	2900kw	1	200kw	9	
				250kw	2	13 units, total 3100kw
				400kw	2	
5#	37600	2300kw	1	150kw	5	
				200kw	6	14 units, total 2800kw
				250kw	2	
				350kw	1	
6#	43068	2600kw	1	150kw	1	
				200kw	1	
				250kw	2	6 units, total 2700kw
				550kw	1	
				1300kw	1	
7#	48773	2700kw	1	100kw	1	
				150kw	3	
				200kw	5	12 units, total 2900kw
				250kw	2	
				900kw	1	
8#	21360	1200kw	1	800kw	1	2 units, total 1300kw
				500kw	1	
Pipes cost		80.28%		58.43%		
Substations cost		19.72%		38.33%		
Total		100%		96.76%		

2.3.2. China

There is a real case in Weihai city in Shandong province of China (Danfoss A/S, 2004), where a total of 343,139 m² heating areas are split into 8 heating zones. Moreover, two scenarios are compared: 8 large substations (Scenario 1) versus 86 building-level substations (Scenario 2), see Table 5. The comparison includes the investment needed for the substations and primary and secondary pipes. This investment calculation does not include the cost of civil works to any extent, nor the cost of electrical facilities (transformers, cubicles, etc.), network valves, or the power connection needed for the group substations. If these expenses were included, it would increase the total investment of Scenario 1 and make Scenario 2 even more favorable. The contrast clearly shows that the investment would be lower when using building-level substations, although the total cost of small substations is double compared to that of large substations. Pipeline routing in the primary side can be done more efficiently and with a greater temperature difference, thus reducing the pipe diameter. From a technical perspective, small pipe size and a high differential temperature are helpful for reducing the heat loss of DH pipelines. In addition, Scenario 2 also gives other additional technical benefits, which remarkably influence the long-term operational costs and the total lifetime of the heating system. These benefits could include, but are not limited to, the following:

* an uncomplicated hydraulic system;
* a reduction of pump operation costs;
* Improvement of heat user comfort level;
* Modular design;
* reduced space requirements;
* The possibility to combine DHW system;
* Ability to charge the heating fee based on actual consumption;
* Flexible and smart control.

According to China's industry standard JGJ173-2009 (Ministry of Housing and Urban-Rural Development P.R.China, 2009), the building-level substation is recommended in 4.2.5 because of obvious technical superiorities, which are mentioned above.

Against the background of the heat reform in China, there is an opportunity to upgrade the DH system of China. Small substations can be in line with current DH industry developments. There is great potential for employing this application in the future.

2.4. End Consumer

Heat reforms are ongoing in China. In July 2003, eight central government ministries and commissions jointly issued a Government Circular calling for each of the 16 Northern provinces (in cold and severe cold climate zones) to implement heating system reforms in several pilot municipalities, according to the specified guidelines in the document "Heat Reform Guidelines." The principles of these Guidelines are the commercialization of urban heating, the promotion of technical innovation in heating systems, the application of energy-saving building construction, and the improvement of living standards. In the section on heat consumers, establishing a heat metering and billing mechanism based on actual consumption and improving building energy efficiency are two main tasks.

2.4.1. Heat Metering and Billing

There are some fundamental differences between DH systems in Denmark and China when it comes to the consumers. Two of the differences are heating billing and metering measurement.

Generally, China uses a fixed heating price based on square meters when charging the heating bill. Heat unit price depends on different factors, such as the type of heat generation (DH plants or HOBs), the type of thermal media (water or steam), building type (residential or commercial). Generally, internal building heating systems follow the constant flow principle, due to the lack of control devices at the end user. The statutory indoor temperature of residential buildings in the heating season is 18 °C. If the temperature is lower than this standard, the customers can refuse to pay the heating fee; if it is higher than this, the heating fee is charged as normal. Under this condition, heat consumers have no incentive to consciously save energy. For this reason, the heat reform aims to install regulation devices at the end of internal building heating systems, thereby making the room temperature adjustable; the heating fee will be charged according to the actual energy consumption. To reach this goal, several technical heat-metering measures have been invented and applied in China. In (Liu et al., 2011), the technical heat metering measures are presented and analyzed according to China's current DH situation. Currently, the heating area in China in 2012 was 4.92 billion m^2; the retrofitted area for heat metering was 0.805 billion m^2 in northern China, which accounted for approximately 66.7% of the total retrofitted area of heat metering devices installed (Ministry of Housing and Urban-Rural Development of China, 2012).

Table 6 lists two households' heating bills from Denmark and China. This seeks to illustrate the differences of heat billing between these two countries, not the price level. As this comparison is not based on the same benchmark, the heating bill of Denmark includes the DHW and SH throughout the year, while the Chinese heating bill contains only the SH fee during the heating season. In Denmark, the cost of DH is split into fixed and variable costs. The fixed cost covers the cost of the distribution network and the variable cost is metered according to actual energy consumption.

In Denmark, regulation devices are mounted at the end of internal heating systems to adjust the thermal flow rate into air heat units, thereby, the heat consumer can take measures to reduce heat consumption, such as closing the thermostatic valve instead of opening the window. In fact, the consumer can have a strong influence over their energy consumption by setting the desired room temperature. Additionally, consumers decide for themselves when their heating season starts or ends. The installation of heat meters and other regulation devices do not, in themselves, save energy. Rather, the energy savings are initiated by the consumers' own awareness. This way of heat billing has shown an average of 20-35% energy savings by the consumers (Drysdale, 2002). Further, modern energy meters are provided with facilities for remote reading. This is not only convenient for the DH companies to monitor the entire heating system, but also facilitates heat users tracking their energy consumption online.

Table 6. Comparison of heating bill between Denmark and China*

Heating bill for a 154 m² one-family house in Denmark for the whole year (365 days)						
variable cost	Heat meter records			Unit price (exl.tax)	DKK	€
	start	end	consumption			
	258.99 GJ	356.75 GJ	97.76 GJ	80 kr./GJ	7820.8	1048.4
fixed cost	effect contribution	heating area 154m²		15 kr./m²/year	2310	309.7
subscription fee					550	73.7
Tax	25%				2670.2	357.9
total					13351	1789.7
Heating bill for a 154m² apartment in a multi-storey building in Beijing for the heating season (125 days)						
fixed cost	Heating generation types		heating area	Unit price (incl.tax)	Total (Yuan)	€
	Gas-fired boiler		154m²	30 Yuan/m2	4620	474.8

*1Euro=8.48 Yuan=7.46 kr.

Table 7. Residential building energy requirements in Denmark

Standard	Building class	Kwh/m²/year	conditions
BR08	Building class 2008	70+2200/HFS*	Minimum requirement in 2006-2010 year
	Low-energy building class 1	35+1100/HFS	Low-energy building class
	Low-energy building class 2	50+1600/HFS	Low-energy building class
BR10	Building class 2010	52.5+1650/HFS	Minimum requirement in 2012 year
	Building class 2015	30+1000/HFS	Low-energy building class
	Building class 2020	20	Low-energy building class

*HFS is the building's heated floor space in m²

2.4.2. Building Energy Efficiency

Building energy efficiency is a key factor influencing the heat load of space heating. Since space heating typically represents a significant share of total building energy consumption, the most beneficial way to implement energy savings is to increase the energy efficiency of stock of houses.

Since the 1970s energy crisis, energy efficiency policies have been implemented in Danish buildings, which is has driven significantly less consumption than is experienced in most other European countries with similar climates (Danish Energy Agency, 2012). Furthermore, Danish authorities' strategy includes announcing future energy efficiency requirements many years in advance. Local municipalities have the power to require new construction to comply with future building requirements. Table 7 lists Danish residential building energy regulations and corresponding heat requirements.

The Chinese residential building sector accounts for approximately 30% of the country's final energy consumption (Richerzhagen et al., 2008). In 2008, heating energy consumption in northern Chinese towns accounted for 23% of total building energy consumption (Tsinghua University building energy research center, 2011). In an effort to reduce heating energy consumption, China began to enforce "Building Energy Efficiency Codes" in 2005 by implementing a three-step approach(Ministry of Housing and Urban-Rural Development of China, 2013).

- Step 1: Residential buildings built in 1991-1999 are required to achieve 30% energy savings compared to average residential buildings built before 1991.
- Step 2: Residential buildings built in 2000-2004 are required to achieve 50% energy savings compared to average residential buildings built before 1991.
- Step 3: Residential buildings built after 2005 are required to achieve 65% energy savings compared to average residential buildings built before 1991 in that location.

Since China's legal heating areas cover different climate zones, building heat consumption index levels vary from case to case. For Beijing, in the 1980s, standard coal consumption per square meter per heating season was 25.2kg. According to the 3-step energy saving approach, this consumption should be reduced to 17.64kg (30%), 12.4kg (50%), and 8.28kg (65%) respectively. In order to reach those levels, the efficiency of the DH distribution pipeline and the efficiency of the boiler are improved accordingly, as well as the building's envelop insulation performance. Consequently, building heat consumption per square meter per heating season is decreasing; see Table 8, with calculations based on equation (2).

$$q_c = 24 * Z * q_H / (H_c * \eta_1 * \eta_2) \qquad (2)$$

The heating season in Beijing (Z) is 125 days and H_c stands for the calorific value of the standard coal equivalent, 8140wh/kg. Table 7 and 8 contain building energy-consumption requirements in Denmark and in China. However, it is illogical to make a simple comparison, since

Danish regulations try to promote long-term thinking concerning energy-efficiency investments. For instance, Danish regulations include requirements for overall building energy demand: SH, ventilation, cooling, DHW, and non-residential lighting. This has encouraged innovation towards more comfortable buildings that have lower overall energy demands. In the case of China, building energy efficiency exclusively focuses on the energy consumption of SH. After the third step energy savings are achieved, the heat consumption of Beijing residential buildings are 43.5kwh/m²/year; this is slightly higher than 40kwh/m²/year in BR10 Building class 2015 (if floor area is 100m²). One could say that Danish buildings have higher energy efficiency than those of China, since overall energy consumption of buildings includes factors others than SH.

The high energy-efficiency of Danish building stock is the result of a consistent effort over many years, relying on strict requirements and standards, an experience that could be an inspiration to China. For China, enhancing energy efficiency could be an effective way to ease the pressure of energy supply security, reduce CO_2 emissions, mitigate the pollution issue, improve the thermal comfort level of building, and so on. There is a significant series of advantages (Richerzhagen et al., 2008).

Table 8. Building energy efficiency codes in China combined with the 3-step approach

	Year and design standard	q_c:Standard coal[1] consumption (kg/m²)	Energy saving ratio	q_H:Building heat consumption index (w/m²)	η_1:Eefficiency of distribution network	η_2:Efficiency of boiler
Datum	1980 Year	25.2	100%	27.4	0.8	0.5
Step 1	1986: JGJ26-86	17.64	30%	22.4	0.85	0.55
Step 2	1995: JGJ26-95	12.4	50%	20.6	0.9	0.68
Step 3	2010: JGJ26-2010	8.28	65%	14.5	0.92	0.7

Table 9. The overview of comparison DH systems used in China and Denmark

Items	Denmark	China	Potentials for China
DH season	Whole year	Winter Only	DHW generation from DH has great potential to expand the market share.
DH system	SH and DHW integrated.	DH is mainly for SH.	
Heat generation	Efficient and flexible heat production system, optimizing the combination of heat generation technologies and mix of fuels. Boilers (biomass, fossil fuel). Heat pump/electric heat boilers. Solar heat. Biomass CHP & geothermal DH plants, Gas CHP. Waste incineration heat/CHP. Surplus heat from industry.	Coal is dominate DH fuel. Large-scale, high-capacity CHP plants are encouraged mostly. Fossil fuel CHP. Fossil fuel heat-only boilers.	Renewable energy, waste energy, clean energy technologies.
Distribution network	Development tendency is LTDH, from 70/40°C ~55/25°C. Reduced heat loss of distribution pipeline based on multiple techniques: directly buried, pre-insulated steel pipe, optimized geometry of service pipes, applied low DH supply, and high temperature differential operation.	High DH supply temperature (130/70°C). Large dimension of distribution pipes due to high heat density. Small temperature difference, hydraulic imbalance and the lack of intelligent control comprise the efficient of DH system.	Improve the efficiency of DH system by achieving the overall hydraulic balance.
DH network control	Building level substation or flat station for each apartment. Single family house and multi-storey buildings are typical.	Large substation for a group of buildings. High-rise and multi-storey buildings are typical.	Building level sub -station, or even flat station concept
End users	Adjustable indoor temperature due to regulation devices at end of internal building heating system. Heat bill is based on actual consumed energy. Government regulates building energy consumption and supervises implementation.	Non-adjustable indoor temperature is min.18°C legally. Heat bill is fixed and charged by floor heating areas. Building efficiency can be improved through reduced consumption of heat.	Retrofit for heat metering and temperature-adjustable heating systems.

1 China typically converts all its energy statistics into "metric tons of standard coal equivalent" (tce), a unit that bears little relation to the heating value of coals actually in use in China. One tce equal 29.31 GJ (low heat) equivalent to 31.52 GJ/tce (high heat).

3. Results

Table 9 gives an overview of comparison of DH systems used in China and Denmark, also states the potentials of the Chinese DH system.

4. Discussion

Energy efficiency permeates the main aspects of the Danish DH system. The fundamental idea of DH, "utilizing local energy otherwise wasted," has been well fulfilled. The idea of heat production is to carry out a wide range of CHP technologies, define according to the corresponding scale, in accordance with local conditions—larger for major cities and smaller for suburban areas. Meanwhile, a diverse range of DH fuels are available, especially renewables, such as biomass, geothermal, and solar energy. Moreover, waste energy is also a well-utilized resource within Denmark's DH system. This utilizes low quality energy in the DH sector, thus reducing the consumption of primary energy. In addition, all kinds of heat storage facilities can adjust the heat supplied from storage systems or heat production units, depending on the price of electricity in different periods, ensuring the economical operation of the DH system. As such, Danish DH systems establish an efficient and flexible heat production system by optimizing the combination of heat generation technologies and a mix of fuels.

As for the distribution network, innovative methods have been explored and utilized, these technologies not only reduce the heat loss of the distribution network, also keep DH competitive in low-heat-density areas. In addition, sophisticated control systems have been implemented, wherein a powerful programmable controller is usually set, with weather compensation and segmentation control. Control valves, working together with sensors, ensure the extract differential pressure in individual branch that the DH system operates under hydraulic balance. Customer substations are even closer to heat users to gain better network control. Additionally, the installation of heat meters and regulation devices at the ends of the heating system enable a heat metering and billing mechanism based on actual consumption. This motivates the heat user to consider means for saving energy. Wide ranging, energy-saving measures and mandatory building energy requirements have improved the energy efficiency of Danish buildings. These facts, together, will accelerate the transition process of Danish DH from 3rd to 4th generation and, as a result, DH will contribute towards realizing the ambitious energy targets.

China has a substantial DH market with large heating areas and high linear heat density. It is full of potential and possibilities. Environmental issues will force China to adjust its energy consumption structure, with less fossil fuel and more sustainable energy as a safer model. Surplus heat from industrial processes presents a valuable resource, which is expected to be utilized properly by combing appropriate

technologies. In addition, the reduction of heat loss and the improvement of hydraulic issues can also greatly enhance the efficiency of DH networks. Applying building-level substations rather than large-scale ones, will allow DH systems to be more flexible and efficient. Heat reform opens the door for establishing wise heat metering and billing mechanisms, and will encourage the heat consumers to consciously save energy. Other focuses of the heat reform are to improve the thermal properties of building envelopes and to upgrade heating systems; these create a platform for applying advanced technologies. One could say that China is in a transition stage of upgrading DH systems and can benefit from the successful experience of other countries. In fact, collaboration and idea exchanges in the DH field between China and Denmark have already started. This does not mean, however, that China should directly copy the experience of other countries. Rather, with sensitivity to national conditions and in compliance with relevant regulations, China can selectively absorb, adopt, and implement best practices in the context of its own heating reforms.

5. Conclusion and Policy Implications

This paper has analyzed the current situation of the DH industry in these two countries. Based on the comparison of the main elements of DH systems used in China and Denmark, it is clear that China can take inspiration from the Danish DH system development and selectively adopt the relevant technologies, based on the real situation.

One experience from Denmark is to establish smart heat production in DH systems by combining different heat generation technologies and a mixture of fuels, as well as the utilization of thermal storage to make the system flexible.

The fundamental idea of DH in Scandinavian countries, "using local energy otherwise wasted," should be propagated in China's DH field. For China, the existing valuable resource could be surplus heat from industrial processes, which is readily available around high-density urban areas, where the DH pipeline infrastructure is available, since DH has developed in these areas for some years.

Improvement of the efficiency of DH networks by enhancing automatic control level into hydraulic balance and achieving higher building energy efficiency would be shortcuts for China's DH system to reach energy-saving and emission-reduction targets.

For China, supply security, pollution, and GHG emissions could be the most important current challenges. Meanwhile, efficiency improvement and modernizing DH with clean energy technologies have the maximum synergy between energy supply security and air pollution abatement. These challenges could also represent other opportunities. Updating DH systems in a sustainable way definitely benefits China in terms of long-term development.

Acknowledgements

The work presented in this paper is one part of an industrial PhD project. This industrial PhD project was jointly funded by Danfoss A/S and DASTI (The Danish Agency for Science, Technology Innovation). I wish to express my sincere gratitude to these institutions, as well as to the other co-authors of this paper.

References

[1] 4DH, Welcome to 4DH. URL http://4dh.dk/

[2] Ajah, A.N., Patil, A.C., Herder, P.M., Grievink, J., 2007. Integrated conceptual design of a robust and reliable waste-heat district heating system. Appl. Therm. Eng. 27, 1158–1164.

[3] Alberg, P., Vad, B., Möller, B., Lund, H., 2010. A renewable energy scenario for Aalborg Municipality based on low-temperature geothermal heat , wind power and biomass. Energy 35, 4892–4901.

[4] Andersen, R.K., Mortensen, J., Copenhagen Cleantech Cluster. Denmark : We Know Waste-Asset mapping of the Danish waste resource management sector. URL https://stateofgreen.com/files/download/446 (access 3.6.2014)

[5] Baeumler, A., Ijjasz-vasquez, E., Mehndiratta, S., 2012. Sustainable Low-Carbon City Development in China. World Bank.

[6] Beijing Heating group, 2011. District Heating Development and Construction Planning of Beijing during 12th Five-Year period.

[7] Beijing shuts coal-fired boilers for clearn air, 2013. URL http://beijing.china.org.cn/2013-05/27/content_28941727.htm (accessed 10.24.13).

[8] Bertelsen, F., Tafdrup, S., n.d. Biomass in the Danish Energy sector. URL http://www.ens.dk/node/2027 (accessed 1.8.14).

[9] Boysen, H., Thorsen, J.E., 2007. Hydraulic balance in a district heating system. Euroheat Power (English Ed). 4, 36 – 41.

[10] Brand, M., Svendsen, S., 2013. Renewable-based low-temperature district heating for existing buildings in various stages of refurbishment. Energy 62, 311–319.

[11] China National Bureau of Statistics, National Bureau of statistics data. URL http://www.stats.gov.cn/ (access 10.10.2013)

[12] Christensen, J.E.S.B., 2008. Why CHP and district heating are important for China. Cogener. On-Site Power Prod. 23–27.

[13] Dalla Rosa, A., Li, H., Svendsen, S., 2011. Method for optimal design of pipes for low-energy district heating, with focus on heat losses. Energy 36, 2407–2418.

[14] Dalla Rosa, A., Li, H., Svendsen, S., 2013. Modeling transient heat transfer in small-size twin pipes for end-user connections to low- energy district heating networks. Heat Transf. Eng. 34, 372–384.

[15] Danfoss A/S, 2004. Real case in Weihai City Shandong Province of China.

[16] Danish Energy Agency, 2012. Energy Efficiency Policies and Measures in Denmark.

[17] Danish Energy Agency, n.d. Energy Policy in Denmark. URL http://www.ens.dk/sites/ens.dk/files/dokumenter/publikationer/downloads/energy_policy_in_denmark_-_web.pdf (accessed 5.28.13).

[18] Dansk Fjernvarme, n.d. Danish district heating association. URL http://www.fjernvarmen.dk/ (accessed 9.15.13).

[19] DBDH, n.d. District heating history. URL http://dbdh.dk/district-heating-history/ (accessed 5.16.13).

[20] Drysdale, A., 2002. Innovative technique for field calibration and inspection of large district heating meters. Euroheat Power/Fernwarme Int. 31, 62 – 65.

[21] Dyrelund, A., 2012. Danish cases to implement the legislation: The future of the energy supply: Smart energy cities. Euroheat Power (English Ed). 9, 12 – 15.

[22] Eldrup, A., 2013. DONG Energy and the Municipality of Fredericia turn waste into valuable resource. URL http://www.dongenergy.com/ (accessed 8.1.13).

[23] EURO HEAT & POWER, n.d. District Heating & Cooling. URL http://www.euroheat.org/Denmark-74.aspx (accessed 1.9.14).

[24] Extranet, P., n.d. Large Scale Solar Heating Plants -Marstal. URL http://www.solar-district-heating.eu/ServicesTools/Plantdatabase.aspx?udt_1317_param_detail=326 (accessed 6.16.13).

[25] Fang, H., Xia, J., Zhu, K., Su, Y., Jiang, Y., 2013. Industrial waste heat utilization for low temperature district heating. Energy Policy 62, 236–246.

[26] Finn Bruus, Halldor Kristjansson, 2004. Principal design of heat distribution. News DBDH July, 14–17.

[27] Gao, Y., Xu, J., Yang, S., Tang, X., Zhou, Q., Ge, J., Xu, T., Levinson, R., 2014. Cool roofs in China: Policy review, building simulations, and proof-of-concept experiments. Energy Policy.

[28] Gebremedhin, A., 2012. Introducing District Heating in a Norwegian town – Potential for reduced Local and Global Emissions. Appl. Energy 95, 300–304.

[29] Gerlach, T., 1991. District heating in Denmark's environmental and energy policy. Fernwaerme Int. 20.

[30] Global Talent, 2013. climate_weather in Denmark. URL http://consortiumforglobaltalent.dk (access 6.10.2013)

[31] IEA, 2007. CHP and DHC in China : An Assessment of Market and Policy Potential Energy and Climate Change Overview. URL http://www.iea.org/media/files/chp/profiles/China.pdf (accessed 4.12.13).

[32] Jørgensen, H., Current status on biorefineries in Denmark. Danish Centre for Forest, Landscape and Planning. University of Copenhagen. URL http://www.iea-bioenergy.task42-biorefineries.com/en/ieabiorefinery.htm (access 3.10.2013)

[33] Li, H., Svendsen, S., 2012. Energy and exergy analysis of low temperature district heating network. Energy 45, 237–246.

[34] Li, J., Colombier, M., Giraud, P.-N., 2009. Decision on optimal building energy efficiency standard in China—The case for Tianjin. Energy Policy 37, 2546–2559.

[35] Li, Y., Fu, L., Zhang, S., Jiang, Y., Xiling, Z., 2011. A new type of district heating method with co-generation based on absorption heat exchange (co-ah cycle). Energy Convers. Manag. 52, 1200–1207.

[36] Liu, L., Fu, L., Jiang, Y., Guo, S., 2011. Major issues and solutions in the heat-metering reform in China. Renew. Sustain. Energy Rev. 15, 673–680.

[37] Lo, K., Wang, M.Y., 2013. Energy conservation in China's Twelfth Five-Year Plan period: Continuation or paradigm shift? Renew. Sustain. Energy Rev. 18, 499–507.

[38] Lund, H., Möller, B., Mathiesen, B.V., Dyrelund, A., 2010. The role of district heating in future renewable energy systems. Energy 35, 1381–1390.

[39] Lund, H., Werner, S., Wiltshire, R., Svendsen, S., Thorsen, J.E., Hvelplund, F., Mathiesen, B.V., 2014. 4th Generation District Heating (4GDH). Energy 68, 1–11.

[40] Lund, H.Ã., Mathiesen, B. V, 2009. Energy system analysis of 100 % renewable energy systems — The case of Denmark in years 2030 and 2050 34, 524–531.

[41] Mahler, A., Røgen, B., Ditlefsen, C., Nielsen, L.H., Pedersen, T.V., 2013. Geothermal Energy Use , Country Update for Denmark, in: Europesn Geothermal Congress 2013. Pisa,Italy.

[42] Mathiesen, B.V., Lund, H., Connolly, D., 2012. Limiting biomass consumption for heating in 100% renewable energy systems. Energy 48, 160–168.

[43] Ministry of Construction of China & General Administration of Quality Supervision Inspection and Quarantine of the P. R. China, 2012. China national standard GB 50019-2012 : Design code of heating ventilation and air conditioning. China Architecture Industry Press, Beijing.

[44] the Ministry of Construction of China & State Bureau of Technical Supervision, 1993. China National Standard GB50176-93:Thermal design code for civil building. China Architecture Industry Press, Beijing.

[45] Ministry of Housing and Urban-Rural Development of China, 2010. JGJ26-2010:Design standard for energy efficiency of residential buildings in severe cold and cold zones. China Architecture Industry Press, Beijing.

[46] Ministry of Housing and Urban-Rural Development of China, 2012. Government work report regarding 2012 heat metering reform special supervision and inspection in North China heating regions.

[47] Ministry of Housing and Urban-Rural Development of China, 2013. JGJ/T 129-2012: Technical specification for energy efficiency retrofitting of existing residential buildings. China Architecture Industry Press, Beijing.

[48] Ministry of Housing and Urban-Rural Development P.R.China, 2009. People's Republic of China Industry Standard JGJ 173-2009: Technical specification for heat metering of district heating system. China Architecture industry Press, Beijing.

[49] Mortensen, H., 1992. CHP DEVELOPMENT IN DENMARK - ROLE AND RESULTS. Energy Policy 20, 1198 – 1206.

[50] Münster, M., Morthorst, P.E., Larsen, H. V., Bregnbæk, L., Werling, J., Lindboe, H.H., Ravn, H., 2012. The role of district heating in the future Danish energy system. Energy 48, 47–55.

[51] Odgaard, O., n.d. Large and small scale district heating plants. URL http://www.ens.dk/en/supply/heat-supply-denmark/large-small-scale-district-heating-plants (accessed 12.2.13).

[52] Orchard, W., 2009. " Carbon footprints of various sources of heat – biomass combustion and CHPDH comes out lowest ". URL http://www.claverton-energy.com/carbon-footprints-of-various-sources-of-heat-chpdh-comes-out-lowest.html (accessed 6.15.13).

[53] Ottosen, P., Gullev, L., 2004. Avedøre unit 2 - the world ' s largest biomass-fuelled CHP plant. DBDH 1.

[54] Persson, U., 2010. Effective width - The relative demand for district heating pipe lengths in city areas. 12th Int. Symp. Dist. Heat. Cool. 128 – 131.

[55] Price, L., Levine, M.D., Zhou, N., Fridley, D., Aden, N., Lu, H., McNeil, M., Zheng, N., Qin, Y., Yowargana, P., 2011. Assessment of China's energy-saving and emission-reduction accomplishments and opportunities during the 11th Five Year Plan. Energy Policy 39, 2165–2178.

[56] Richerzhagen, C., Hansen, N., Netzer, N., 2008. Energy Efficiency in Buildings in China Policies , Barriers and Opportunities. German Development Institute / Deutsches Institut für Entwicklungspolitik (DIE), Bonn.

[57] Runager, J.M., Nielsen, J.E., 2009. LARGE SOLAR THERMAL SYSTEM-DEVELOPMENT AND PROSPECTIVES. ENERGY Environ. 8–11.

[58] SDH, n.d. Ranking List of European Large Scale Solar Heating Plants. URL http://www.solar-district-heating.eu/ServicesTools/Plantdatabase.aspx (accessed 6.16.13).

[59] Thomson, E., 2014. Introduction to special issue: Energy issues in China's 12th Five Year Plan and beyond. Energy Policy 73, 1–3.

[60] Thorsen, J.E., 2010. Analysis on flat station concept. Preparing dhw decentralised in flats. 12th Int. Symp. Dist. Heat. Cool. 16 – 21.

[61] Thorsen, J.E., Gudmundsson, O., 2012. Danfoss district heating application handbook. Danfoss A/S, Nordborg.

[62] Tsinghua University building energy research center, 2011 Annual Report on China Building Energy Efficiency.

[63] U.S.Environmental Protection Agency, Combined Heat and Power Partnership, Asia Pacific Partnership on Clean Development and Climate, 2008. Facilitating Deployment of Highly Efficient Combined Heat and Power Applications in China:Analysis and Recommendations. URL http://www.epa.gov/chp/documents/chpapps_china.pdf (accessed 6.20.13).

[64] Vestforbrænding, 2013. Why incineration. URL http://www.vestfor.com/why-incineration (accessed 6.12.13).

[65] WADE, 2010. The Potential for Clean DE and CHP in China – Executive Summary. URL http://www.localpower.org/ (access: 5.10.2013)

[66] Werner, S., Frederiksen, S., 2013. District Heating and Cooling. Studentlitteratur AB, Lund.

[67] Wikipedia, 2014. Avedøre Power Station. URL http://en.wikipedia.org/wiki/Aved%C3%B8re_Power_Station (accessed 2.10.14).

[68] Xu, zhongtang, 2010. 60 years development of urban heating. Dist. Heat. China 1–10.

[69] Xu, B., Fu, L., Di, H., 2009. Field investigation on consumer behavior and hydraulic performance of a district heating system in Tianjin, China. Build. Environ. 44, 249–259.

[70] Xu, Z., 2000. The development of Chinese urban central heating. City Dev. Res. China 51–55.

[71] Yan, D., Zhe, T., Yong, W., Neng, Z., 2011. Achievements and suggestions of heat metering and energy efficiency retrofit for existing residential buildings in northern heating regions of China. Energy Policy 39, 4675–4682.

20

Bioethanol Production from *Eucalyptus camaldulensis* Wood Waste Using *Bacillus subtilis* and *Escherichia coli* Isolated from Soil in Afaka Forest Reserve, Kaduna State Nigeria

Usman Yahaya[1, *], Umar Yahaya Abdullahi[2], Denwe Samuel Dangmwan[2], Muhammad Muktar Namadi[3]

[1]Forestry Research Institute of Nigeria, Trial Afforestation Research Station, Kaduna, Nigeria
[2]Department of Biological Sciences, Nigerian Defence Academy, Kaduna, Nigeria
[3]Department of Chemistry, Nigerian Defence Academy, Kaduna, Nigeria

Email address:

usmanyahayaks@yahoo.com (U. Yahaya), umaryahaya09@gamil.com (Y. A. Umar), dangmwansamuel@gmail.com (S. D. Denwe),
ammimuktar@yahoo.com (M. M. Namadi).

Abstract: The economic and ecological problems associated with fossil fuel have raised interest in biofuel research in recent times in different parts of the world. The use of *Eucalyptus* forest waste biomass with no appreciable value to industries or for food as alternative and cost effective feedstock for bioethanol production was evaluated in this study. *E. camaldulensis* biomass (bark and leaves) were pretreated separately with acid (2M H_2SO_4) and Microwave irradiation (250V, 50Hz) prior to fermentation with *Bacillus subtilis* and *Escherichia coli* isolated from surrounding soil. Higher yield of reducing sugar were obtained from bark (43 %) and leaves (38.5 %) pretreated by microwave irradiation as compared with acid treated plant biomass. Similarly, Bioethanol volume and concentration of 34.89 g/l and 0.51 % respectively were higher in Microwave irradiated bark of *E. camaldulensis* at 21 days of fermentation when *E. coli* and *B. subtilis* were used in synergy The least bioethanol volume yield of 18.79 g/l and concentration of 0.12 % when bark and leaves of *E. camaldulensis* were combined was obtained on day 7 of fermentation using *E. coli*. The study concludes that the amount of dried wastes generated (37.8 kg) from one average stand of *Eucalyptus* tree could yield significant volume (131,884.2 g/l) of bioethanol when *B. subtilis* and *E. coli* are used in synergy.

Keywords: *Eucalyptus*, Biomass, Bioethanol, Fermentation

1. Introduction

The world is facing the crisis of global warming and environmental degradation which mainly has been associated with excessive use of fossil fuels. Alternative sources of energy are being explored the world over in order to reduce oil dependence and increase energy production [1]. Among the various sources been explored, biofuels offer one of the best alternative options as they have much lower life cycle Green House Gas (GHG) emissions compared to fossil fuels [2]. Biofuels which could be solid, liquid or gaseous fuel derived from biological materials can be used to generate energy [3]. Energy produced through these processes could help to reduce world's dependence on oil and therefore cut

CO_2 emission, thus mitigating global warming. In addition, bi-products of biofuel production can provide new income and employment opportunities in rural areas [4].

Bioethanol which is one of the biofuel derived from plant biomass is an ethyl alcohol, grain alcohol, $CH_3–CH_2–OH$ or ETOH. It is a liquid biofuel which is produced from several different biomass feedstocks and conversion technologies. Bioethanol has been reported as an attractive alternative fuel because of its renewable bio-based resource and its oxygenation which provides the potential to reduce particulate emissions in compression–ignition engines [5]. It is one of the promising future energy alternatives that could contribute to the reduction of negative environmental impacts generated by the use of fossil fuels [6]. Bioethanol has been produced from a variety of raw materials containing fermentable sugars.

Eucalyptus species is one of the commercially important fast-growing trees in Nigeria. It provides raw material for papermaking and is widely used in the construction industries, although large amounts of wood residue, such as bark, leaves, cork residue, cross-cut ends, edgings, grinding dust and saw have not been efficiently utilized [7].

Considering the high cellulose content, fast growth of Eucalyptus trees and the fact that waste generated during wood processing has no human and animal food values, the plant could serve to provide the much needed feedstock for bioethanol production in Nigeria.

2. Methodology

2.1. Study Area

Afaka Forest Reserve occupies about 7,093.1366 hectares of land (Fig. 1). It lies on latitudes 10^0 33'N and 10^0 42'N; Longitudes 7^0 13'E and 7^0 24'E. The Forest Reserve provides a mixture of both natural and man-made vegetation characteristic of guinea savannah vegetation. Some of the indigenous and exotic plants in the forest reserve include *Pakia biglobolsa, Ceiba petandra, Azadirachta indica, Mangifera indica, Eucalyptus spp, Tectona grandis, Pinus caribae, Gmelina arborea* among others

The forest reserve is the main source of electricity pole for most part of the state in addition to providing wood for the construction industry.

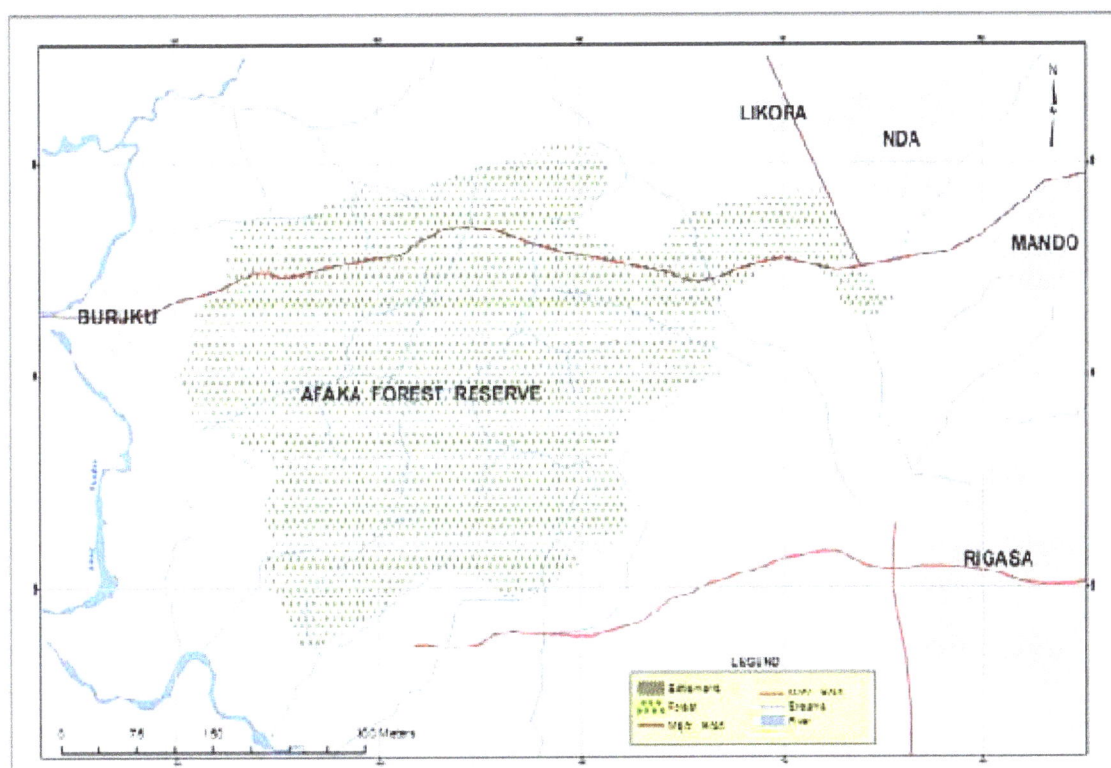

Figure 1. Map of Afaka Forest Reserve and Immediate communities.

Source: Department of Geographgy, NDA Kaduna.

2.2. Sample Collection and Processing

One kilogram (Kg) each of Bark and leaves of *E. camaldulensis* were collected separately in clean polyethene bags from Afaka Forest Reserve, Kaduna State and transported immediately to the Centre for Energy and Environment, Nigerian Defense Academy - Kaduna. Samples collected were washed several times to remove adhering dirt and later chopped into small pieces using a sharp knife. Chopped samples were oven dried in an oven at 150 ^{0}C for 6 hours, pulverized to powder using mortar and pestle, and stored in capped wide mouthed plastic containers until needed [8].

2.3. Isolation and Identification of Fermenting Bacteria from Soil

2.3.1. Collection and Preparation of Soil Sample for Serial Dilution

Collection and preparation of soil sample for serial dilution was carried out in accordance with the standard method described by [9]. Briefly, five (5) grams of upper soil layer were collected at 5 different locations within Afaka Forest Reserve using clean, dry plastic sample tubes with the aid of a sterile spatula. Soil samples collected were mixed together to form composite soil from which 1g was suspended in 10ml of sterile water in a ratio of 1:10 (10^{-1}). Further dilution of

10^{-2}, 10^{-3}, 10^{-4} and 10^{-5} were prepared from the stock (10^{-1}) preparation.

2.3.2. Preparation of Media

Preparation of nutrient agar media was carried out in accordance with the standard procedure described by [9]. Twenty eight (28) grams of nutrient agar was added to 1000ml of distilled water in a beaker, stirred vigorously and dissolved by heating on a hot plate. This was later sterilized by autoclaving for 15 minutes at 121 ^{0}C and allowed to cool before dispensing into petri-dishes. The preparations were allowed to solidify at room temperature.

2.3.3. Inoculation of Media

Media inoculation was done by streaking different dilutions (10^{-2}, 10^{-3}, 10^{-4} and 10^{-5}) on solidified nutrient agar in petri-dishes and incubated at 37 ^{0}C for 24 hours (9). Each petri-dish was later observed for appearance of colonies.

2.3.4. Fermentation Test

Each bacteria isolated from soil was screened for fermentation ability by Carbohydrate Fermentation Test using Triple Sugar Iron agar (TSI) prepared as agar slope [10]. Test organisms were inoculated by stabbing and streaking the medium with the aid of a sterilized straight wire loop and then incubated at 37 ^{0}C for 24 hours. Gas production was determined by cracking of the medium while H_2S formation was determined by blackening at the slant butt junction. Determination of glucose fermentation was achieved by yellowing of the butt. Fermentation of lactose or sucrose or both was determined by yellowing of both the butt and the slant, while motility was determined by observing the line of inoculation. Sharp defined line of inoculation indicates positive motility.

2.4. Morphological and Biochemical Characterization of Bacteria Isolates from Soil Samples

Colonies of bacteria with fermenting ability were characterized and identified based on morphological and biochemical characteristics using standard techniques described by [11] and [10] respectively.

2.5. Pretreatment of E. camaldulensis Wood Waste

2.5.1. Microwave Irradiation

Ten grams each of dried bark and leaves of E. camaldulensis were taken in separate glass beakers and microwaved (model no-QMWO-25L) for 3 minutes at 250V, 50Hz [12]. To the content in each of the beakers, 100 ml of distilled water was added and autoclaved at 121 ^{0}C for 15 minutes. The mixtures were then filtered through No1 Whatman filter paper into a conical flask and the hydrolysate collected for further analysis.

2.5.2. Acid (2M H_2SO_4) Pretreatment

Ten grams each of dried bark and leaves of E. camaldulensis were soaked separately in 100 ml of 2M H_2SO_4 in a beaker. The mixtures were allowed to stand for 4 hours and later autoclaved at 121 ^{0}C for 15 min. The

mixtures in each beaker were then filtered into a conical flask through a Whatman No.1 filter paper. Hydrolysates collected were subjected to further analysis.

2.6. Hydrolysate Detoxification

The hydrolysate from bark and leaves of E. camaldulensis collected from both microwave irradiated and acid treated biomass were separately heated to 60 ^{0}C and basified by adding at intervals 0.5g solid NaOH until a pH of 5.5 was achieved. To the solution, 1g of $Ca(OH)_2$ was added to detoxify harmful materials present in the hydrolysate and then filtered through a Whatman No.1 filter paper to remove insoluble residues. The filtrates containing fermentable sugars were then stored in capped plastic containers for determination of reducing sugar [13].

2.7. Determination of Reducing Sugar

The reducing sugar content of the hydrolysates was assayed by adding 3ml 0f 3, 5 - dinitrosalicylic acid (DNS) to 3 ml of each hydrolysate sample. The mixture was heated in hot water bath for 10 minutes until red-brown color was observed. To the mixture, 1 ml of 40 % potassium sodium tartrate solution was then added to stabilize the color and the mixture cooled to room temperature under running tap. Absorbance of each sample was measured at 491 nm using UV-VIS spectrophotometer. The reducing sugar content was subsequently determined by reference to a standard curve of known glucose concentration [14].

2.8. Fermentation of the Hydrolysate

Fifty (50) milliliters of bark, leaves and combination of bark and leaves of E. camaldulensis hydrolysates were separately dispensed into three 100 ml capacity conical flasks and each flask replicated three times. The flasks were then covered with cotton wool, wrapped in aluminium foil, and autoclaved at 120 ^{0}C for 15 minutes. The flasks were allowed to cool at room temperature and aseptically inoculated with the fermentative organisms (6.00×10^2cfu/ml) isolated from soil as follows:
   a) *Bacillus subtilis*
   b) *E.coli*
   c) *Bacillus subtilis + E.coli*
All flasks were incubated anaerobically at 30 ^{0}C and each examine at seven days interval for 3 weeks. The fermented broth was distillated at 78 ^{0}C and the distillate collected for determination of bioethanol concentration in the fermented medium.

2.9. Determination of Concentration of Bioethanol

The concentration of bioethanol in distillates was carried out by the method described by [15] using UV-VIS quantitative analysis of alcohols. This involves taking 1 ml of standard ethanol and diluting with 100 ml of distilled water to produce 1% stock solution. To obtain 0%, 0.2%, 0.4%, 0.6% and 0.8% of the stock ethanol solution, 0 ml, 2ml, 4ml, 6ml and 8ml of the stock solution was diluted in 10ml of distilled water. To each of the varying ethanol concentrations, 2 ml of chromium reagent was added and allowed to stand for an hour and the absorbance of each concentration measured at 588 nm using UV-VIS spectrophotometer. Readings obtained were used to develop standard ethanol curve. To determine the concentration of bioethanol produced, 4 ml of each bioethanol sample was transferred into a test tube and treated with 2 ml of the chromium reagent. The mixture was allowed to stand for an hour and the absorbance measured at 588 nm using the UV-VIS spectrophotometer.

2.10. Quantification of Ethanol

To determine the quantity of ethanol produced, distillate from hydrolysate of bark, leaves and a combination of bark and leaves fermented with *B. subtilis*, *E. coli* and combination of the two bacteria were collected over a slow heat at 78 ^{0}C. The quantity of ethanol produced in g/l was then obtained by multiplying the volume of distillate collected at 78°C by the density of ethanol (0.8033 g/ml) [16].

2.11. Statistical Analysis

Data obtained were statistically analyzed by one-way analysis of variance. Comparison of means were done by the New Duncan's multiple range test (P = 0.05).

3. Result

The amount of waste generated from one average fell stand of *Eucalyptus* plant in Afaka forest Reserve is presented in Fig. 2. Fresh Bark, leaves and unused branches of felled *Eucalyptus* plant produced 38.0 kg, 7.0 kg and 26.0 kg respectively while the dried bark, leaves and unused branches produced 32.0 kg, 5.8 kg and 20.8 kg respectively.

The reducing sugar yields from the hydrolysates of bark, leaves and, bark and leaves of *E. camaldulensis* pretreated with acid (2M H_2SO_4), microwave irradiation and untreated biomass are presented in figure 3. Microwave irradiation produce the highest (38.5% - 43%) yield of reducing sugar compared with untreated (07% - 08%) and acid (2M H_2SO_4) (35.5% - 38%) pretreated biomass. Although there was no significant difference in reducing sugar yields of acid (2M H_2SO_4) treated and microwave irradiated biomass (P>0.05), the difference in reducing sugar yield between treated and untreated biomass is statistically significant (P<0.05).

Table 1 shows the quantity in volume (g/l) of ethanol produced over a 3 weeks period from microwave treated *E. camaldulensis* wood waste biomass using bacteria isolated from soil sample. Although the performance of individual fermenting bacteria is lower than when both are used in synergy, the volume of ethanol produced by these organisms irrespective of plant biomass used increased steadily over the 3 wks period.

However, highest volume (31.22 ± 0.54g/l – 34.89 ± 0.07g/l) of ethanol production was achieved when microwave irradiated bark was fermented by *B. subtilis* + *E. coli* than leaves (25.95 ± 0.04g/l – 33.95 ± 0.02g/l) or bark and leaves (26.18 ± 0.50g/l – 29.05 ± 0.02g/l) using the same fermenting organisms.

Similarly, there was an increase in percentage concentration of bioethanol produced from day 7 to day 21 irrespective of plant part or fermenting organism used. However, microwave irradiated bark biomass when fermented by *B. subtilis* + *E. coli* had higher percentage concentration of bioethanol produced at day 7 (0.42 ± 0.02), day 14 (0.47 ± 0.03) and day 21 (0.51 ± 0.02) than either of the plant parts fermented by individual bacteria (Table 2).

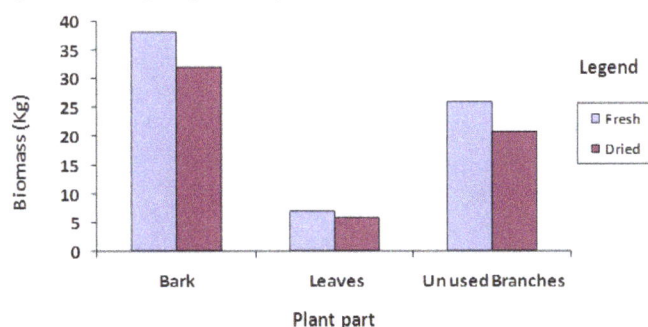

Figure 2. *Biomass by weight generated from an average fell stand of E. camalendulensis in Afaka Forest Reserve, Kaduna*

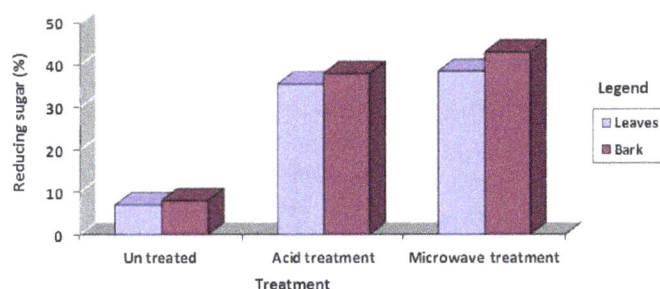

Figure 3. *Percentage yield of reducing sugar from E. camaldulensis waste using different treatment methods.*

Table 1. The volume of bioethanol (g/l) produced from E. camaldulensis biomass using bacteria isolated from soil at Afaka Forest Reserve Kaduna, Nigeria.

Plant part	Fermenting Organisms	Fermentation days 7	14	21
Bark and leaves	B. subtilis	$19.28^a \pm 0.02$	$21.13^a \pm 0.01$	$25.18^a \pm 0.05$
	E. coli	$18.79^a \pm 1.10$	$20.09^b \pm 0.04$	$22.88^b \pm 0.51$
	B.subtilis+ E.coli	$26.18^b \pm 0.50$	$27.70^c \pm 0.06$	$29.05^c \pm 0.02$
Leaves	B. subtilis	$21.09^a \pm 0.04$	$25.28^a \pm 0.07$	$27.82^a \pm 0.01$
	E. coli	$19.80^b \pm 0.06$	$25.72^a \pm 0.07$	$27.28^a \pm 0.09$
	B.subtilis+ E.coli	$25.95^c \pm 0.04$	$30.58^b \pm 0.01$	$33.93^b \pm 0.02$
Bark	B. subtilis	$25.84^b \pm 0.06$	$27.50^c \pm 1.04$	$28.93^b \pm 0.05$
	E. coli	$25.26^b \pm 0.02$	$26.25^{bc} \pm 0.08$	$27.55^b \pm 0.09$
	B.subtilis+ E.coli	$31.22^c \pm 0.54$	$31.77^d \pm 0.41$	$34.89^c \pm 0.07$

a,b,c, means within a column in each plant part with different superscripts are significantly different (P<0.05). Values are means ± standard deviation of three replicates

Table 2. Percentage concentration of bioethanol produced from E. camaldulensis wood waste biomass at Afaka Forest Reserve Kaduna, Nigeria.

Plant part	Fermenting Organisms	Fermentation days 7	14	21
Bark and leaves	B. subtilis	$0.13^a \pm 0.01$	$0.16^a \pm 0.01$	$0.25^a \pm 0.02$
	E. coli	$0.12^a \pm 0.03$	$0.15^a \pm 0.02$	$0.22^b \pm 0.02$
	B.subtilis+ E.coli	$0.23^b \pm 0.02$	$0.30^b \pm 0.02$	$0.39^c \pm 0.01$
Leaves	B. subtilis	$0.16^a \pm 0.01$	$0.25^a \pm 0.02$	$0.33^a \pm 0.02$
	E. coli	$0.14^a \pm 0.02$	$0.23^a \pm 0.02$	$0.31^a \pm 0.02$
	B.subtilis+ E.coli	$0.27^b \pm 0.01$	$0.42^b \pm 0.03$	$0.47^b \pm 0.01$
Bark	B. subtilis	$0.25^a \pm 0.03$	$0.34^a \pm 0.02$	$0.38^a \pm 0.01$
	E. coli	$0.22^a \pm 0.03$	$0.32^a \pm 0.02$	$0.36^a \pm 0.03$
	B.subtilis+ E.coli	$0.42^b \pm 0.02$	$0.47^b \pm 0.03$	$0.51^b \pm 0.02$

a,b,c, means within a column in each plant part with different superscripts are significantly different (P<0.05). Values are means ± standard deviation of three replicates

4. Discussion

The basic structural framework of plants consists of cellulose, hemicellulose and lignin. The close and complex association between these three lignocellulosic materials causes physical and chemical barriers that have to be broken to release fermentable sugars for bioethanol production. Various pretreatment techniques have been developed for various biomass feedstocks [7]. The differences observed in the yield of reducing sugar in microwave and acid pretreatment in this study is indicative of the differences in efficiency of these techniques to release fermentable sugars from E. camaldulensis biomass. Yields of reducing sugar from lignocellulose biomass processed by different pretreatment methods including microwave irradiation, concentrated and dilute acids are well documented [7, 17, 18].

Microwave irradiation is known to enhance the digestibility of cellulosic biomass, increases surface area, decreases the degree of polymerization and crystallinity of cellulose, enhances hydrolysis of hemicelluloses and results in partial depolarization of lignin [7]. This is in contrast to acids which though are powerful agents for cellulosic hydrolysis results in the formation of degradation products and releases natural inhibitors which affect the yields of fermentable sugars [19].

The high volume (34.89 ± 0.07g/l) of bioethanol obtained from bark of E. camaldulensis in the present study using B. subtilis + E. coli in synergy at days 21 is higher than the volume of ethanol obtained for other plant biomass such as guinea corn husk (26.31g/l), sawdust (12.30g/l)[15]; Sweet potato peels (16.47g/l), rice husk (06.22g/l)[20]; and empty fruit branches of palm oil tree (10.32g/l)[21]. These differences in volume of bioethanol obtained from the different plant biomass could be associated with the major composition of the various feedstocks in addition to the fermenting organisms involved in the production process. B. subtilis and E. coli are excellent organisms that are resistant to salt/ toxic inhibitors in addition to being good anaerobic fermenters. According to [22], B. subtilis and E. coli are organisms that grow very fast and can utilize pentose (C_5) and hexoses (C_6) including glucose, xylose, mannose, cellobiose among other simple sugars. They are also reported to possess native hemicellulases. Although both B. subtilis and E. coli have excellent fermentation ability, B. subtilis is reported to produces larger number of polysaccharide degrading enzymes such as α amylase, pullulanase, endo β-1-4 mannase, levenase, pectate lyases, β-1-4-endogluconase, β-1,3-1,4-endogluconase, and endo-1,4- β- xylanases. The ability of B. subtilis to efficiently break down polysaccharides into soluble carbohydrates is reflected in the relatively high volume of bioethanol produce when these organisms are used individually in the present study.

5. Conclusion

This study revealed that *B. subtilis* and *E. coli* isolated from soil have a great potential in the fermentation of *E. camaldulensis* biomass into bioethanol. The amount (34.89 g/l) of bioethanol produced from 10g of *E. camaldulensis* biomass after 21 days of fermentation using *B. subtilis* and *E.coli* in synergy translates into about 131,884.2g/l of bioethanol derivable from waste generated from an average stand (37.80 kg) of *Eucalyptus spp*. Further studies would however be needed to enhance the performance of these organisms through genetic manipulation to achieve higher yields thus reducing over dependence on fossil fuel.

Acknowledgement

The Authors acknowledge the support received from staff and management of Forestry Research Institute of Nigeria and the Nigerian Defense Academy Kaduna. We are also grateful to the Department of Geography for providing the sketch map of the study area and to the Centre for Energy and Environment, NDA for allowing the use of their facilities.

References

[1] G. Berndes, M. Hoogwijk and R.Van den Broek. "The contribution of biomass in the future global energy supply: a review of 17 studies". *Biomass and Bioenergy.* vol. 25, pp. 1-28, 2003.

[2] A. Kumar. "Next generation bio-fuels for greenhouse gas mitigation and role of biotechnology". [*Proc. Natl Conf. Emerging trends in biotechnology and pharmaceutical research,* Mangalayatan University, Aligarh Feb. 18-19, pp 38, 2012].

[3] A.M. Agba, M.E. Ushie, F.I. Abam, M.S. Agba and J. Okoro. "Developing the Biofuel Industry for Effective Rural Transformation". *European Journal of Scientific Research* vol. 40, pp. 441-449, 2010.

[4] C.V. Stevens and R. Verhe. Renewable bioresources scope and modification for nonfood applications. John Wiley and Sons Ltd, England, 2004, pp 310.

[5] A.C. Hansen, Q. Zhang and P.W.L. Lyne. "Ethanol–diesel fuel blends: a review". *Bio resource Technology*, vol. 96, pp. 277–285, 2005.

[6] J. D. McMillan. "Bioethanol production: Status and prospects". *Renewable Energy*, vol. 10(2), pp. 295-302, 1997.

[7] Y. Zheng, Z. Pan, R. Zhang and D. Wang. "Overview of biomass pretreatment for cellulosic ethanol production". *International Journal of Agriculture and Biol Eng*, vol. 2(3), pp. 51 – 68, 2009.

[8] M. Galbe and G. Zacchi. "Pretreatment of lignocellulose materials for efficient bioethanol production". *Advances in Biochemical Engineering / Biotechnology*, vol. 108, pp. 41-65, 2007.

[9] A. Musliu and W. Salawudeen. "Screening and isolation of soil bacteria for ability to produce antibiotics". *European Journal of applied Science*, vol. 4(5), pp. 211-215, 2012.

[10] B.S Manga and S.B. Oyeleke. Essentials of Laboratory Practical's in Microbiology 1st ed., Tobes Publishers, 2008, pp. 56-76.

[11] B. Chitra, P. Harsha, G. Sadhana and R. Soni. "Isolation and characterization of bacterial isolates from agricultural soil at Durg district". *Indian journal of Science Research*, vol. 4(1), pp. 221-226, 2014.

[12] S. Nivedita, R. Shelly and P. Shruti. "Production, Purification and Characterization of cellulose free-xylanase by Bacillus coagulans B30 using lignocellulosic forest wastes with different pretreatment methods". *Journal of Agroalimentary processes and technologies*, vol. 19(1), pp. 28-36, 2013.

[13] A. Martinez, M.E. Rodriques, S.W. York, J.F. Preston and L.O. Ingram. "Effect of $Ca(OH)_2$ treatments on the composition and toxicity of bagasse Hemicellulose Hydrolysates". *Biotechnology and Bioengineering*, vol. 6, pp. 526-36, 2000.

[14] G.L. Miller. "Use of dinitrosalicylic acid reagent for the determination of reagent for the determination of reducing sugar". *Analytical Chemistry*, vol. 31(3), pp. 426-428, 1959.

[15] S.B. Oyeleke and N.M. Jibrin. "Production of bioethanol from guinea corn husk and millet husk" *.African journal of Microbiology Research*, vol. 3(4), pp. 147-151, 2009.

[16] C.N. Humphrey and U.O. Caritas. "Optimization of ethanol production from *Garcinia kola* (bitter kola) pulp agro waste". *African Journal of Biotechnology*, vol. 6(17), pp. 2033-2037, 2007.

[17] R. Arumugam and M. Manikandan. "Fermentation of pretreated hydrolyzates of banana and mango fruit wastes for ethanol production". *Asian Journal of Experimental Biological Science* vol. 2(2), pp. 246-256, 2011.

[18] M. Lima, B. Gabriela, K. Hana, B. Juliano, A. Camila, D. Oigres, R. Euduardo, D. Leonardo, J. Simon, A. Carlos and P. Igor. "Effect of pretreatment on morphology, chemical composition and enzymatic digestibility of *Eucalyptus* bark: a potentially valuable source of fermentable sugars for bioethanol production-part 1". *Biotechnology for biofuel*, vol. 6, pp. 75, 2013.

[19] V. Chaturvedi and P. Verma. "An overview of key pretreatment processes employed for bioconversion of lignocellulosic biomass into biofuels and value added products". *3 Biotechnology*, vol. 3(5), pp. 415-431, 2013.

[20] M. Nikzad, K. Movagharnejad, G.D. Najafpour and F. Talebnia. "Comparative studies on effect of pretreatment of rice husk for enzymatic digestibility and bioethanol production". International journal of engineering. Transactions B: Applications, vol. 26(5), pp. 455 – 464, 2013.

[21] A.K. Mohd, S.K. Loh, A. Nasrin, A. Astimar and M.S. Rosnah. "Bioethanol production from empty fruit bunches hydrolysate using *Saccharomyces cerevisiae*". *Research Journal of Environmental Science*, vol. 5(6), pp. 573-586, 2011.

[22] j. Deutscher, A. Galinier and I. Martin-Verstraete. In: *Carbohydrate uptake metabolism. Bacillus subtilis* and its Closest Relatives: from Genes to Cells. A.L. Sonenshein, J.A. Hoch, R. Losick, Eds. American Society for Microbiology Press, Washington, DC, 2002, pp. 129–150.

[23] A.A. Brooks. "Ethanol production potential of local yeast strains isolated from ripe banana peels". *African journal of Biotechnology*, vol. 7(20), pp. 3749-3752, 2008.

[24] B.S. Dien, M.A. Cotta and T.W. Jeffries. "Bacteria engineered for fuel ethanol production current Status". *Applied Microbiology and Biotechnology*, vol. *63*, pp. 258-266, 2003.

[25] R.C. Dubey. A text book of Biotechnology, 4[th]ed. Published in Ram Nagar, New Delhi, India, 2012, pp 545-547.

[26] Z. Hu and Z. Wen." Enhancing enzymatic digestibility of switch grass by microwave-assisted alkali pretreatment," *Biochemical Engineering Journal*, vol. 38, pp. 369-378, 2008.

[27] S.A. Nikolic. "Microwave-assisted liquefaction as a pretreatment for the bioethanol production by the simultaneous saccharification and fermentation of corn meal". *Chemical Industry and Chemical Engineering*, vol. 14(4), pp. 231–234, 2008.

[28] A. Rabah, S. Oyeleke, S. Manga and L. Hassan. "Dilute acid pretreatment of millet and guinea corn husks for bioethanol production". *International Research Journal of Microbiology,* vol. 2 (11), pp. 460-465, 2011.

[29] R. Templer and R.J. Murphy. "Environmental sustainability of bioethanol production from waste paper: sensitivity to the system boundary". *Energy and Environmental Science*, vol. 2, pp. 8281-8293, 2012.

[30] J. Xiang, J. Ye, W.Z. Liang and P.M. Fan. "Influence of microwave on the ultrastructure of cellulose". *Journal of South China University of Technology*, vol. 28, pp. 84-89, 2000.

[31] Q. Yu, X. Zhuang, Z. Yuan, Q. Wang, W. Qi, W. Wang, Y. Zhang, J. Xu and H. Xu. "Two step liquid hot water pretreatment of *Eucalyptus grandis* to enhance sugar recovery and enzymatic digestibility". *Bioresource Technology*, vol. 101, pp. 4895-4899, 2010.

21

Basic Guidelines for LED Lamp Package Design

Song Jae Lee

Electronics Engineering Department, Chungnam National University, Daejeon, Korea

Email address:

sjlee@cnu.ac.kr

Abstract: Even though significant amount of researches has been done to develop LED lamp packages for improved performance especially in terms of output power, it is believed that no standard theories or guidelines have been established yet for designing LED lamp packages. In this paper, both the InGaN/Sapphire LED chip structure and its Epi-Up or Epi-Down chip-mounting scheme have been analyzed by using Monte Carlo photon simulation method. Based on the analysis, we have established guidelines for designing LED lamp packages.

Keywords: LED, LED Lamp, LED Lamp Package Design

1. Introduction

Even though visible light-emitting diodes (LEDs) have been commercially very successful, it is believed that the basic principles or guidelines for designing LED lamp packages may not have been well established yet.

Conventional LED lamps in general may fall into either the leaded type or SMD (surface mount design) type as schematically described in Fig. 1. The leaded type has relatively long leads that are often to be inserted through the holes made in the printed circuit board (PCB) and soldered at the bottom. On the other hand, the SMD type has structures that are more suited to be attached directly on the top surface of the PCB. Even though the leaded type LED lamps are still used extensively, for instance, in applications such as traffic lights and outdoor displays, the general trend seems to favor ever more the SMD type LED lamps.

A serious problem in the SMD structure shown in Fig. 1(b) may be that the heat generated in the LED chip is not dissipated easily. In general, both the molding compounds encapsulating both the LED chip and the dielectric chip mount have relatively poor thermal conductivities. Thus, in order to prevent severe heating in the junction of the chip, the LED driving current should be kept relatively small leading to very limited light output. In general, the more the junction temperature rises, the more likely the injected electrons and holes recombine nonradiatively or overflow to the carrier confinement layers and lead to significantly decreased internal quantum efficiency.[1] In addition, the rise of the junction temperature would in general shrink the bandgap energy of the active layer, which in turn shifts the spectrum to the red and changes the color of the light output.

The key component of the LED lamp is obviously the LED chip. Thus it would be a first task in LED lamp design to accurately analyze LED chips. Fig. 2 shows a typical structure of the InGaN/Sapphire LED chip, which currently enjoys a commanding position in visible LEDs. They have excellent reliability and brightness. Furthermore their emission spectrum from ultraviolet to amber is much broader than that of the AlInGaP/GaAs system that emits mostly in the red spectral region.[2] It is noted also that the blue emission from InGaN/Sapphire LEDs is exploited to implement the white LEDs that are essential for the general lighting. As well known, blue LEDs can also be fabricated by using the InGaN/SiC system. However, they may hardly compete with the InGaN/Sapphire LEDs especially in terms of maximum light output and reliability.

InGaN/Sapphire LEDs are grown on the non-conducting sapphire substrate and, as a result, their chip structures are quite unique in that the two electrodes both formed on the epitaxial side are displaced from each other in the lateral direction. Another important feature of the electrode design is that a very thin semi-transparent p-ohmic material is deposited on the top surface of the p-GaN carrier confinement layer outside the p-electrode pad. The thin ohmic material is to compensate the very low conductivity of the p-GaN carrier confinement layer and prevents the driving voltage from rising too much. The deposition of the thin ohmic material outside the p-electrode would make the isolation distance between two electrodes as small as typically $10{\sim}20~\mu m$. It is noted that this small separation between the two electrodes would make it crucial to align the chip very accurately when the LED chip is to be attached on the chip mount in the Epi-down (flip chip) mode. We will discuss this problem in more detail later.

In this paper we first analyze both the InGaN/Sapphire LED chip structure and chip-mounting schemes of either Epi-up or Epi-down. Then, based on the analysis, the basic guidelines for designing LED lamp packages would be made.

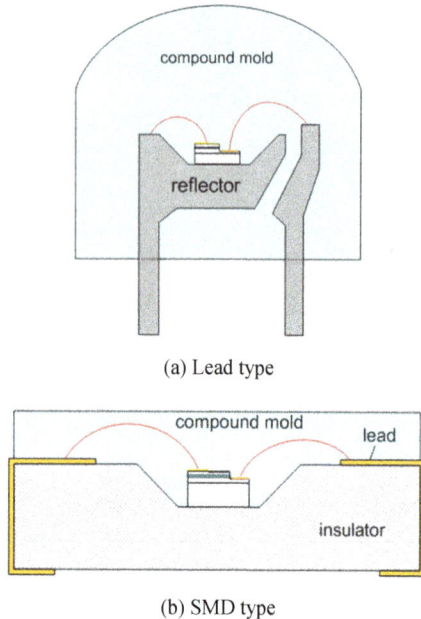

(a) Lead type

(b) SMD type

Fig. 1. Conventional LED lamps.

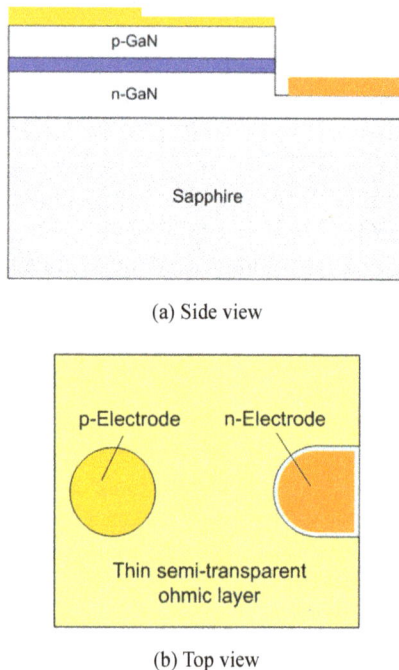

(a) Side view

(b) Top view

Fig. 2. Typical InGaN/Sapphire LED structure.

2. Different Types of InGaN/Sapphire LED Chips

The first task in the LED lamp package design may be to determine the type of the chip to be employed in the lamp. In the viewpoint of LED lamp design, the InGaN/Sapphire LED chip may be classified into the three types illustrated in Fig. 3. The structure in Fig. 3(a) is a regular chip, which is commonly used in practical LED lamps and has a size of typically $300\,\mu m \times 300\,\mu m$. With such standard-sized chips, the maximum light output power achievable would be seriously limited. Thus many researchers have been tempted, for the purpose of enhancing the maximum light output per unit LED lamp, to try large-sized chips as in Fig. 3(b). The structure in Fig. 3(c) is the so-called vertical structure, in which the sapphire substrate with a relatively poor thermal conductivity is removed and the p-electrode and n-electrode formed on top and bottom, respectively are displaced in the vertical direction.

The above rather arbitrary classification of the InGaN/Sapphire LED chips may be valid if each structure shows its own distinctive characteristics. For instance, the vertical structure in Fig. 3(c) has a very small thermal resistance as resulting from the removal of the sapphire substrate of a poor thermal conductivity. Another important feature that should be taken into account in choosing the LED chip would be the photon extraction efficiency (or photon output coupling efficiency) η_{cpl}. The external quantum efficiency η_{ext}, which is the most important parameter to determine the overall light-emitting efficiency of LED, is related to the photon extraction efficiency η_{cpl} as

$$\eta_{ext} = \eta_{in}\eta_{cpl} \tag{1}$$

where η_{int} is the internal quantum efficiency. It is noted that the photon extraction efficiency η_{cpl} is determined mostly by the detailed structure of the chip, whereas the internal quantum efficiency η_{int} is determined mostly by the quality of the epitaxial layers in the chip. Thus, it is crucial in lamp package design to maximize the photon extraction efficiency η_{cpl}.

In this work the photon extraction efficiency η_{cpl} was calculated by using the Monte Carlo photon simulation method that has been described elsewhere and used extensively to analyze the LED chips and lamps.[3,4] In the calculation we have assumed that all the chips are mounted on the surface with the photon reflectivity of 80% and encapsulated by the molding compound with reflective index of 1.50. Other important chip simulation parameters are basically the same as those used in the previous works.[3, 4]

Fig. 4 shows the photon extraction efficiency η_{cpl} calculated as a function of the chip size. As the chip size increases, the average generated photons have to travel a longer distance to escape the chip. And the probability for the photons to be absorbed increases exponentially with the traversal distance. Consequently, the photon extraction efficiency η_{cpl} should decrease with the chip size.

In order to compensate the degraded photon extraction efficiency in enlarged chips, the driving current should be increased by more than proportionally to the increased chip area. Consequently, the heating per unit chip area would be in general more severe in enlarged chips than in small area chips. Without significantly improved heat sinking, the junction temperature rise would be larger in enlarged chips than in small area chips. As discussed already, the internal quantum efficiency η_{int} decreases in general with the junction temperature. Taking into account all these effects, it would be reasoned that the external quantum efficiency η_{ext} in general degrades more rapidly with the chip size than the photon extraction efficiency η_{cpl}.

Based on the above reasoning, we may reach an agreement that the approach to increase the chip size too much in order to increase the per-unit light output would seriously compromise one of the most important advantages of LEDs, i.e., the high efficiency. Thus, a rather practical approach to increase per-unit light output may be to combine an LED chip of relatively small size with a good heat sink and thereby increase both the driving current and light output.

One of the important issues regarding the vertical structure in Fig. 3(c) is how the substrate removal affects the photon extraction efficiency η_{cpl}. Even though the removal of the sapphire substrate is certainly beneficial in terms of heat dissipation, it would in general degrade the photon extraction efficiency. In other words, the sapphire substrate in the regular structure helps to increase photon extraction efficiency. The refractive index of the sapphire substrate is about 1.77 and is between 2.48 and 1.50 that are approximate refractive indices for the GaN carrier confinement layer and the molding compound, respectively. Such index-matched substrate is to help enhance the photon extraction efficiency. It is noted, however, that the index-matching effect of the substrate occurs only to the photons transmitted into the substrate and reach the sidewall of the substrate either directly or by way of reflection off the bottom of the substrate. The thinner the substrate, the larger fraction of the photons that are transmitted into the substrate would be reflected off the bottom of the substrate and transmitted back into the GaN confinement layer without touching the sidewall of the substrate. As a result, as shown in Fig. (5), the photon extraction efficiency η_{cpl} in the regular InGaN/Sapphire structure decreases as the thickness of the sapphire substrate decreases.

Following the above reasoning, the photon extraction efficiency in the vertical structure with the sapphire substrate completely removed should similarly be poorer than in the regular structure with the substrate remained intact. For instance, the photon extraction efficiency in the vertical structure is estimated to be about 0.334 and is quite smaller than the photon extraction efficiency 0.564 estimated of the regular structure with the sapphire substrate of thickness 100 μm. A rather interesting point to note here may be that the photon extraction efficiency in the vertical structure is even smaller than the photon extraction efficiency 0.435 estimated of the regular chip with the sapphire substrate completely removed as in the vertical structure. The poor photon extraction efficiency of the vertical structure compared to the substrate-removed regular structure is due to the difference in the electrode design. In the case of the vertical structure, the photons would be generated more likely in the central region

of the actively layer and they have to travel a longer distance to reach the edge of the chip. On the other hand, in the regular structure, the photons would be generated more likely in the region near the edge of the chip and they will couple out of the chip more easily. Even though the photon extraction efficiency in the vertical structure could be improved somewhat, for instance, by optimizing the electrode pattern design5, it may not easily surpass that of the regular structure with the substrate completely removed. Based on the above reasoning we may reach an agreement that when the vertical structure is chosen instead of the regular structure, the benefit of its very low thermal resistance should be more than enough to compensate the degraded photon extraction efficiency.

3. Chip Mounting Scheme

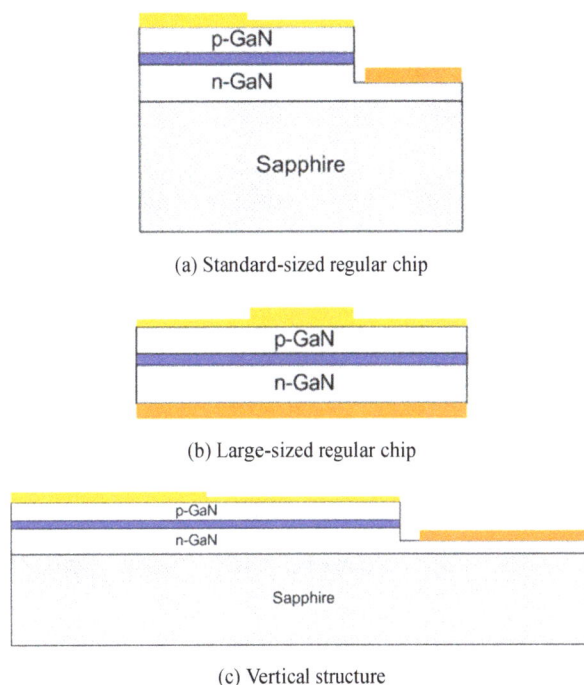

Fig. 3. *Classification of InGaN/Sapphire LED structures in the viewpoint of LED lamp package design.*

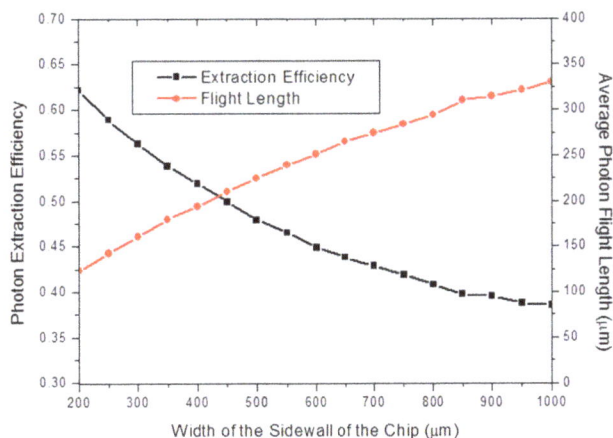

Fig. 4. *The photon extraction efficiency and the average flight length of the output photons as a function of the chip size in regular chip.*

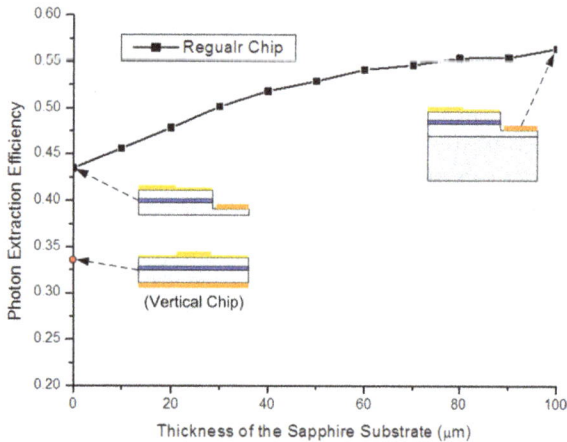

Fig. 5. *Photon extraction efficiency as a function of the sapphire substrate thickness in regular chip.*

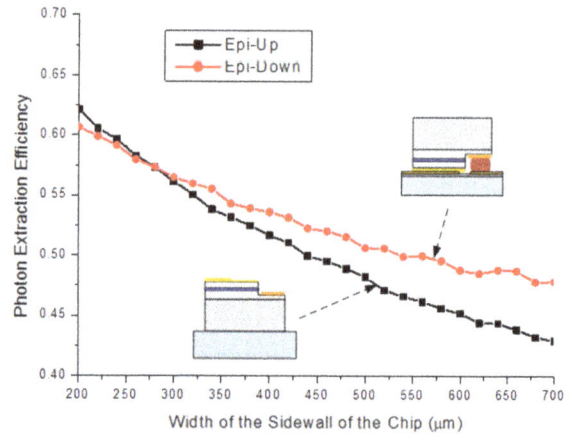

Fig. 7. *Photon extraction efficiency as a function of the chip size, depending on chip-mounting schemes.*

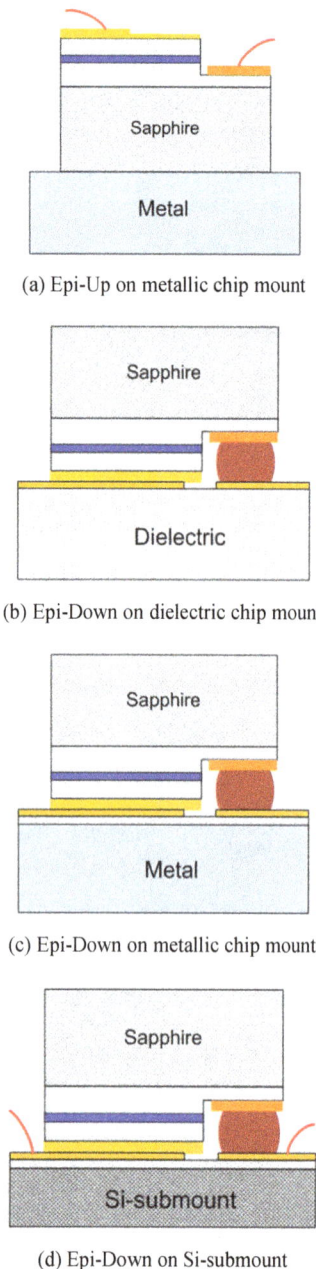

(a) Epi-Up on metallic chip mount

(b) Epi-Down on dielectric chip mount

(c) Epi-Down on metallic chip mount

(d) Epi-Down on Si-submount

Fig. 6. *Chip mounting schemes.*

Once the type of the LED chip is chosen to be employed in the lamp packaging, the next important task may be to decide how to attach the chip on the chip mount. In general the performance of the chip, in terms of both the heat dissipation and the photon extraction efficiency η_{cpl}, would depend significantly on the chip mounting scheme. Fig. 6 shows the regular chips attached on the chip mount either the epitaxial side up or down. It is noted that, in the case of the vertical chip with the sapphire substrate completely removed, the distinction between up and down will not have much meaning.

In the Epi-up mounting scheme, in which the insulating sapphire substrate of the chip is made contact directly with the chip mount, a plate of bare metal of high thermal conductivity can be utilized as the chip mount as shown in Fig. 6(a). In this configuration, the two electrodes of the chip will be connected by the bond wire to the current leads that would be formed separately somewhere else in the package. In the case of Epi-down chip mounting scheme, in which the epitaxial side of the chip is made contact with the chip mount, the chip mount should be equipped on top with two electrically isolated current leads that are to be soldered to the respective electrodes of the chip. The electrically isolated current leads may be formed either on a bare dielectric plate as in Fig. 6(b) or on a dielectric layer deposited on a metallic plate as in Fig 6(c). It is noted that either the dielectric plate or layer in the Epi-down chip mount would seriously hamper the heat dissipation from the LED chip junction to the chip mount, which would also function as a heat sink, and thus somewhat compromise the thermal advantage unique in the Epi-down scheme.

However, the most critical disadvantage of the Epi-down scheme may be that it can hardly be applied to mass production, due to the difficulties in attaching the LED chip on the chip mount. As mentioned in the previous section, the lateral separation between the two electrodes in the regular InGaN/Sapphire LED chip is as small as typically $10\sim20^{\mu m}$, and thus it is critical to align the chip on the chip mount very accurately so that the two electrodes on the chip are in proper contact with the respective current leads on the chip mount. It is noted, however, that the margin of positioning error of typical die bonders that are often employed in the LED manufacturing industry is usually much larger than $10\sim20^{\mu m}$, the typical separation between two electrodes.

One of the approaches to overcome the alignment problem in Epi-down scheme may be, for instance, to attach the LED epitaxy wafer, before being diced into separate chips, on a Si wafer that has current lead patterns formed on top, by using a mask aligner that is widely used in semiconductor device fabrication. The resulting epitaxy-Si integrated wafer is then diced into separate chips as shown in Fig. 6(d). It is noted, however, that in chip dicing process a portion of the current lead patterns on the Si wafer should be exposed for the bond wire connection. The overall processes may be too difficult or expensive for them to be practically employed in the LED industry.

Another important issue regarding the chip mounting scheme is how it affects the photon extraction efficiency η_{cpl}. Fig. 7 shows the photon extraction efficiency calculated as a function of the chip size both in the Epi-up and Epi-down chip mounting schemes. A rather intuitive prediction may be that the photon extraction efficiency would be generally higher in the Epi-down scheme in which the sapphire substrate is completely cleared of the interference from the chip mount. However, in our calculation in Fig. 7, when the width of the chip is relatively small, i.e., below about 260 μm, the photon extraction efficiency in the Epi-down scheme is actually poorer than in the Epi-up scheme.4

In order to properly understand how the chip mounting scheme really affects the photon extraction efficiency, it is necessary to consider the effect of photon trapping inside the epitaxial region as a result of the total internal reflection off the sapphire substrate interface. As discussed in the previous section, the refractive index of the sapphire substrate is about 177 and is considerably smaller than the refractive index of the GaN carrier confinement layer, about 2.48. Thus a significant fraction of the emitted photons, i.e. the ones travelling roughly parallel to the epitaxial layers, would be reflected total-internally off the sapphire substrate boundary and be trapped inside the epitaxial region. And the rest of the emitted photons, i.e. the ones travelling rather perpendicularly to the epitaxial layers would avoid the total internal reflection and mostly transmit into the substrate region.

An important point is that the two categories of the photons escape the chip through different windows in each chip mounting scheme. First, for the photons trapped inside the epitaxial region, only the sidewall is opened in the Epi-down scheme, whereas in the Epi-up scheme the top surface of the chip where the thin semitransparent p-ohmic material is deposited is also partially opened in addition to the sidewall. Thus, for the photons trapped inside the epitaxial region to escape the chip, the Epi-up scheme is preferred over the Epi-down scheme. Next, for the photons transmitted into the sapphire substrate, only the substrate sidewall is opened in the Epi-up scheme, whereas in the Epi-down scheme both the substrate sidewall and substrate bottom surface that is now directed upward and therefore cleared of the chip mount are opened. Thus, for the photons transmitted into the sapphire substrate to escape the chip, the Epi-down scheme is preferred over the Epi-up scheme.

Each chip mounting scheme, as reasoned above, having its

own unique advantage but enjoyed by only a fraction of the photons, i.e., the photons trapped inside the epitaxial region or the photons transmitted into the sapphire substrate, it is not simple to judge which chip mounting scheme is preferred for the average photons to couple out of the chip. However, our calculation in Fig. 7 shows that when the chip size is relatively large the Epi-down scheme has higher photon extraction efficiency than the Epi-up scheme. It is noted that, in this regime of large area chip, most of the output photons would have passed through the upward windows, which have much larger area than the sidewall windows. Consequently, in the Epi-up scheme of this regime, the photon extraction efficiency would be determined largely by the amount of the photons trapped inside the epitaxial region as well as the transmittance of the thin ohmic material deposited on the top surface of the chip. Similarly, in the Epi-down scheme the photon extraction efficiency will be determined largely by the amount of the photons transmitted into the substrate. It is noted that the approach to grow the epitaxial layers on the roughened surface of the sapphire substrate would increase significantly the number of photons transmitted into the substrate and thereby improve significantly the photon extraction efficiency in the Epi-up scheme.[4]

Another important point in Fig. 7 is that when the chip size is below about 260 μm the Epi-up scheme has higher photon extraction efficiency than the Epi-down scheme. It is noted that, in the Epi-up scheme of this regime of small area chip, even the photons transmitted into the substrate are able to couple out of the chip through the widely opened substrate sidewalls.

4. Combinations of LED Chip and Chip Mount

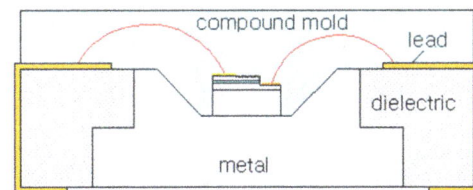

(a) Regular chip on metallic chip mount

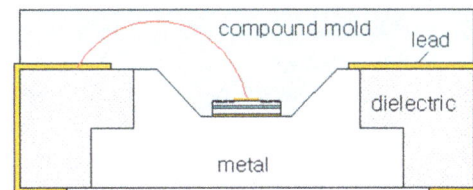

(b) Vertical chip on metallic chip mount

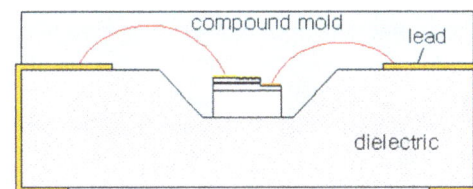

(c) Regular chip on dielectric chip mount

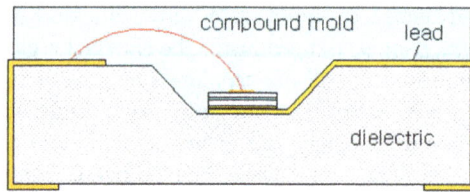

(d) Regular chip on dielectric chip mount

Fig. 8. Four basic LED lamp packaging structures.

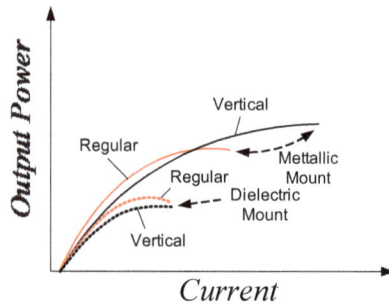

Fig. 9. Schematic current-output power characteristics depending on chip/chip-mount combinations.

Based on the previous discussions regarding both the type of LED chip and chip mounting scheme, we may reach the following agreements. First, the large area chip in Fig. 3(b) with the photon extraction efficiency seriously degraded may not be used better, except for special applications requiring large light output of short duration as in flash lamps for digital cameras. On the other hand, the vertical chip in Fig. 3(b) also with the photon extraction efficiency seriously degraded has a very low thermal resistance and thus it would be possible, when combined with a good heat sink, to increase significantly the driving current and thereby the light output. Regarding the chip mounting methods, the Epi-down scheme requiring too complicated chip attaching process may hardly be adopted in the LED manufacturing industry. Furthermore, the Epi-down scheme may not be preferred much over the Epi-up scheme even in terms of the photon extraction efficiency, especially when the chip size is relatively small as is the case in most practical LED lamps. Thus, both the large area chip and the Epi-down chip mounting scheme not being allowed in LED lamp package design, most practical LED lamp structures would then be obtained by attaching, in Epi-up mode, the two types of the chip, i.e., the regular chip or the vertical chip on the two types of the chip mount, i.e., the metallic mount or the dielectric mount. The resultant four types of LED lamp structures based on chip-chip mount combination are schematically described in Fig. 8.

One of the important output characteristics of LED lamps may be their light out power vs. current curve. In general the output power of an LED lamp would depend very complicatedly on various parameters of the chip and chip mount employed, and thus it is very difficult to generalize the current-output power characteristics for various LED lamp structures.[6] Thus, here we may better present the

schematic current-output power curves that are expected typically of the four types of LED lamp structures in Fig. 8.

The LED lamp structure in Fig. 8(a) consists of the regular chip attached on the metallic chip mount. In this case, the overall thermal resistance form the junction to the bottom of the chip mount would be relatively large, due to the relatively thick sapphire substrate with relatively poor thermal conductance in the heat dissipation path. Thus, when the driving current is relatively low and therefore the heating in the junction of the chip is not very severe, the output power would increase approximately linearly with the current. However, when the driving current is increased enough, the large amount of heat generated would not be dissipated easily and as a result the output power would start to saturate with the current.

In the LED lamp structure in Fig. 8(b), which consists of the vertical chip attached on the metallic chip mount, the overall thermal resistance form the junction to the bottom of the chip mount would be very small. Thus, the saturation of the output power would occur at a much larger current level than in the structure in Fig. 8(a). It is noted, however, that the output power level in the linear region is actually smaller than in the structure in Fig. 8(a), due to the photon extraction efficiency seriously degraded in the chosen vertical structure. The structure in Fig. 8(b), however, has a significantly extended linear region compared to the structure in Fig. 8(a), and as a result, a power-crossover point is often observed between the two structures.

The LED lamp structure in Fig. 8(c) consists of the regular chip attached on the dielectric chip mount. In this case, the overall thermal resistance form the junction to the bottom of the chip mount would be extremely large due to the added thermal resistance coming from the dielectric chip mount. As a consequence, the output power would start to saturate at an even smaller current level than in the structure in Fig. 8(a). Another important point is that the output power level is also usually smaller than in the structure in Fig. 8(a). In general the dielectric chip mount has poor photon reflectivity compared to metallic chip mount, and thus the more output photons incident on the chip mount surface would be absorbed leading to reduced output power level.

Lastly, the LED lamp structure in Fig. 8(d) consists of the vertical chip attached on the dielectric chip mount. In this case, even though the sapphire substrate is eliminated in the heat dissipation path, the overall thermal resistance form the junction to the bottom of the chip mount would be still very large. Thus, the output power, which is already compromised as result of the degraded photon extraction efficiency in the chosen vertical structure, would start to saturate at almost the same current level as in the structure in Fig. 8(c). Thus, the power-crossover point would in general not be observed between the two structures.

5. About Chip-On-Board Packaging

In real applications of LED lamps, they would be installed

often on a special printed circuit board (PCB) called the metal core printed circuit board (MCPCB). The MCPCB with improved thermal characteristics would help not only increase per-unit output power but also stabilize the various LED output characteristics, such as the spectrum distribution and reliability. Fig. 10(a) shows a unit LED lamp put on the MCPCB. Considering only the thermal characteristics, however, it would be much more helpful to install LED chips directly on PCB (COB: chip on board) as shown in Fig. 10(b). Since the PCBs on which LED chips are installed are usually much thicker and wider than the chip mounts in unit LED lamps and thus have much larger heat capacity compared to the chip mount. Thus the direct attachment of the LED chip on PCBs would lead to drastically improved output power and stabilized output characteristics.

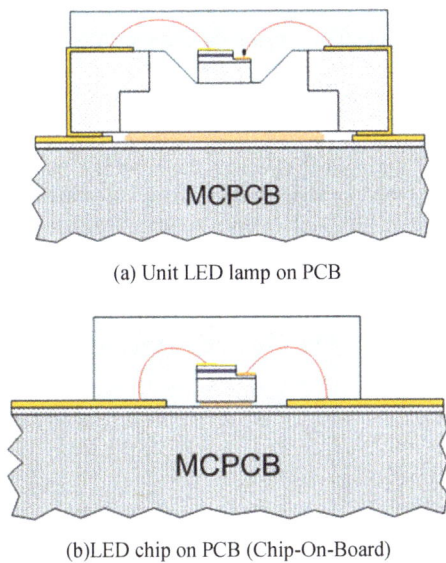

(a) Unit LED lamp on PCB

(b)LED chip on PCB (Chip-On-Board)

Fig. 10. *Two different approaches to implement LED lamps on PCB.*

It is noted, however, that the COB scheme may not be practical enough to be adopted extensively in the LED industry for various reasons. First the PCB on which LED chips are installed, possibly with other electrical or mechanical components, will have a shape, size, or design that may in general be widely different depending on the particular application. As a consequence, it would be extremely difficult or expensive to automatize, on this type of nonstandard PCBs, the various LED lamp manufacturing processes, such as chip attachment and encapsulation with molding compound. Similarly the testing of the each LED lamp for the important output characteristics, such as the output power, beam pattern, spectrum, and reliability cannot be easily automatized on the nonstandard PCB. Furthermore, if any of the LED lamps on the PCB fails to meet any of the various specifications required in the particular application, it would be very difficult or nearly impossible to repair or replace the failed one that has been molded strongly on the PCB.

On the other hand, in the case of the unit LED lamp installed on the PCB as shown in Fig. 10(a), all the manufacturing processes would be performed on standard lead frames, whose shape, size, or design would be determined independently of the particular applications, and would easily be automatized. The resultant uniform unit LED lamps of the same shape and design would then be tested using the standard and automatized testing procedures. Lastly only the unit LED lamps meeting all the required specifications would be installed on the PCB and be replaced quite easily, if any problem occurs after installment.

6. Conclusion

(a) Chip mount without isolating inter-walls

(b) Chip mount with isolating inter-walls

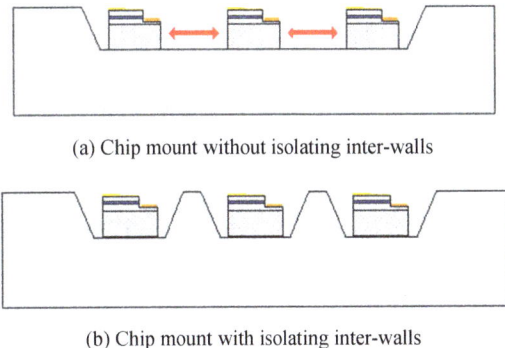

Fig. 11. *Multi-Chip integration schemes.*

Using Monte Carlo photon simulation method, the performance of the regular chip, vertical chip, and large area chip are analyzed and compared in terms of the photon extraction efficiency. In addition, the Epi-Up and Epi-Down chip-mounting schemes are analyzed and compared in terms of both the photon extraction efficiency and the difficulties of implementation. Based on the results of the analysis, we have established the basic guidelines for the LED lamp package design as follows:

- LED chips with chip area considerably larger than about $300\mu m \times 300\mu m$ have the photon extraction efficiency seriously degraded. Increasing the driving current by more than proportionally to increased chip area in order to compensate degraded photon extraction efficiency, would inevitably lead to junction temperature rise, which in turn would tend to decrease even the internal quantum efficiency too. Thus, the large area chips are in general not desired, except for special applications requiring large output power for a short duration of time.
- The approach to increase the output power or photon flux density by integrating multiple chips in one package as shown in Fig. 11(a) would also suffer from the degradation of the photon extraction efficiency as a result of the significant optical coupling between neighboring chips. In order to eliminate the in-between optical coupling, the chip mounts with isolating inter-walls as shown in Fig. 11(b) should be used.
- The regular chip has higher photon extraction efficiency than the vertical chip, as a compensation for the increased thermal resistance. It is always preferred over the vertical chip in applications requiring low output power that is to be achieved with minimal heating.
- The vertical chip has a much lower thermal resistance than the regular chip, as a compensation for the

degraded photon extraction efficiency. In order to fully exploit its thermal potential for increasing the driving current and thereby output power, it should be mounted on metallic chip mounts with improved heat sinking capacity.

- Dielectric chip mounts with a larger thermal resistance also have usually a poor photon reflectivity leading to significant optical loss, and therefore they would not in general be suited for applications requiring large output power. Especially it is not very sensible to attach the vertical chip on the dielectric chip mount since the approach leads to both poor efficiency and poor thermal behavior.

- The Epi-down chip mounting scheme requires a chip attachment process that is so delicate that it can hardly be adopted in the LED industry. Furthermore, it is not preferred over the more conventional Epi-up scheme even in terms of the photon extraction efficiency, when the LED chip has a smaller area than about $300 \mu m \times 300 \mu m$.

- The COB scheme, despite the advantages of increasing output power and better stabilizing output characteristics, cannot be used extensively in the LED industry due to the difficulties both in implementing it and in replacing the failed LED lamps.

References

[1] Y. Li, W. Zhao, Y. Xia, M. Zhu, J. Senawiratne, T. Detchprohm, E. F. Shubert, and C. Wetzel, "Temperature dependence of the quantum efficiency in green light emitting diode dies", Phys. Stat. Sol. (c) vol. 4, pp. 2784-2787, 2007.

[2] T. Mukai, M. Yamad, and S. Nakmura,, "InGaN-based uv/blue/green/amber/red LEDs", Proc. SPIE, vol. 3621, pp. 2-14, San Jose, January 1999.

[3] S. J. Lee, "Analysis of light-emitting diodes by Monte Carlo photon simulation", Appl. Opt., vol. 40, pp. 1427-1437, 2001.

[4] S. J. Lee, "Study of photon extraction efficiency in InGaN light-emitting diodes depending on chip structures and chip-mount schemes", SPIE. Opt. Eng., vol. 45, pp. 1/14601-14/14601, 2006.

[5] S. J. Lee, "Electrode design for InGaA/Sapphire LED's based on multiple thin ohmic metal patches", Proc. SPIE, vol. 5530, pp. 339-346, Denver, Aug. 2004.

[6] D. W. Hong, J. K. Yoo, J. M. Kim, M. J. Yoon, and S. J. Lee, "Analysis of the optical effect of the substrate removal and chip-mount type on light output power characteristics in InGaN/Sapphire LEDs," Hankook Kwanghak Hoeji (Korean J. of Opt. Photon.) vol 19, pp. 381-385, 2008.

22

Knowledge on Household Biodegradable Waste Management in Bangalore City

Asha Jyothi U. H.[1, *], Mamatha B.[1], H. S. Surendra[2]

[1]Department of Resource Management, Smt. V.H.D. Central Institute of Home Science, Bangalore University, Bangalore, India
[2]Department of Statistics, University of Agricultural Sciences, GKVK, Bangalore, India

Email address:
sa_uh@yahoo.com (Asha J. U. H.)

Abstract: Waste is not seen as a resource that can be refined or recycled and thereby generate wealth. Instead, it is often treated as the evil leftover that needs to be eliminated. This indiscriminate disposal with little concern leads to many health and environment problems. The objectives of the study were to elicit information on quantity and composition of household waste generated by households in Bengaluru city, to identify the influencing factors on waste generation, to elicit information on the knowledge of waste management at household level from selected homemakers. The sample households for the study were identified through multistage selection. 20 households each from the 40 wards identified were selected randomly for the study. Thus, 800 households were selected for the baseline data collection of the study with the help of questionnaire. Intervention programme was conducted for a period of one month to 80 homemakers through posters, lectures, power point presentation and group discussion regarding role of individuals in waste management. Education has been known to be an empowering tool for people at both the household and society levels. Intervention programme increased the knowledge of the homemakers from moderate to adequate level. Adequate knowledge on the influence of improper waste disposal may encourage people to adopt positive waste management practices.

Keywords: Biodegradable Waste, Waste Management, Knowledge, Demographic Characteristics, Association, Intervention Programme

1. Introduction

The status of development of a country may be categorized in several ways. With respect to its impact on solid waste management, the status of development is categorized on the basic management adapted to the nature and quantities of waste generated and to the availability of technology for handling and processing it UNEP, (2005).

Indian cities will be the locus and engine of economic growth over the next two decades, and the realization of an ambitious goal of 9 percent–10 percent growth in GDP depend fundamentally on making Indian cities much more liveable, inclusive, bankable, and competitive. The income level of the people is the most important factor for waste generation rates (Medina, 2010; World Bank, 2012). Suggested other factors of importance for the waste generation rate is the degree of industrialisation, public habits, local climate and level of urbanization (World Bank, 2012). Solid waste generation is suggested by UNEP to reflect the lives of people and the activities in the country. With that perspective the waste generation would be a combined function of the living standards of the inhabitants and the region's natural resources (United Nations Environment Program, 2005). Our city is littered with uncollected solid waste and no public place or street is free of litter. Though much recycling takes place by rag pickers and waste collectors, a lot is left to be disposed off. To keep cities clean, citizen involvement is essential to sort waste at source and minimize waste that needs to be collected and disposed. Programme should be implemented to obtain citizens' cooperation planning commission (2007). Whenever the community has been involved from planning stage, the programme has always become sustainable. While our programme have elaborate guidelines for community involvement, it is obvious that field-level adoption is far from satisfactory planning commission (2007).

The degree of community sensitization and public awareness is low. There is no system of segregation of organic, inorganic and recyclable wastes at household level. Hence, this study was taken up to create community awareness about the likely imperilment of poor waste management and the rudiments of handling the waste through segregating of biodegradable waste material like kitchen & garden waste from other waste generated in the household. This will help in promoting effective management of solid waste generated through proper practices of storing in a separate bag or a bin installed at their respective houses. Also the biodegradable fraction of the waste could be recycled through composting.

2. Objectives

The objectives of the study were to elicit information on quantity and composition of household waste generated by households in Bengaluru city, to identify the influencing factors on waste generation, to elicit information on the knowledge of waste management at household level from selected homemakers.

3. Methodology

The sample households for the study were identified through multistage selection. First four zones namely east, west, south and Bommanahalli were selected from the total of eight zones of Bengaluru. 20 households each from the 40 wards identified were selected randomly for the study. Thus, 800 households were selected for the baseline data collection. Questionnaire was the tool used to collect the required information. A handout on various methods of waste management at household level was developed and handed over to the participating respondents. Intervention Programme for a period of one month was conducted on 10 percent of the total sample comprising of 80 homemakers through Posters, Lectures, PowerPoint presentation and group Discussion regarding role of individuals in waste management

4. Results and Discussion

The major portion of the municipal solid waste of Bengaluru city consists of organic or biodegradable waste (60%). This was followed by recyclable waste in the form of plastics (14%), and paper (12%). Small quantities of other recyclable waste generated are glass, metal, card board, rubber, bio medical waste and miscellaneous. The Biodegradable waste which when mixed with other types of neither waste neither decomposes completely nor can the recyclables can be recycled as it is contaminated with organic waste.

Waste management is a complex process that requires a lot of information from various sources such as factors on waste generation and waste quantity forecasts. It was observed that 35.9 percent of the homemakers are in the age group of 26 – 35 years while 26 percent of them are in the age group of 36 - 45 years (Table 1a).

Table 1a. Socio-Demographic Characteristics of the Homemakers (N = 800)

Category	Respondents	
	Number	Percent
Age group (years)		
16-25	167	20.9
26-35	287	35.9
36-45	209	26.1
46+	137	17.1
Educational level		
Up to 5th Std	80	10.0
6-10th Std	182	22.7
PUC/Diploma	201	25.1
Graduate	249	31.1
PG/Professional	88	11.1
Number of Children		
None	183	22.9
One	197	24.6
Two	297	37.1
Three	123	15.4
Occupational Status		
Government	30	3.7
Private	200	25.0
Self Employed	75	9.4
Professional	31	3.9
Homemaker	464	58.0

Source: Field Study

Table 1b. Socio Demographic Characteristics of the Homemakers (N = 800)

Category	Respondents	
	Number	Percent
Household Size (Members)		
2 - 3	206	25.7
4 – 5	520	65.0
6+	74	9.3
Family Income/Month		
Rs. 2,001 – 5,000	201	25.1
Rs. 5,001 – 15,000	236	29.5
Rs. 15,001 – 25,000	169	21.11
Above 25,001	194	24.3
Type of Family		
Nuclear	592	74.0
Joint	156	19.5
Extended	52	6.5
Type of House		
Independent house	290	36.2
Compound House	180	22. 5
Row	131	16.4
Storied	123	15.4
Apartment	76	9.5
Type of Ownership		
Own	418	52.2
Rented	253	31.6
Leased	107	13.4
Quarters	22	2.8

Source: Field Study.

With regards to the educational qualification of the samples, all the respondents had some form of formal education. Occupational status of the surveyed families revealed that in more than half of the households, the women were full time homemakers. The numbers of children per household in 37.1 percent of the families were two children and one child in 24.6 percent of the households.

When operations related to promotion of waste

management systems are considered it is observed that generation of waste and planning was found to be influenced by different factor. The household size was found to be relatively small across the sample with 4 -5 members in 65 percent of the households followed by 2 -3 members in 25.7 percent of the households (Table 1b).

The income of the families ranged from Rs 2,000 to above Rs 25,000. The sample households were almost equally distributed between the different income groups. It was revealed that 36.2 percent of the families were residing in independent houses. With regards to ownership of the house, it was observed that a little more than half of the surveyed households lived in own houses followed by tenants (31.6%).

It was found that, kitchen waste mostly consisting of biodegradable waste like vegetable peels, spoiled food and fruits, and food remains after consumption, are generated daily in 90.7 percent of the households (table 2). Narayana (2009), observed that unlike in western countries, the solid waste of Asian cities is often comprised of 70-80% organic matter, dirt and dust. Sivakumar (2010) observed that the food waste is usually the predominant component in the waste stream due to the habit of fresh food consumption and composition of all

other types of waste are low in all households

Table 2. *Type of Waste Generated in the House (N = 800)*

No	Type of Waste @	Daily	Monthly	Occasional	Total
1	Kitchen	90.7	2.0	7.3	100
2	Plastic	46.0	26.0	28.0	100
3	Garden	13.8	34.9	51.3	100
4	Metal	5.0	21.8	73.2	100
5	Tins	4.5	21.9	73.6	100
6	Cans	7.9	19.6	72.5	100
7	Glass	11.0	30.5	58.5	100
8	Ceramics	6.1	15.8	78.1	100
9	Paper	40.9	30.8	28.3	100
10	Books	19.5	29.4	51.1	100
11	Newspaper	31.3	35.6	33.1	100
12	Textiles	7.5	23.8	68.7	100
13	Electronic Items	5.4	12.5	82.1	100
14	Others	6.3	5.9	87.8	100

Source: Field Study, @ Multiple Responses.

Table 3. *Source of Knowledge on Segregation of Waste (N = 800).*

Type of Media @	Full Extent	Partial Extent	Not at all	Total	Average	Preferential Ranking
Electronic media	50.4	40.4	9.2	100	70.6	1
Print media	23.6	52.1	24.3	100	49.7	8
Meeting/Lectures/Talks	23.4	52.9	23.7	100	49.9	7
Family/Relatives/Members	35.3	52.5	12.2	100	61.6	5
Friends/Neighbours	40.9	46.6	12.5	100	64.2	4
Self Motivation	46.0	48.8	5.2	100	70.4	2
Health Personnel	47.9	38.5	13.6	100	67.2	3
Others	16.6	70.1	13.3	100	51.7	6

Source: Field Study, @ Multiple Responses.

It was found from table 3 that, 70.6 percent of the households got information from electronic media in the form of internet, radio and audio tapes played by BBMP waste carrier vehicles. 50.4 percent of the households felt that the information given is complete enough for them to understand about waste segregation and its uses. The results corroborates with the findings of Afroz, Hanaki and Tuddin, (2010) that the majority obtained their knowledge about recycling from newspaper and television

Table 4. *Knowledge about Waste and its Disposal (N = 800).*

Aspects @	Response	Yes	No
Knowledge about Waste disposal	Waste generation can be reduced	17.4	82.3
	Willing to know about household waste management	27.1	72.9
	Want to learn composting waste at home	22.3	77.7
	Waste can cause Environmental Problems	10.3	89.7
	Waste can cause health problems	13.1	86.9
Awareness on how BBMP disposes waste	Landfills	51.4	48.6
	Incineration	77.0	23.0
	Composting	68.3	31.7
	Others	86.0	14.0

Source: Field Study, @ Multiple Responses.

Majority (89.7%) of the households did not know that waste can cause environmental problems (table 4). Negative response was obtained from 86.9 percent of the households with regards to health problems caused due to improper waste disposal techniques. This finding contradicts with Yoada, Chirawurah and Adongo (2014) study that 83% of the respondents were aware that improper waste management contributes to disease causation like malaria and diarrhoea.

Table 5a. Association between Demographic characteristics and Knowledge level on Waste Management (N= 800).

| Characteristics | Category | Sample (n) | Knowledge level | | | | χ^2 Value |
| | | | Moderate | | Adequate | | |
			N	%	N	%	
Age group (years)	16-25	167	63	37.7	104	62.3	4.76 NS
	26-35	287	128	44.6	159	55.4	
	36-45	209	74	35.4	135	64.6	
	46+	137	56	40.9	81	59.1	
Educational level	Up to 5th Std	80	45	56.3	35	43.7	45.55*
	6-10th Std	182	98	53.9	84	46.1	
	PUC/Diploma	201	73	36.3	128	63.7	
	Graduate	249	65	26.1	184	73.9	
	PG/Professional	88	40	45.5	48	54.5	
Occupational status	Government	30	10	33.3	20	66.7	2.40 NS
	Private	200	75	37.5	125	62.5	
	Self employed	75	29	38.7	46	61.3	
	Professional	31	15	48.4	16	51.6	
	Home maker	464	192	41.4	272	58.6	
Number of Children	None	183	64	35.0	119	65.0	10.64*
	One	197	76	38.6	121	61.4	
	Two	297	116	39.1	181	60.9	
	Three	123	65	52.9	58	47.1	
Combined		800	321	40.1	479	69.9	

Source: Field Study, * Significant at 5% Level, NS : Non-Significant.

It can be inferred from table 5a that whatever the age and occupation of the homemakers knowledge does not have any effect on the waste management at household level. This findings to certain extends negates the findings of Nguyen (2010) were Negative correlation was found between household size and positive correlation between Age, education, family income and number of children in the household.

It was found that education and number of children has statistically significant influence on knowledge associated with waste management. This finding corresponds with the finding of Samuel (2006), found that level of education had statistical significant influence on the knowledge of environmental sanitation. Jatau (2013) implied that level of education has statistically significant influence on the knowledge of improper waste management on health. Otitoju (2014), disclose education is a powerful tool that should be used towards building a more sustainable society. The way humans respond and co-operate on waste management issues is influenced by their education. Individuals need to be given the necessary knowledge in the scheme in order to ensure maximum participation.

Table 5b. Association between Demographic characteristics and Knowledge level on Waste Management (N= 800).

| Characteristics | Category | Sample (n) | Knowledge level | | | | χ^2 Value |
| | | | Moderate | | Adequate | | |
			N	%	N	%	
Household size (members)	2-3	206	87	42.2	119	57.8	3.53 NS
	4-5	520	198	38.1	322	61.9	
	6+	74	36	48.7	38	51.3	
Family Income/ month	Rs.2,001-5,000	201	108	53.7	93	46.3	25.49*
	Rs.5,001-15,000	236	96	40.7	140	59.3	
	Rs.15,001-25,000	169	51	30.2	118	69.8	
	Above Rs.25,000	194	66	34.0	128	66.0	
Type of Family	Nuclear	592	221	37.3	371	62.7	11.24*
	Joint	156	81	51.9	75	48.1	
	Extended	52	19	36.5	33	63.5	
Type of House	Apartment	76	33	43.4	43	56.6	4.58 NS
	Storied	123	46	37.4	77	62.6	
	Row	131	58	44.3	73	55.7	
	Compound house	180	79	43.9	101	56.1	
	Independent	290	105	36.2	185	63.8	
Type of Ownership	Own	418	170	40.7	248	59.3	14.03*
	Rented	253	93	36.8	160	63.2	
	Leased	107	41	38.3	66	61.7	
	Quarters	22	17	77.3	5	22.7	
Combined		800	321	40.1	479	69.9	

Source: Field Study, * Significant at 5% Level, NS : Non-Significant.

It can be established that as the family income increases better the knowledge the homemakers will possess on waste management (table 5b). Ezebilo and Animasaun (2011) imply that respondents who have more money were more likely to pay more for private solid waste management services.

People who have more money often have more capacity to consume food that are packaged in quick to disposal containers are more likely to be affected when solid waste services are ineffective.

Table 6. *Knowledge level of Homemakers on Waste Management (N = 80).*

Knowledge Level	Category	Pre test		Post test		χ² Value
		Number	Percent	Number	Percent	
Inadequate	≤ 50 % Score	29	36.3	0	0.0	
Moderate	51-75 % Score	41	51.2	21	26.3	70.25**
Adequate	> 75 % Score	10	12.5	59	73.7	
Total		80	100.0	80	100.0	

Source: Intervention programme, ** Significant at 1% level, χ² (0.01, 2df) = 15.086.

Knowledge level of the homemakers before and after participation in the intervention programme was assessed (table 6). It can be observed that, in pre interventions, 36.3 percent of the homemakers were having score of less than 50% indicating that the homemakers possessed inadequate knowledge. Post intervention revealed that the knowledge level increased to moderate level (26.3%) and adequate knowledge (73.7%). Chi-square analysis revealed that in the post intervention; the homemakers had enhanced information on household waste management.

5. Conclusion

Various kinds of waste are generated by the households on a daily basis. It was found that kitchen waste comprising of biodegradable waste was generated every day in majority of the houses. Electronic media was the main source for disseminating knowledge on waste segregation followed by self motivation with a concern for environment. The overall message received through the different sources for waste segregation was reduction and recycling of waste. Age, education and occupation of the homemakers, household size, type of family and type of ownership had positive significant influence on the knowledge level of homemakers on waste management. Post intervention programme, the knowledge on household waste management increased from moderate to adequate level.

Acknowledgements

Grateful acknowledgement to University Grants Commission of India for providing study leave through Faculty development programme for completing the study.

References

[1] Afroz Rafia, Keisuke Hanaki and Rabbah Tuddin, (2010), The Role of Socio-Economic Factors on Household Waste Generation: A Study in a Waste Management Program in Dhaka City, Bangladesh, Research Journal of Applied Sciences, Volume: 5, Issue: 3, pp 183-190 http://www.medwelljournals.com/fulltext/?doi=rjasci.2010.183.190 dated 05.01.2015

[2] Ezebilo Eugene E and Emmanuel D. Animasaun (2011), Economic Valuation of Private Sector Waste Management Services, Journal of Sustainable Development Vol. 4, No. 4, Pp 38-46 http://www.ccsenet.org/journal/index.php/jsd/article/viewFile/10504/8227 dated 06.01.2015.

[3] Jatau Audu Andrew (2013), Knowledge, Attitudes and Practices Associated with Waste Management in Jos South Metropolis, Plateau State, Mediterranean Journal of Social Sciences, Vol 4 No 5, Pp 121 -127. http://mcser.org/journal/index.php/mjss/article/viewFile/667/690 dated 04.01.2015.

[4] Medina Martin, (2010), Solid Wastes, Poverty and the Environment in Developing Country Cities Challenges and Opportunities, Working Paper No. 2010/23, UNU-WIDER http://www.rrojasdatabank.info/2010-23.pdf dated 18.05.2014

[5] Narayana Tapan, (2009). Municipal Solid Waste Management in India: From Waste Disposal to Recovery of Resources. Waste Management, Journal of Waste Management, 29(3), PP-1163-1166. http://www.ncbi.nlm.nih.gov/pubmed/18829290 retrieved on 30.12.2014

[6] Nguyen, Phuc Thanh, (2010), A study on evaluation methodologies for household solid waste management toward a sustainable society in Vietnam, Journal of Vietnamese Environment. https://oa.slub-dresden.de/ejournals/jve/thesis/view/6 dated 09.09.2014

[7] Otitoju Tunmise A, (2014), Individual Attitude toward Recycling of Municipal Solid Waste in Lagos, Nigeria, American Journal of Engineering Research (AJER), Volume-03, Issue-07, pp-78-88. http://www.ajer.org/papers/v3%287%29/L0377888.pdf dated 05.01.2015

[8] Planning Commission of India, (2007), Report of The Steering Committee on Water Resources For Eleventh Five Year Plan (2007-2012).

[9] Top of Form. http://planningcommission.nic.in/aboutus/committee/strgrp11/str11_wtr.doc

[10] Samuel, E.S (2006), Environmental Sanitation Knowledge among Primary School pupils in Idemili North Local Government Area of Anambra State, Journal of Environmental Health (3) 1, PP 5 – 12.

[11] Sivakumar. K, and M. Sugirtharan, (2010), Impact of Family Income and Size on Per Capita Solid Waste Generation: A Case Study in Manmunai North Divisional Secretariat Division of Batticaloa, J Sci.Univ.Kelaniya 5 (2010): 13-23.file:///C:/Documents%20and%20Settings/Asha/My%20Documents/Downloads/4087-14603-1-PB.pdf dated 10.08.2014.

[12] UNEP, (2005), Solid Waste Management (Vol. I), United Nations Environment Programme International Environmental Technology Centre. CalRecovery, Inc.

[13] The World Bank, 2012. What a Waste - a global review of solid waste management, Washington: The World Bank, and How we classify countries. [Online] http://data.worldbank.org/about/country-classifications Accessed 06 08 2012.

[14] Yoada Ramatta Massa, Dennis Chirawurah and Philip Baba Adongo (2014), Domestic waste disposal practice and perceptions of private sector waste management in urban Accra, BMC Public Health 2014, 14, Pp 697http://www.biomedcentral.com/1471-2458/14/697 dated 04.01.2015.

23

Microcontroller Based Automatic Solar Tracking System with Mirror Booster

Protik Kumar Das[1], Mir Ahasan Habib[1], Mohammed Mynuddin[2]

[1]Dept of Mechanical Engineering, Rajshahi University of Engineering & Technology (RUET), Rajshahi, Bangladesh
[2]Dept of EEE, Atish Dipankar University of Science and Technology, Dhaka, Bangladesh

Email address:
protikdasruet09@gmail.com (P. K. Das), kanan_092086@yahoo.com (M. A. Habib), myn101eee@gmail.com (M. Mynuddin)

Abstract: This paper is designed solar tracking system with mirror booster using microcontroller. Solar energy is rapidly becoming an alternative means of electrical source all over the world. To make effective use of solar energy, its efficiency must be maximized. A feasible approach to maximizing the power output of solar array is by sun tracking. This paper deals with the design and construction of solar tracking system by using a stepper motor, gear motor, photo diode. Mirror is used as booster to maximize the efficiency. The whole frame will travel circularly and the mirror will travel from south to north and vice-versa. The prototype is considered around a programmed microcontroller which controls the system by communicating with sensors and motor driver based on movement of the sun. The performance and characteristics of the solar tracker are experimentally analyzed.

Keywords: Types of Renewable Energy, Types of Tracking Mechanism, Gear Motor, Maximize the Panel Efficiency

1. Introduction

Energy is a key to the advancement and prosperity of humans. With civilization development the energy consumption has increased steadily. To sustain the current human development more energy is required in near future. The primary solution is to burn more fossil fuels [1] .But burning more and more fossil fuels will cause environmental problems such as water pollution, Air pollution, soil contamination and most dangerously greenhouse gases. The increased use of fossil fuels has significantly increased greenhouse gas emission, particularly carbon dioxide, creating an enhanced greenhouse effect known as global warming. Fossil fuels required millions of year to generate and going to finish in very short time. However, the world supplies of fossil fuels which is our main source of electricity will start to run out from the years 2020 to 2060.The increasing demand for energy, the continuous reduction in existing sources of fossil fuels and the growing concern regarding environment pollution, have pushed mankind to explore new technologies for the production of electrical energy using clean, renewable sources, such as solar energy, wind energy, etc [2]. Renewable energy resources will be an increasingly important part of power generation in the new millennium. Besides assisting in the reduction of the emission of greenhouse gases, they add the much- needed flexibility to the energy resource mix by decreasing the dependence on fossil fuels. Among the renewable energy resources, solar energy is the most essential and prerequisite resource of sustainable energy because of its ubiquity, abundance, and sustainability. Regardless of the intermittency of sunlight, solar energy is widely available and completely free of cost. Recently, photovoltaic (PV) [3]-[6] system is well recognized and widely utilized to convert the solar energy for electric power applications. Among the non-conventional, renewable energy sources, solar energy affords great potential for conversion into electric power, able to ensure an important part of the electrical energy needs of the planet.

The sun is the prime source of energy, directly or indirectly, which is also the fuel for most renewable systems. Among all renewable systems, photovoltaic system is the one which has a great chance to replace the conventional energy resources. Solar panel directly converts solar radiation into electrical energy. Solar panel is mainly made from semiconductor materials. Si used as the major component of solar panels, which is maximum 24.5% efficient. Unless high efficient solar panels are invented, the only way to enhance the performance of a solar panel is to increase the intensity of light falling on it. Solar trackers are the most appropriate and proven technology to increase the efficiency of solar panels through keeping the panels aligned with the sun's position. Solar trackers get popularized around the world in recent days to harness solar energy in most efficient way. This is far more cost effective solution than purchasing additional solar panels [7]-[8]. In this paper the design methodology of a microcontroller based simple and easily programmed automatic solar tracker is presented. A prototype of automatic solar tracker ensures feasibility of this design methodology.

Solar tracking is necessary for most of the solar systems to collect maximum amount solar radiation. Concentrators require a high degree of accuracy to ensure that the reflected sunlight is directed to the absorber, which is at the focal point of the reflector. Sun Tracker can help increase overall efficiency of a solar installation by over 40%. The applications of solar tracking system are given below: i) Satellite dish ii) Concentrating collectors (cylindrical parabolic collector, parabolic dish collector, power tower concentrating collector etc.), iii) PV cell and iv) Flat plate collector etc.

1.1. Renewable Energy

Any energy resource that is naturally regenerated over a short time scale and derived directly from the sun (such as thermal, photochemical, and photoelectric), indirectly from the sun (such as wind, hydropower, and photosynthetic energy stored in biomass), or from other natural movements and mechanisms of the environment (such as geothermal and tidal energy) is called renewable energy [9].

1.2. Types of Renewable Energy

i) Solar energy [10], ii) Hydroelectric power [11] iii) Biomass [12] iii) From hydrogen [13] iv) Geothermal energy [14] v) Wind power [15] vi) Ocean energy [16]-[17] vii) Waste renewable energy [18]-[19]

Solar energy: Most renewable energy comes either directly or indirectly from the sun. Sunlight, or solar energy, can be used directly for heating and lighting homes and other buildings, for generating electricity, and for hot water heating, solar cooling, and a variety of commercial and industrial uses. Solar energy has many advantages: a) Need no fuel b) Has no moving parts to wear out c)Non-polluting & quick responding d) Adaptable for on-site installation e) Easy maintenance f) Can be integrated with other renewable energy sources g) Simple & efficient and so on.

1.3. Application of Renewable Energy

Renewable energy is of many uses and it can support small as well as large applications. Renewable energy from wind, sun and geothermal is used to produce electricity and heat for use. The solar power plants are used to generate electricity and steam for industrial projects. The energy form the geothermal heat is used to heat radiators in the homes. Thus the renewable energy sources can viably help users to their

heat homes. Some other applications of renewable energy sources include heating space, to generate electricity, solar oven, ventilation, day lighting, space cooling, water heating, mechanical energy to cut woods and grinding grains. The renewable energy sources and the technologies associated with them are equally important to households and industry. Renewable energy sources are of many uses to domestic users.

1.4. Solar Energy

Solar energy is the term used for the heat and light which the sunlight contains. Sunlight reaches to earth in the form of photons. Photons are energy packets that contain light in it. Solar energy is considered as a renewable energy source because it does not destroy our eco system and is present naturally in the environment.

1.5. Electricity from Solar Power

Solar power is the form of energy that helps in generation of electricity from the sunrays. There are many methods to generate electric current using sunlight but the most common methods are photovoltaic and concentrated solar power. Photovoltaic contains an array of solar cells which are pressed in solar panels. These solar panels are protected and framed by a glass sheet. This sheet does not allow any impurities to pass in. Hence only sunrays can make their way in. These solar panels are made up of conductive materials like impure silicon and copper indium mostly. These conductors help and support the flow of electrons, thus the heat present on the solar panels is able to generate direct electric current. This electric current cannot support the electrical devices. Therefore it is converted to alternative current by using inverter and battery. Photovoltaic energy is growing rapidly and it is so far the only rapidly progressing renewable energy technology. Concentrating solar power (CSP) [20] systems work on the principle of converging the sunlight form many kilometers to single focal point. The concentrated energy stored in this form is converted to thermal energy which is utilized to support photovoltaic cell. The solar energy stored by CSP is also helpful in running steam turbines.

1.6. Types of Solar Energy System

Solar energy systems are of many types dependent upon their use. These include concentrating solar power systems, parabolic dishes, sunlight Stirling dishes, updraft towers, photovoltaic and solar ponds. All these systems are used to generate electricity in an economical and environment friendly manner. Photovoltaic solar panels are also used to heat water. The steam produced in this process is used for running the industrial machinery. Concentrating solar systems uses lenses, tracking system and glass to convert the sunlight into single beam. The heat stored in this way is used to support the conventional power houses. Parabolic dishes are another system which is used save solar energy. These parabolic antennas work on the mechanism of tracking through single axis. Stirling solar dish systems contain a parabolic reflector that gathers the light to single focal point onto as a receiver. The updraft solar systems work by running wind turbines connected to it. These wind turbines produce electricity which is stored in the form of direct current in collectors at one end. Solar ponds were constituted to perform an experiment over the layers of salt present in red sea. The heat from the sun is stored into the lower bed of salt. This heat is then used to heat water in collectors.

1.7. The Development in Solar Power System

The development in solar power systems took place when human beings realized that the bio fuels like coal and oil will not be sufficient to produce electricity in future. The development of solar energy systems faced many hindrances because of the unavailability of solar heating system and its equipment's. It was in 1973 again when the fuel and oil crisis reached to the peak. The desire to find alternative energy sources again became the need of time. Hence form that time developed countries like USA star deploying solar panels on barren land to cater the expected deficiency of bio fuels in future. Soon after wards in 1985 when the prices of oil feel again the development and attention toward solar systems declined once again. It is now again that we need renewable energy sources to fight the problem of global warming and pollution. Therefore sunlight is the only energy source which provides electricity for industrial consumption without producing any harmful waste. However the commercial production of solar cells has resulted in its deployment even in residents. The governments throughout the world are giving great incentives to industry and individuals to switch to this major energy sources. Solar energy systems are great way to utilize abundant solar energy to support unlimited commercial and domestic applications.

1.8. Advantages of Solar Power Energy

The importance and advantages of solar energy and its uses were not even declined in prehistoric times. Sunlight helps plants to generate food for them during the process of photosynthesis. Solar power energy free of cost energy source helped people store their food for longer when refrigerators were not in use, people used it for killing germs in clothes and most importantly this useful star provides us vitamin D to support healthy growth of our bones. Nowadays the practices of using solar energy are changing as it has been identified as an inexpensive way to produce electricity.

1.9. Storage of Solar Energy

Producing electricity from sun is also termed as a renewable solar energy source. Renewable energy source refers to all those energy sources other than traditional bio fuels. Solar energy is of many uses. When the process of generating electricity was under consideration, the major challenge was to store solar energy to be used later in night, storms and rains. Hence useful devices like solar panels, solar heating systems and solar cells supported this immense

challenge. Solar energy can be used unless we have sun. Hence all the solar power applications can help us utilize solar energy rays to produce electricity for supporting personal, domestic and industrial applications. Now it is possible to store solar energy in batteries which are attached to the solar powers panels. The electricity generated through solar energy process can be used with or without traditional utility grids. Another experiment carried at the red sea demonstrated that solar energy can also be stored in the beds of salt to support solar heating systems.

2. Photovoltaic Cell and Solar Tracking System

2.1. Solar Panel

A solar panel is a set of solar photovoltaic modules electrically connected and mounted on a supporting structure. A photovoltaic module is a packaged, connected assembly of solar cells. The solar module can be used as a component of a larger photovoltaic system to generate and supply electricity in commercial and residential applications. Each module is rated by its DC output power under standard test Condit Solar modules use light energy (photons) from the sun to generate electricity through the photovoltaic effect. The majority of modules use wafer-based crystalline silicon cells or thin-film cells based on cadmium telluride or silicon. The structural (load carrying) member of a module can either be the top layer or the back layer. Cells must also be protected from mechanical damage and moisture. Most solar modules are rigid, but semi-flexible ones are available, based on thin-film cells.

Fig. 1. A Typical Solar Panel.

2.2. Principle of Photovoltaic Cell

Photovoltaic (PV) [21] system is well recognized and widely utilized to convert the solar energy for electric power applications. It can generate direct current (DC) electricity without environmental impact and emission by way of solar radiation. The DC power is converted to AC power with an inverter, to power local loads or fed back to the utility. Being a semiconductor device, the PV systems are suitable for most operation at a lower maintenance costs.

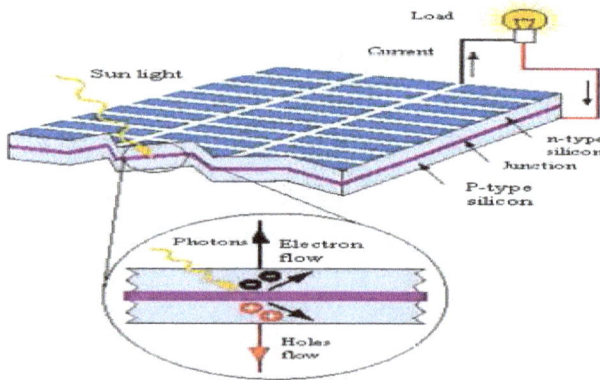

Fig. 2. Principle of photovoltaic cell.

2.3. Solar Tracker

Solar tracker is rack for photovoltaic modules that move to point at or near the sun throughout the day. Trackers add to the efficiency of the system, reducing its size and the cost per KWH.A solar tracker is a generic term used to describe devices that orient various payloads toward the sun. Payloads can be photovoltaic panels, reflectors, lenses or other optical devices.

2.4. Tracking Technique

There are several forms of tracking currently available, these vary mainly in the used are fixed control algorithms and dynamic tracking. The inherent difference between the two methods is the manner in which the path of the sun is determined. In the fixed control algorithm systems, the path of the sun is determined by referencing an algorithm that calculates the position of the sun for each time period. That is, the control system does not actively find the sun's position but works it out given the current time, day, month, and year. The dynamic tracking system, on the other hand, actively searches for the sun's position at any time of day or night. Control system is common for both tracking techniques. This system consists of some method of direction control, such as DC motors, stepper motors, and servo motors, which are directed by a control circuit, either digital or analog.

2.5. Types of Solar Collector

Different types of solar collector and their location (latitude) require different types of tracking mechanism. Solar collectors may be: i) non-concentrating flat-panels, usually photovoltaic or hot-water, ii) Concentrating systems, of a variety of types.

Solar collector mounting systems may be fixed (manually aligned) or tracking. Tracking systems may be configured as: i) Fixed collector / moving mirror ii) Moving collector

Fixed mount: Domestic and small-scale commercial photovoltaic and hot-water panels are usually fixed, often flush-mounted on an appropriately facing pitched roof. Advantages of fixed mount systems (i.e. factors tending to indicate against trackers) include the following :i) Mechanical simplicity: and hence lower installation and ongoing maintenance costs ii) Wind-loading: it is easier and

cheaper to provision a sturdy mount; all mounts other than fixed flush-mounted panels must be carefully designed having regard to their wind loading due to their greater exposure iii) Indirect light: approximately 10% of the incident solar radiation is diffuse light, available at any angle of misalignment with the direct sun and iv) Tolerance to misalignment: effective collection area for a flat-panel is relatively insensitive to quite high levels of misalignment with the sun – see table and diagram at Accuracy Requirements section below – for example even a 25° misalignment reduces the direct solar energy collected by less than 10%. Fixed mounts are usually used in conjunction with non-concentrating systems, however an important class of non-tracking concentrating collectors, of particular value in the 3rd world, are portable solar cookers. These utilize relatively low levels of concentration, typically around 2 to 8 Suns and are manually.

Fixed collector or moving mirror: Many collectors cannot be moved, for example high-temperature collectors where the energy is recovered as hot liquid or gas (e.g. steam). Other examples include direct heating and lighting of buildings and fixed in-built solar cookers, such as Schaffer reflectors. In such cases it is necessary to employ a moving mirror so that, regardless of where the Sun is positioned in the sky, the Sun's rays are redirected onto the collector. Due to the complicated motion of the Sun across the sky, and the level of precision required to correctly aim the Sun's rays onto the target, a heliostat mirror generally employs a dual axis tracking system, with at least one axis mechanized. In different applications, mirrors may be flat or concave.

Non-Concentrating Photovoltaic (PV) Trackers: Photovoltaic panels accept both direct and diffuse light from the sky. The panels on a Standard Photovoltaic Trackers always gather the available direct light. The tracking functionality in Standard Photovoltaic Trackers is used to minimize the angle of incidence between incoming light and the photovoltaic panel. This increases the amount of energy gathered from the direct component of the incoming light.

2.6. Concentrated Photovoltaic (CPV) Trackers

The optics in CPV modules accept the direct component of the incoming light and therefore must be oriented appropriately to maximize the energy collected. In low concentration applications a portion of the diffuse light from the sky can also be captured. The tracking functionality in CPV modules is used to orient the optics such that the incoming light is focused to a photovoltaic collector. CPV modules that concentrate in one dimension must be tracked normal to the sun in one axis. CPV modules that concentrate in two dimensions must be tracked normal to the sun in two axes.

2.7. Single Axis Trackers

The axis of rotation of single axis trackers is typically aligned along a true North meridian. It is possible to align them in any cardinal direction with advanced Single axis trackers have one degree of freedom that acts as an axis of rotation tracking algorithms. There are several common implementations of single axis trackers. These include horizontal single axis trackers (HSAT), vertical single axis trackers (VSAT), tilted single axis trackers (TSAT) and polar aligned single axis trackers (PSAT). The orientation of the module with respect to the tracker axis is important when modeling performance.

Fig. 3. Single axis tracker.

2.7.1. Horizontal Single Axis Tracker (HSAT)

The axis of rotation for horizontal single axis tracker is horizontal with respect to the ground. The posts at either end of the axis of rotation of a horizontal single axis tracker can be shared between trackers to lower the installation cost. Field layouts with horizontal single axis trackers are very flexible. The simple geometry means that keeping the entire axis of rotation parallel to one another is all that is required for appropriately positioning the trackers with respect to one another. Appropriate spacing can maximize the ratio of energy production to cost, this being dependent upon local terrain and shading conditions and the time-of-day value of the energy produced. Backtracking is one means of computing the disposition of panels. Horizontal Trackers typically have the face of the module oriented parallel to the axis of rotation. As a module tracks, it sweeps a cylinder that is rotationally symmetric around the axis of rotation. Several manufacturers can deliver single axis horizontal trackers. In these, a long horizontal tube is supported on bearings mounted upon pylons or frames. The axis of the tube is on a North-South line. Panels are mounted upon the tube, and the tube will rotate on its axis to track the apparent motion of the sun through the day.

Fig. 4. Horizontal Single Axis Tracker in California.

2.7.2. Vertical Single Axis Tracker (VSAT)

The axis of rotation for vertical single axis trackers is vertical with respect to the ground. These trackers rotate from East to West over the course of the day. Such trackers are more effective at high latitudes than are horizontal axis trackers. Field layouts must consider shading to avoid unnecessary energy losses and to optimize land utilization. Also optimization for dense packing is limited due to the nature of the shading over the course of a year. Vertical single axis trackers typically have the face of the module oriented at an angle with respect to the axis of rotation. As a module tracks, it sweeps a cone that is rotationally symmetric around the axis of rotation.

2.7.3. Tilted Single Axis Tracker (TSAT)

All trackers with axes of rotation between horizontal and vertical are considered tilted single axis trackers. Tracker tilt angles are often limited to reduce the wind profile and decrease the elevated end's height off the ground. Field layouts must consider shading to avoid unnecessary losses and to optimize land utilization. With backtracking, they can be packed without shading perpendicular to their axis of rotation at any density. However, the packing parallel to their axis of rotation is limited by the tilt angle and the latitude. Tilted single axis trackers typically have the face of the module oriented parallel to the axis of rotation. As a module tracks, it sweeps a cylinder that is rotationally symmetric around the axis of rotation.

Fig. 5. Tilted single axis tracker.

2.7.4. Polar Aligned Single Axis Trackers (PASAT)

One scientifically interesting variation of a tilted single axis tracker is a polar aligned single axis tracker (PASAT). In this particular implementation of a Tilted Single Axis Tracker the tilt angle is equal to the latitude of the installation. This aligns the tracker axis of rotation with the earth's axis of rotation. These are rarely deployed because of their high wind profile.

2.8. Dual Axis Trackers

Dual axis trackers [30] have two degrees of freedom that act as axes of rotation. These axes are typically normal to one another. The axis that is fixed with respect to the ground can be considered a primary axis. The axis that is referenced to the primary axis can be considered a secondary axis.

Fig. 6. Dual axis tracker.

There are several common implementations of dual axis trackers. They are classified by the orientation of their primary axes with respect to the ground. Two common implementations are tip-tilt dual axis trackers (TTDAT) and azimuth-altitude dual axis trackers (AADAT).The orientation of the module with respect to the tracker axis is important when modeling performance. Dual axis trackers typically have modules oriented parallel to the secondary axis of rotation. Dual axis trackers allow for optimum solar energy levels due to their ability to follow the sun vertically and horizontally. No matter where the sun is in the sky, dual axis trackers are able to angle themselves to be in direct contact with the sun.

2.8.1. Tip–Tilt Dual Axis Tracker (TTDAT)

A tip–tilt dual axis tracker has its primary axis horizontal to the ground. The secondary axis is then typically normal to the primary axis. The posts at either end of the primary axis of rotation of a tip–tilt dual axis tracker can be shared between trackers to lower installation costs. Field layouts with tip–tilt dual axis trackers are very flexible. The simple geometry means that keeping the axes of rotation parallel to one another is all that is required for appropriately positioning the trackers with respect to one another. In addition, with backtracking, they can be packed without shading at any density The axes of rotation of tip–tilt dual axis trackers are typically aligned either along a true North meridian or an east west line of latitude. It is possible to align them in any cardinal direction with advanced tracking algorithms. Manufacturers include Patriot Solar Group.

2.8.2. Azimuth-Altitude Dual Axis Tracker (AADAT)

An azimuth–altitude dual axis tracker has its primary axis vertical to the ground. The secondary axis is then typically normal to the primary axis. Field layouts must consider shading to avoid unnecessary energy losses and to optimize land utilization. Also optimization for dense packing is limited due to the nature of the shading over the course of a year.

This mount is used as a large telescope mount owing to its structure and dimensions. One axis is a vertical pivot shaft or horizontal ring mount that allows the device to be swung to a compass point. The second axis is a horizontal elevation pivot mounted upon the azimuth platform. By using combinations of the two axis, any location in the upward

hemisphere may be pointed. Such systems may be operated under computer control according to the expected solar orientation, or may use a tracking sensor to control motor drives that orient the panels toward the sun. This type of mount is also used to orient parabolic reflectors that mount a Sterling engine to produce electricity at the device.

Fig. 7. Azimuth-altitude dual axis tracker - 2 axis solar tracker, Toledo, Spain.

2.9. Tracker Type Selection

The selection of tracker type is dependent on many factors including installation size, electric rates, government incentives, land constraints, latitude, and local weather. Horizontal single axis trackers are typically used for large distributed generation projects and utility scale projects. The combination of energy improvement and lower product cost and lower installation complexity results in compelling economics in large deployments. In addition the strong afternoon performance is particularly desirable for large grid-tied photovoltaic systems so that production will match the peak demand time. Horizontal single axis trackers also add a substantial amount of productivity during the spring and summer seasons when the sun is high in the sky. The inherent robustness of their supporting structure and the simplicity of the mechanism also result in high reliability which keeps maintenance costs low. Since the panels are horizontal, they can be compactly placed on the axle tube without danger of self-shading and are also readily accessible for cleaning. A vertical axis tracker pivots only about a vertical axle, with the panels either vertical, at a fixed, adjustable, or tracked elevation angle. Such trackers with fixed or (seasonably) adjustable angles are suitable for high latitudes, where the apparent solar path is not especially high, but which leads to long days in summer, with the sun travelling through a long arc. Dual axis trackers are typically used in smaller residential installations and locations with very high government Feed in Tariffs.

2.10. Multi-Mirror Concentrating PV

This device uses multiple mirrors in a horizontal plane to reflect sunlight upward to a high temperature photovoltaic or other system requiring concentrated solar power. Structural problems and expense are greatly reduced since the mirrors are not significantly exposed to wind loads. Through the employment of a patented mechanism, only two drive systems are required for each device. Because of the configuration of the device it is especially suited for use on flat roofs and at lower latitudes. The units illustrated each produce approximately 200 peak DC watts. A multiple mirror reflective system combined with a central power tower is employed at the Sierra Sun Tower, located in Lancaster, California. This generation plant operated by e Solar is scheduled to begin operations on August 5, 2009. This system, which uses multiple heliostats in a north-south alignment, uses pre-fabricated parts and construction as a way of decreasing startup and operating costs.

2.11. Advantage of Solar Tracker

The main reason to use a solar tracker is to reduce the cost of the energy we want to capture. A tracker produces more power over a longer time than a stationary array with the same number of modules. This additional output or "gain" can be quantified as a percentage of the output of the stationary array. Gain varies significantly with latitude, climate, and the type of tracker you choose—as well as the orientation of a stationary installation in the same location. (The energy required to move the tracker is insignificant in these calculations.) Climate is the most important factor. The more sun and less clouds, moisture, haze, dust, and smog, the greater the gain provided by trackers. At higher latitudes gain will be increased due to the long arc of the summer sun.

3. Methodology

3.1. Project Design Methodology

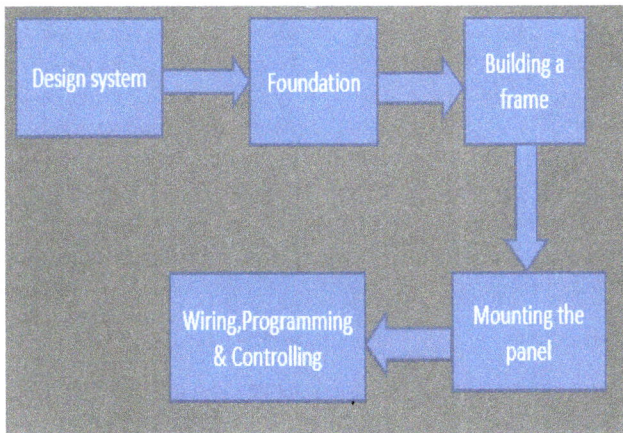

Fig. 8. Block diagram of project design methodology.

Design System:
Firstly to build a solar tracker designing model is a basic need. The solar panel is fastened to the frame, as well as the mirror to improve the design. The mirror will help to increase the fall of the sun rays upon the panel. The factors in which design criteria depends are:i) Dimension of the panel ii) Capacity of the panel iii)Selection of structural material iv) Selection of gear motor and v) Space available.

Foundations: For balancing the solar tracker a good foundation is a very important thing. Secondly basic

Foundation is needed to implement this project. The foundation should have adequate strength to hold the total system.

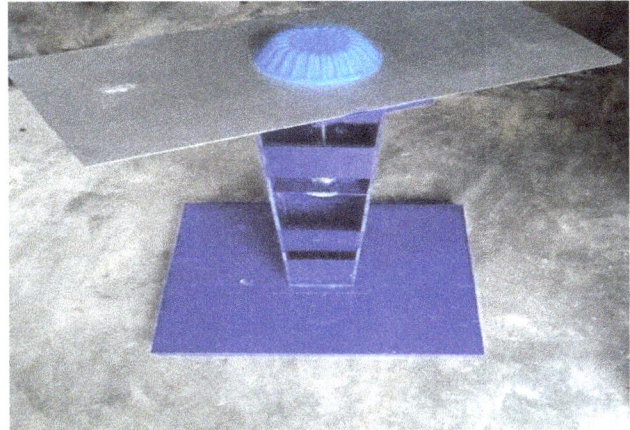

Fig. 9. Foundation of the solar tracker.

Building Frame: To set the solar system building a frame is also an important thing. The panel & mirror are stay in the frame.

Mounting the Solar Panel: Here to mount the solar panel, fasten the panel to the frame has done.

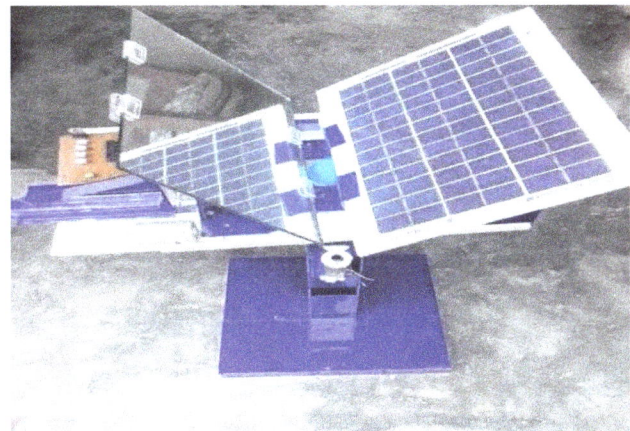

Fig .10. Mounting the panel on the frame.

Wire and Program Controller: Proper wiring system, to control the motor program is set & also to control the overall system.

3.2. Boosting the Solar Panel

Solar panels are a great way to make some green electricity for our home or workplace but they're kind of expensive and sometimes the wattage produced can be a bit disappointing. If we use a sun tracking system to keep our panels facing the sun we can considerably improve the watt yield but these are not cheap and on a small system they can add considerably to the cost. Here's a really cost effective and simple way to get 75% more power from any ordinary solar panel. Most of the time a solar panel is working well below peak power, on hazy days and when the sun is lower in the sky, early morning, and late afternoon for example. The light levels are just not high enough, so to boost the light level aligning a mirror will reflect more light onto solar panel. It worked really well and after a bit of experimentation it was found that placing a mirror at least twice the size of the solar panel on the ground in front of the panel could boost the output by as much as 75%.

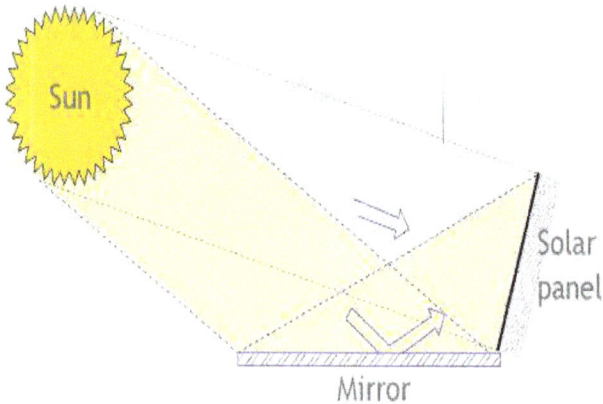

Fig. 11. Solar panel boosting with mirror [22].

Using a bigger mirror can reflect light onto panel over a longer period during the day so tracking the sun is not needed, just face the panel and mirror due south[23]. This is probably one of the cheapest and easiest ways to boost the power of a small solar panel, but this method does have some limitations. Using more mirrors to reflect more light onto the solar panel increase its power further but on a sunny summer's day the extra light can build up a lot of heat that may damage the panel. Placing mirrors either side of the panel to reflect doesn't work well because as the sun moves west it will cast a shadow across the panel. The only place that the mirror won't cast a shadow at any time in the day is on the ground in front of the solar panel. On a dull day the mirror doesn't give much of a power boost at all.

4. Microcontroller

4.1. Microcontroller

A microcontroller is a small computer on a single integrated circuit containing a processor core, memory, and programmable input/output peripherals. The Program memory in the form of NOR flash or OTP ROM is also often included on chip, as well as a typically small amount of RAM. Microcontrollers are designed for embedded applications, in contrast to the microprocessors used in personal computers or other general purpose applications. Microcontrollers [31] are used in automatically controlled products and devices, such as automobile engine control systems, implantable medical devices, remote controls, office machines, appliances, power tools, toys and other embedded systems. By reducing the size and cost compared to a design that uses a separate microprocessor, memory, and input/output devices, microcontrollers make it economical to digitally control even more devices and processes. Mixed

signal microcontrollers are common, integrating analog components needed to control non-digital electronic systems [25]. ATmega 8 microcontroller [23]-[25] is used for controlling the direction of DC Gear motor.

Advantage used of Microcontroller

Microcontrollers, as stated, are inexpensive computers .The microcontroller has ability to store and run a unique program makes it extremely versatile. For instance one can program a microcontroller to makes decisions based on predetermined situations and selections .The microcontroller has ability to perform math and logic functions allows it to mimic sophisticated logic and electronic circuit.

4.2. Photodiode Sensor

The construction of the Photodiode light sensor is similar to that of a conventional PN junction diode except that the diodes outer casing is either transparent or has a clear lens to focus the light onto the PN junction for increased sensitivity. The junction will respond to light particularly longer wavelengths such as red and infra-red rather than visible light. This characteristic can be a problem for diodes with transparent or glass bead bodies such as the 1N4148 signal diode. LED's can also be used as photodiodes as they can both emit and detect light from their junction. All PN-junctions are light sensitive and can be used in a photo-conductive unbiased voltage mode with the PN-junction of the photodiode always "Reverse Biased" so that only the diodes leakage or dark current can flow.

The current-voltage characteristic (I/V Curves) of a photodiode with no light on its junction (dark mode) is very similar to a normal signal or rectifying diode. When the photodiode is forward biased, there is an exponential increase in the current, the same as for a normal diode. When a reverse bias is applied, a small reverse saturation current appears which causes an increase of the depletion region, which is the sensitive part of the junction. Photodiodes can also be connected in a current mode using a fixed bias voltage across the junction. The current mode is very linear over a wide range.

4.3. Gear Motor

The DC Gear motor, consisting of a DC electric motor and a gearbox, is at the heart of several electrical and electronic applications. Precision Micro drive's have been designing and developing such high quality mini DC gear motors in an easy-to-mount package for a range of products and equipment. The miniature gear motor work smoothly and efficiently, supporting these electrical and electronic applications. These geared motors have reduction gear trains capable of providing high torque at relatively low shaft speed or revolutions per minute (RPM). Precision Micro drive's DC geared motors reduce the complexity and cost of designing and constructing applications such as industrial equipment, actuators, medical tools, and robotics. The voltage is regulated by using LM7805 [27] voltage regulator.

Fig. 12. *Geared motor.*

5. Experimental Implementation

5.1. Total Arrangement of the System

Fig. 13. *Total arrangement of the system.*

5.2. Working Principle

Photodiode is used as sensors. Two photodiode are used to sensing the position of the panel and two photodiode are used to sensing the mirror position for boosting the panel. If voltage drop between the two sensors of panel is not same a signal will go to the microcontroller, Microcontroller calculates the ADC [28] value. Then the ADC value it goes to motor driver A. Motor A will rotate the panel. Again when the voltage drop of two sensors of the mirror is not same, a signal will go to the microcontroller, microcontroller calculates the ADC value. Then the ADC value goes to the motor driver B and motor B control the mirror position to boosting the panel efficiency.

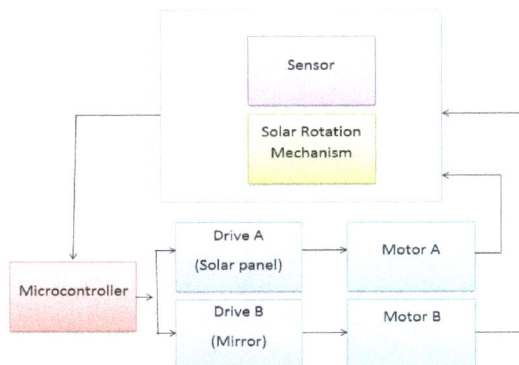

Fig. 14. *Block diagram of the system.*

5.3. Circuit Diagram of the System

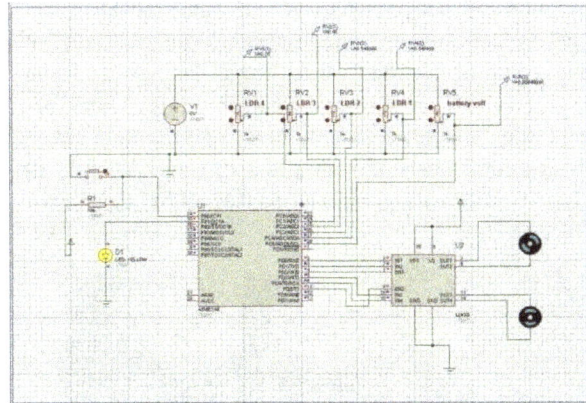

Fig. 15. *Circuit diagram of the system.*

5.4. Experimental Procedure

The experimental result is depends on i) Solar intensity ii) Area of the PV cell iii) Voltage iv) Current. The total system set on the direction of the sun. If the panel does not get its position perfect, it will track automatically the maximum light of sun by its sensors. Thus a current & voltage drop is found. Then by using millimeter , the readings of voltage and current is noted down. The photovoltaic trainer instruments give the solar intensity for different period of time. By measuring theses values, the power output of the solar panel and its efficiency can be calculated.

Fig. 16. *Automatic solar tracker.*

Fig. 17. *Multimeter.*

5.5. Calculation of Intensity

The relationship between radiant sun light and electrical power generated can be demonstrated with the aid of measuring equipment called photovoltaic Trainer. The sun intensity i (w/m2) can be calculated by this There are following things in photovoltaic trainer: a) Tilting Support on Caster b) Variable Resistor with slider c) Solar Module d) Temperature Sensor e) I luminance Sensor and f) Connector of the Solar Module.

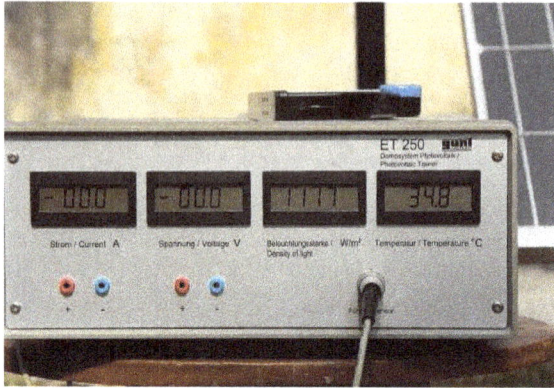

Fig. 18. *Photovoltaic trainer showing intensity and temperature.*

6. Results and Discussions

6.1. Description

From Fig 19 to Fig 23 shows the tracking results using microcontroller and mirror. These figures are based on intensity, current, power and efficiency on the basis of Appendix-A.

6.2. Intensity Vs. Day Time Curve

Intensity increase with respect to the day time and after 1 pm the intensity decreases. The graph shows right fulfill the all condition of Power vs. Day time graph.

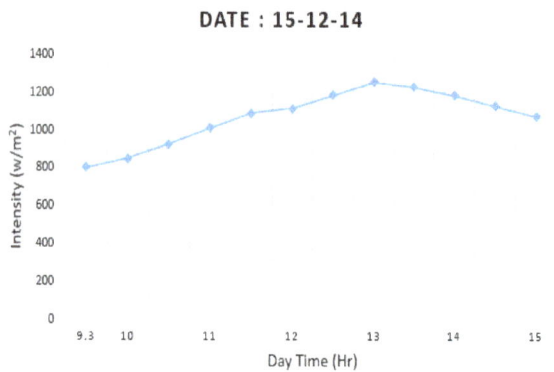

Fig. 19. *Intensity vs. Day time curve.*

6.3. Current Vs. day Time Curve

This graph shows that at first current increases and then decrease with the increase of day time. As the intensity increase the current also increase and as the intensity decrease the current also decrease.

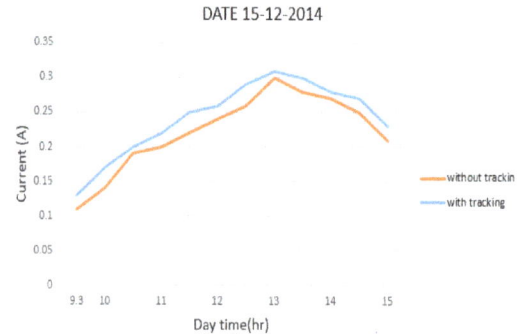

Fig. 20. *Current vs. Day time curve.*

6.4. Power Vs. Day Time Curve

Power increase with respect to the day time because intensity increases. After 1 pm the intensity decreases so the powers automatically decrease. This graph shows difference of power output between with tracking and without tracking.

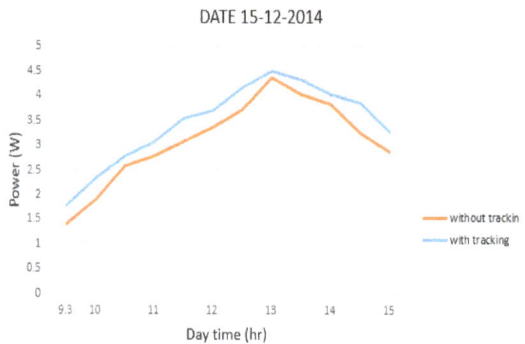

Fig. 21. *Power vs. Day time curve.*

6.5. Efficiency Vs. Day Time Curve

This graph represents that panel efficiency increase with respect to the day time because voltage, current and intensity increase. After 1pm the voltage, current and intensity decreases so the efficiency automatically decrease.

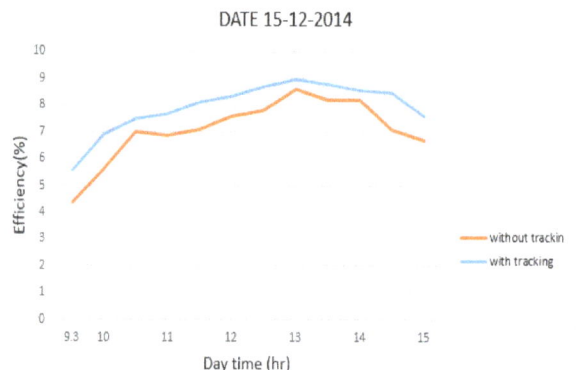

Fig. 22. *Efficiency vs. Day time curve.*

6.6. Power Vs. Day Time Curve

Power increase with respect to the day time because intensity increases. After 1 pm the intensity decreases so the powers automatically decrease. The graph shows right fulfill the all condition of Power vs. Day time graph. From the graph we observed that the power out-put of solar panel with mirror is greater than without mirror.

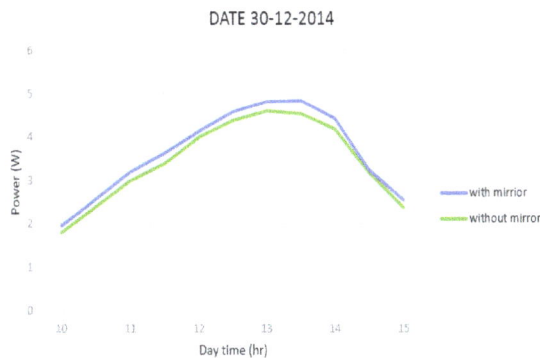

Fig. 23. *Power vs. Day time curve.*

7. Conclusions

The microcontroller based solar tracking system with mirror booster has been fabricated. The efficiency of solar panel is 9%. The paper has been presented a novel and a simple control implementation of a Sun tracker that employed to follow the Sun and produce electricity. The proposed two-motor design is simple and self-contained, and do not require any programming and a computer interface. The proposed methodology is an innovation so far. It achieves the following attractive features are simple and cost-effective control implementation, bulky but it is effective for large power generation, Battery will not take any charge after it is fully charged. This will increase battery's life time. Photo-diode is used instead of LDR sensor because; it can only take infra-red. Gear motor is used because it gives more power than stepper motor.

Sample calculations 1:
Solar panel length=0.25 m
Width=0.16 m
Area of the panel=Length x width =0.25 x 0.16=0.04 m^2
V=14.57v
I=0.31A
Power, P=14.57 x 0.31
=4.51 watt
Efficiency= (V x I)/ (A x i)
 = (14.57 x 0.31)/ (0.04 x 1256)
=8.99 %
Sample calculations 2:
V= 14.49v
I=0.30A
Power, P=V x I=14.49 x 0.30=4.377
Efficiency= (V x I)/ (A x i)
 = (14.49 x 0.30)/ (1256 x 0.04)
=8.6%

References

[1] A. J. Sangster. Engineering the Early Demise of Fossil Fuels "International Journal of Sustainable and Green Energy" Vol. 3, No. 6, 2014, pp. 115-122. doi: 10.11648/j.ijrse.20140306.11

[2] Feltrin, A., Freundlich, A., Material considerations for terawatt level deployment of photovoltaics. Renewable Energy, volume 33, 2008, Pages 180-185.

[3] K. H. Hussein et al, "Maximum Photovoltaic Power Tracking: An Algorithm for rapidly changing atmospheric conditions," Proc. Inst. Elect. Eng. volume 142, pt. G, no.1, January 1995, Pages 59–64.

[4] C.S. Chin, A. Babu, W. McBride. Design,modeling & testing of a standalone single axis active solar tracker using MATLAB/Simulink.

[5] C.R. Sullivan and M.J. Powers,"A High-Efficiency Maximum Power Point Tracking for Photovoltaic Arrays in a Solar-Power Race Vehicle", IEEE PESC'93, 1993, Pages 574-580.

[6] Xuejun Liu and A.C.Lopes,,"An Improved Perturbation and Observe Maximum Power Point Tracking Algorithm for PV Arrays" IEEE PESC '2004, Pages 2005-2010.

[7] Rashmi Swami"Solar Cell" International Journal of Scientific and Research Publications, Volume 2, Issue 7, July 2012,pp 1-5

[8] Sayran A. Abdulgafar 1, Omar S. Omar 2 ,Kamil M. Yousif 3 "Improving The Efficiency Of Polycrystalline Solar Panel Via Water Immersion Method"International Journal of Innovative Research in Science,Engineering and Technology,pp8127-8132.

[9] http://www.geo.lu.lv/fileadmin/user_upload/lu_portal/projekti/gzzf/videunilgtspejigaattistiba/VidZ1000/7.LECTURE-Energy_resources.pdf

[10] http://aerostudents.com/files/solarCells/solarCellsTheoryFullVersion.pdf

[11] Kumar, A., T. Schei, A. Ahenkorah, R. Caceres Rodriguez, J.-M. Devernay, M. Freitas, D. Hall, Å. Killingtveit, Z. Liu, 2011: Hydropower. In IPCC Special Report on Renewable Energy Sources and Climate Change Mitigation [O. Edenhofer, R. Pichs-Madruga, Y. Sokona, K. Seyboth, P. Matschoss, S. Kadner, T. Zwickel, P. Eickemeier, G. Hansen, S. Schlömer, C. von Stechow (eds)], Cambridge University Press, Cambridge, United Kingdom and New York, NY, USA.

[12] http://www.energy.gov/eere/bioenergy/bioenergy-technologies-office

[13] http://www.nrel.gov/docs/fy08osti/39146.pdf

[14] http://energy.gov/eere/geothermal/geothermal-basics

[15] http://web.mit.edu/windenergy/windweek/Presentations/Wind%20Energy%20101.pdf

[16] http://energy.gov/eere/energybasics/articles/wave-energy-basics

[17] http://folk.ntnu.no/falnes/teach/wave/JF_introduction2010-06-28.pdf

[18] http://www.e-renewables.com/documents/Waste/Waste%20to%20Energy%20-%20The%20Basics.pdf

[19] http://www.r-e-a.net/pdf/energy-from-waste-guide-for-decision-makers.pdf

[20] Yingxue Yao, Yeguang Hu, Shengdong Gao, Gang Yang, Jinguang. A multiple dual-axis solar tracker with two tracking strategies. Renewable Energy, Volume 72, December 2014, Pages 88-98.

[21] Kroposki B, DeBlasio R, (2000), Technologies for the New Millennium: Photovoltaics as a Distributed Resource, IEEE Power Engineering Society Summer Meeting, pages 1798 – 1801.

[22] http://www.geo-dome.co.uk/article.asp?uname=solar_mirror

[23] http://www.engineersgarage.com/microcontroller

[24] Electronic Devices and Circuit Theory-9th edition by Robert L. Boylestad and Louis Nashelsky.

[25] Datasheet of 2011 Atmel Corporation 8159D–AVR–02/11.

[26] Microcontroller, Sahil Kapoor, Narendra Singh, Shiv Chauhan. Vth Semester,Department of Electronic Communication & Engg. Dronacharya College of Engineering, Gurgaon, India123506. 2014 IJIRT , Volume 1 Issue 6 , ISSN : 2349-6002

[27] https://www.sparkfun.com/datasheets/Components/LM7805.pdf

[28] http://www.atmel.com/Images/Atmel-8456-8-and-32-bit-AVR-Microcontrollers-AVR127-Understanding-ADC-Parameters_Application-Note.pdf

[29] Laughlin Barker, Matthew Neber, Hohyun Lee.Design of a low-profile two-axis solar tracker .Solar Energy, Volume 97, November 2013, Pages 569-576.

[30] S. Ozcelik, H. Prakash, R. Challoo. Two axis solar tracker analysis and control for maximum power generation. Procedia Computer Science, Volume 6, 2011, Pages 457-462.

[31] B.K. Bose, P.M. Szczesny and R.L. Steigerwald,," Micro-computer Control of a Residential Photovoltaic Power Conditioning System", IEEE Trans. On Industry Applications, volume IA-21, no. 5, September 1985, Pages ll82-1191.

24

Three-phase Matrix Converter Applied to Wind Energy Conversion System for Wind Speed Estimation

Alaa Eldien M. M. Hassan, Mahmoud A. Sayed, Essam E. M. Mohamed

Department Electrical Engineering, South Valley University, Qena, Egypt

Email address:

alaaeldien@eng.svu.edu.eg (A. E. M. M. Hassan), mahmoud_sayed@ieee.org (M. A. Sayed),
essam.mohamed@eng.svu.edu.eg (E. E. M. Mohamed)

Abstract: With continuous increasing concerns of the energy issues, renewable energy sources are getting much attention worldwide. This paper presents a full description of the grid-tie Wind Energy Conversion System (WECS) based on interfacing a Permanent Magnet Synchronous Generator (PMSG) to the utility grid by using the direct AC/AC matrix converter. Due to the random variation of wind velocities, wind speed estimation control technique is used to estimate the wind velocity and extracts the maximum power at all wind velocities. The matrix converter controls the maximum power point tracking MPPT by adjusting the PMSG terminal frequency, and hence, the shaft speed. In addition, the matrix converter controls the grid injected current to be in-phase with the grid voltage for the unity power factor. Space Vector Modulation is used to generate the PWM signals of the matrix converter switches. The MPPT algorithm is included in the speed control system of the PMSG. The system dynamic performance is investigated using Matlab/Simulink.

Keywords: Permanent Magnet Synchronous Generator (PMSG), Matrix Converter (MC)

1. Introduction

Due to the continuous increased demand of the electrical energy, and the lack in reserve stocks of non-renewable energy sources such as fossil fuels, huge efforts are being made to generate the electrical power from renewable energy sources. Renewable energy sources, such as wind and solar, provide clean and cheap energy sources, [1-3]. Wind power is now one of the most energy sources that expand rabidly in the industry due to its abundance. Wind Energy is also advantageous over traditional methods of generating energy, in the sense that it is getting cheaper and cheaper to produce energy. Different types of generators are being used with wind turbines such as induction machine, double fed induction machine and PMSG [4-7]. PMSG has become the most familiar type of synchronous generators because of its good features that it has High efficiency and low maintenance cost. In addition, it has Small size and simple construction. There is no need for a separate DC excitation source. On the other hand, PMSG has some drawbacks that it has high initial cost compared with other generators at the same rating. And Overloading, short circuit and high temperate reduce, and weaken the magnetization of the

permanent magnets [8-10]. The work in this paper is based on PMSG.

Because of the random variation on the wind velocities, the electrical power generated from the PMSG will vary also and unable to connect to the grid. Therefore, MPPT has been emerged and becoming an essential part in the variable speed wind turbine. Many methods have been proposed to locate and track the maximum power point. tip speed ratio (TSR) technique was presented, which depends mainly on measuring the wind velocity, then using the optimal value for the tip speed ratio in order to estimate the optimal value for the shaft speed[11, 12]. Power signal feedback (PSF) control technique was presented, which depends upon lookup tables for the shaft speed and its maximum output power at each wind velocity [13-15]. Also the optimal control technique was presented in some researches [16, 17]. The main theory of this method is to adjust the generator torque in order to obtain wind turbine torque for each wind velocity. Perturbation and observation (P&O) method that is independent of the wind velocity detections has been presented in [15, 18, 19]. This method is based on perturbing the shaft generator speed in small step-size and observing the resulting changes in the output power until the slope becomes zero. The P&O is a robust, simple and reliable technique [15]. Wind speed estimation control technique presents in [20]. This method uses the efficiency curve of the blade to estimate the wind velocity without using wind velocity sensor. The work in this paper based on this technique.

In order to connect the electrical power to the grid so it converted to DC power then it enter an inverter to convert it to AC power compatible with the grid, this converter called traditional back-to-back converter. but this converter have some drawbacks including: operation on two stages reduces the reliability of the system, the bulky short life-time capacitor on the rectification stage, high power losses, and high amount of harmonics [21]. To overcome the previous mentioned drawbacks of traditional converters the AC/AC matrix converters are applied. Matrix converter is a single-stage AC/AC bi-directional power flow converter that takes power from AC source and converts it to another AC system with different amplitude and frequency. The reduction in the number of switches provides more compact than the traditional converter. Since it has only one power stage, there is no need for the bulky and lifetime limited energy-storing electrolytic capacitor that is considered an essential part in the conventional converters based two stages. It also has the ability to control the output voltage magnitude and frequency in addition to operation at unity power factor for any load. Moreover, it provides sinusoidal input and output waveforms, with minimal higher order harmonics and no sub-harmonics[21, 22]. The first attempt for modulation strategy technique of matrix converters is the Venturini method that depends mainly on the product of the input voltage and the modulation index matrix to generate the output voltage. However, this method has a poor voltage transfer ratio of 50% [23] . A modification to the strategy proposed in [24] was presented to enhance the maximum transfer ratio to its maximum limit of 86.7%. The scalar modulation strategy is used to utilizes the input voltage to produce the active and non-active states of the converter switches[25-27].

This paper presents a full description of grid-tie wind energy conversion system based on PMSG interfaced to the grid with the direct AC/AC matrix converter. The matrix converter controls the PMSG speed in order to track the maximum power point at all wind speeds based on wind velocity estimation technique. In addition, the matrix converter controls the grid side power factor to be unity in order to inject only active power to the grid. Section I represents the introduction. Section II gives a description of wind turbine and its mathematical model. Section III describes the MPPT control technique based on wind velocity estimation method. Section IV gives a full description of the matrix converter and its PWM using indirect SVM technique. Section V presents a control technique of the WECS. The result of this work is presented in section VI. Finally, the conclusions are presented in section VII.

2. Wind Turbine Performance

The wind kinetic energy is converted into mechanical power on the PMSG shaft through the wind turbine. This PMSG converts the mechanical power into electrical power. The mechanical power can be formulated as follows:

$$P_{Mech} = 0.5 \rho A C_p V_w^3 \qquad (1)$$

where: P_{mech} is the output mechanical power from the turbine in Watts, ρ is the air density in Kg/m^3, A is the turbine rotor area in m^2 ($A=\pi R_r^2$) and R_r is the rotor blade radius, C_p is the power coefficient and V_w is the wind velocity in m/s. If the air density ρ and the area A are constant, the power coefficient C_p can be formulated as follows:

$$C_p = \frac{p_{turbine}}{p_{wind}} \qquad (2)$$

$$C_p = 0.5176[\frac{116}{\lambda_i} - (0.4\theta - 5)]e^{\frac{-21}{\lambda_i}+(0.0086\lambda)} \qquad (3)$$

$$\lambda_i = \frac{1}{\frac{1}{\lambda + 0.08\theta} - \frac{0.035}{\theta^3 + 1}} \qquad (4)$$

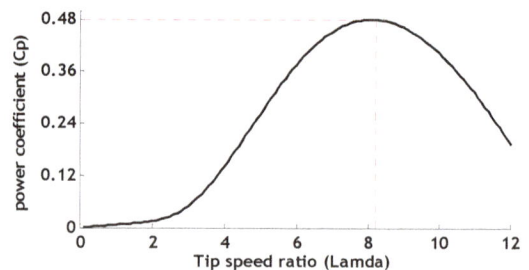

Fig. 1. *Wind turbine power coefficient versus tip speed ratio.*

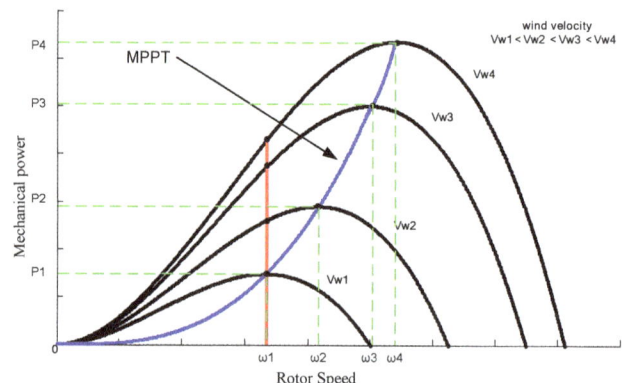

Fig. 2. *MPPT control scheme at different wind velocities.* where θ is the pitch angle, which is considered zero. Therefore, the tip speed ratio (λ) is formulated as follows:

$$\lambda = \frac{Tip_{speed}}{Wind_{speed}} = \frac{\omega_r R_r}{V_w} \qquad (5)$$

The optimum value of C_p is about 0.48 for a tip speed ratio of 8.1. In order to keep the system operates at the MPP, the tip speed ratio should be always at this value at operational conditions [28], as shown in Fig. (1).

3. Maximum Power Point Tracking Control (MPPT)

Fig. 2 shows the relationship between the output mechanical power and shaft speed at different values of wind velocity. According to Fig. 2, for each wind speed there is a unique rotational speed that record maximum mechanical power. Maximum Power Point Tracking (MPPT) is used to extract the maximum mechanical power from the wind turbine at each wind velocity.

It is clear that the maximum power extraction occurs at different rotational speeds. For example, at wind speed (V_{w1}), the maximum output power is P_1, which is obtained at a rotational speed ω_1. If the wind speed increases to V_{w2}, V_{w3}, or V_{w4}, applying a constant shaft speed controller to keep the rotational speed constant at ω_1 results in a mechanical power less than the maximum available power at the corresponding wind speed. Therefore, in order to extract the maximum power at each wind speed, the rotational speed should be controlled to follow the change in wind speed. If MPPT control is applied, the shaft speed will increase to ω_2 at wind speed of

V_{w2} to extract the maximum power available for this point P_2. The same concept applies for the other velocities, i.e. V_{w3} and V_{w4} [5].

To estimate the value of the wind velocities at each change the power coefficient (C_p) equations in (2, 3), this nonlinear equation depend on TSR (λ) that formulate in (5). It approximately rewritten in third order polynomial form to be formulated as in (6)[20]:

$$C_p(\lambda) = a_0 + a_1\lambda + a_2\lambda^2 + a_3\lambda^3 \qquad (6)$$

Where: a_0 through a_3, are constant which can be found numerically. The numerical solution generates values:

$a_0 = 0.00715814, a_1 = -0.04454063, a_2 = 0.02899277,$

$a_3 = -0.00202519.$

Referring to the mechanical power that formulated in (1), after equation (6) substitute the power coefficient it can be formulated as follow:

$$P_{Mech} = 0.5\rho A(a_0 + a_1\lambda + a_2\lambda^2 + a_3\lambda^3)V_w^3 \qquad (7)$$

Also equation (5) substitutes the TSR (λ) in (7):

$$P_{Mech} = 0.5\rho A(a_0 V_w^3 + a_1\omega_r R_r V_w^2 + a_2\omega_r^2 R_r^2 V_w + a_3\omega_r^3 R_r^3) \qquad (8)$$

$$V_w^3 + \frac{a_1}{a_0}\omega_r R_r V_w^2 + \frac{a_2}{a_0}\omega_r^2 R_r^2 V_w + \frac{a_3}{a_0}\omega_r^3 R_r^3 - \frac{P_{Mech}}{a_0 0.5\rho A} = 0 \qquad (9)$$

The numerical solution for equation 9 generates three values

for the wind velocity. The second answer value is the more accurate empirical solution[29, 30].

4. Matrix Converter Switching Scheme

Matrix converter is a one stage converter consists of nine bi-directional switches that establish a 3x3 matrix as shown in Fig. 3. In order to avoid short circuit that might be occurred between the input phases and open circuit on the output phases, one and only one switch per column must be ON and the other switches still OFF.

$$S_{m1} + S_{m2} + S_{m3} = 1 \ m \in \{1,2,3\} \qquad (10)$$

The input three-phase voltage can be formulated as follows:

$$\begin{bmatrix} e_r \\ e_s \\ e_t \end{bmatrix} = E_m \begin{bmatrix} \cos(\omega t + \phi_i) \\ \cos(\omega t + \phi_i - 2\pi/3) \\ \cos(\omega t + \phi_i + 2\pi/3) \end{bmatrix} \qquad (11)$$

Where ϕ_i is the phase angle of the input voltages waveform.

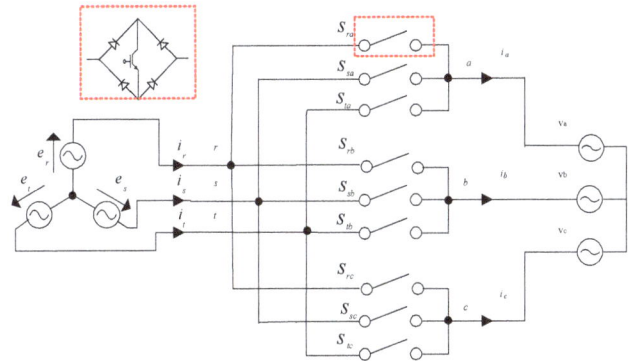

Fig. 3. *Matrix Converter.*

The output three-phase voltage can be formulated as follows as follow:

$$\begin{bmatrix} V_a \\ V_b \\ V_c \end{bmatrix} = V_m \begin{bmatrix} \cos(\omega t + \phi_o) \\ \cos(\omega t + \phi_o - 2\pi/3) \\ \cos(\omega t + \phi_o + 2\pi/3) \end{bmatrix} \qquad (12)$$

Where ϕ_o is the phase angle of the output voltages waveform.

The relation between the input and the output waveforms can be formulated as follows:

$$\begin{bmatrix} V_a \\ V_b \\ V_c \end{bmatrix} = \begin{bmatrix} S_{ra} & S_{sa} & S_{ta} \\ S_{rb} & S_{sb} & S_{tb} \\ S_{rc} & S_{sc} & S_{tc} \end{bmatrix} \begin{bmatrix} e_r \\ e_s \\ e_t \end{bmatrix} \qquad (13)$$

The input current equation is obtained as follows:

$$\begin{bmatrix} i_r \\ i_s \\ i_t \end{bmatrix} = \begin{bmatrix} S_{ra} & S_{sa} & S_{ta} \\ S_{rb} & S_{sb} & S_{tb} \\ S_{rc} & S_{sc} & S_{tc} \end{bmatrix}^T \begin{bmatrix} i_a \\ i_b \\ i_c \end{bmatrix} \qquad (14)$$

Each element in the 3x3 modulation index matrix represents the duty cycle of the switches. These duty cycles of switches are determined by the space vector modulation control.

Space-vector modulation (SVM) is better than conventional PWM techniques because of its advantages as it generates controlled output voltage magnitude and frequency, it generates lower THD and it is suitable for digital controllers [31]. However, the implementation of space vector modulation control technique is complex, and need more switching states.

In order to simplify the states of direct matrix converter, indirect space vector modulation method is considered.

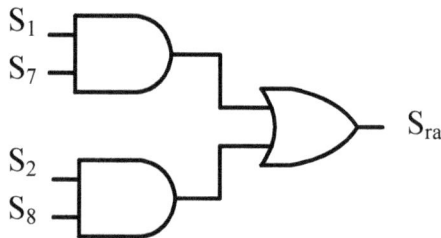

Fig. 4. Implementation of the duty cycle of the switch Sra.

The indirect space-vector modulation technique depends on two virtual stages, i.e., the rectification and the inversion stages[32, 33]. The modulation index for the matrix that formulated in (13) is subdivided into the product of two matrices, the rectifier matrix and the inverter matrix (15).

$$A = I_{inverter} \cdot R_{rectifier}$$

$$\begin{bmatrix} S_{ra} & S_{sa} & S_{ta} \\ S_{rb} & S_{sb} & S_{tb} \\ S_{rc} & S_{sc} & S_{tc} \end{bmatrix} = \begin{bmatrix} S_7 & S_8 \\ S_9 & S_{10} \\ S_{11} & S_{12} \end{bmatrix} \bullet \begin{bmatrix} S_1 & S_3 & S_5 \\ S_2 & S_4 & S_6 \end{bmatrix} \quad (15)$$

The combination of the rectifier and the inverter duty cycles is used to generate the duty cycles of the nine switches of the Matrix Converter, so that $S_{ra} = S_1 * S_7 + S_2 * S_8$. This equation is represented using the logic circuit as shown in Fig. 4 [32]. The rectification matrix (S_1-S_6) represents the duty cycles of the rectification stage switches, and the inversion matrix (S_7-S_{12}) represents the duty cycles of the inversion stage switches.

5. Control Scheme of Wind Energy Conversion System.

Fig. 5 shows the overall system of the grid-tie wind energy conversion system including PMSG, Matrix converter and its controller circuit, an input filter, and finally the grid side. The wind turbine developed torque T_m is applied to the generator

shaft; the wind velocity estimation block is used to estimate the value of the wind velocity which multiplying with the optimum value of the Tip Speed Ratio to generate the MPP speed that achieves the maximum power extracted at this wind velocity. The reference generator speed is compared with the actual generator speed and the error signal is applied to conventional PI controllers to generate the reference value for the q-axis component; in the other hand the reference value for the d-axis component (Id^*) is zero in order to keep the d-axis component flux equals zero. The actual three-phase generator currents are detected and converted to the d-q axis component (i_{ds}-i_{qs}) using Park/Clark Transformation. The actual load current in the d-q axis components are compared with their reference values and the error signal is applied to conventional PI controllers to generate the reference d-q axis voltage components V_d^* and V_q^*. The gains of the PI controllers have been manually tuned in order to achieve acceptable transient response. The d-q reference voltage components are converted to the three-phase axis using the invers Park/Clark Transformation in order to obtain the relevant three-phase voltage of matrix converter voltage. Fig.6 Illustrates the block diagram of control scheme of wind energy conversion system.

6. Results

The grid-tie wind turbine model based on PMSG and three-phase-to-three-phase matrix interface converter has been carried out in Matlab/Simulink environment. The parameters of the whole system are listed in Table I. Fig. 7 shows the wind velocity profile applied in the simulation. Fig. 8 illustrates the estimated and actual wind speed. Fig. 9 shows the effectiveness of the controller as the feedback speed tracks well the reference speed. Fig. 10 shows the value of C_p that remains constant at its optimal value regardless of the variation in the wind and generator speed. Fig. 11 shows the Tip Speed Ratio (λ) that remains constant at its optimal value at all values of the wind speed. The power generated from the machine varies according to the wind speed variation, as shown in Fig. 12. The generated three-phase currents of the PMSG change due to the wind speed variation as shown in Fig. 13. Fig. 14 shows the variation in the mechanical and electromagnetic torque of the PMSG shaft. It is clear that both torques are in good agreement. Fig. 15 shows the utility grid three-phase currents and voltages at all values of wind velocities. fig.16 shows the grid side phase voltage and its corresponding line current at all wind speeds. It is clear that the grid current and voltage are in-phase at all wind speeds. Therefore, the wind turbine injects active power only to the utility grid. Fig. 17 provides the FFT analysis of the three-phase currents of the PMSG, which shows a decrease in the THD of the output current waveform to 1.96% which can be neglected compared to the other control techniques. Fig. 18 analysis of the grid current waveform which shows that the THD of the grid current has a good level of 9.4%.

7. Conclusions

This paper presents a study of a grid connected wind energy conversion system for electrical power extraction from wind energy based on PMSG to compensate the shortage of non-renewable energy sources. In order to extract the maximum available power at each wind speed the excellent tracking of the Maximum Power Point based on wind velocity estimation method is applied. The electrical

power generated by the PMG is tied to the grid through a matrix converter, which solve all the problems of the traditional converters. The Matrix Converter, controlled by SVM, enables excellent transient response while sinusoidal current waveforms is dominant with grid currents in-phase with the grid voltage for unity power factor. Simulation results prove that the wind turbine system based three-phase PMSG tracks the MPP and injects only active power to the grid at all wind speeds. Fig. 5 WECS modeling

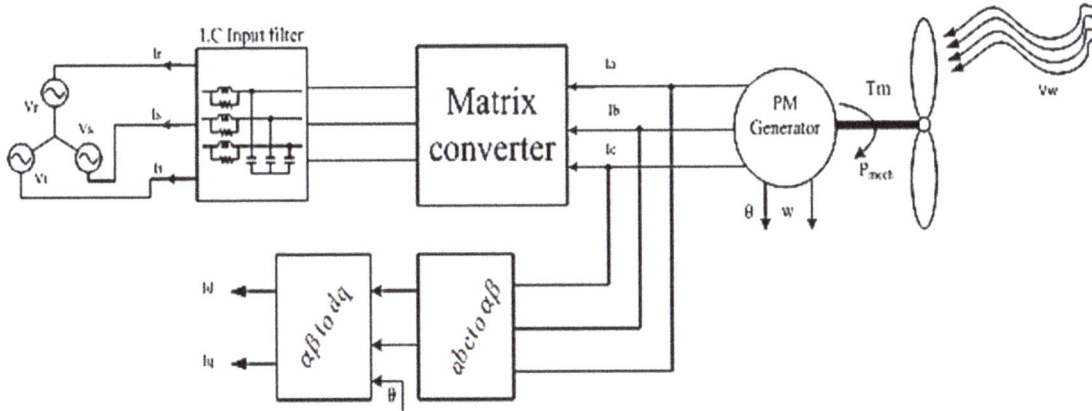

Fig. 5. *WECS modeling.*

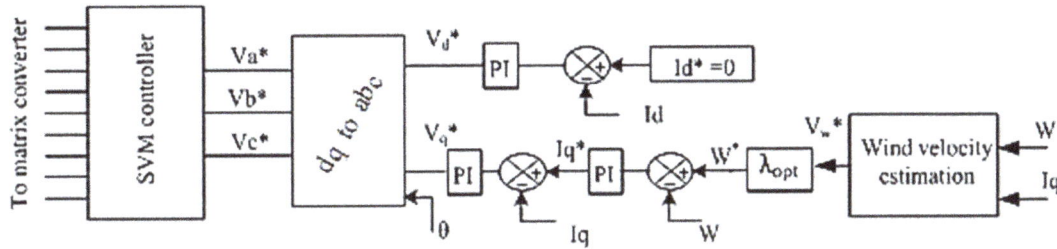

Fig. 6. *Block diagram of Matrix Converter Controller.*

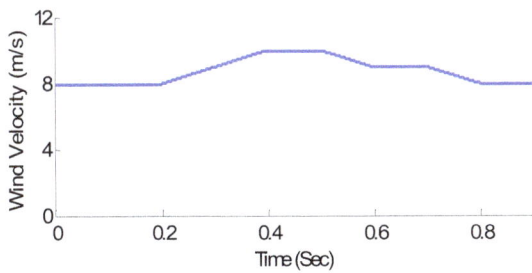

Fig. 7. *wind velocity profile.*

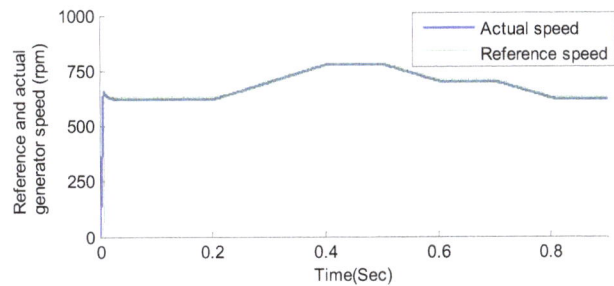

Fig. 9. *actual and reference generator speed.*

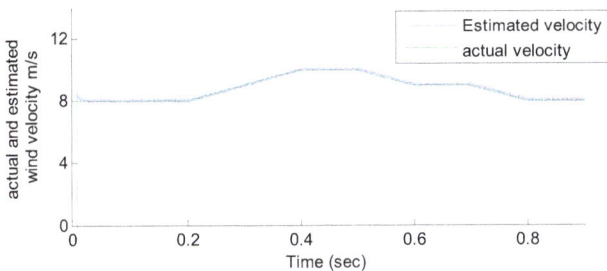

Fig. 8. *The actual and estimated wind velocity.*

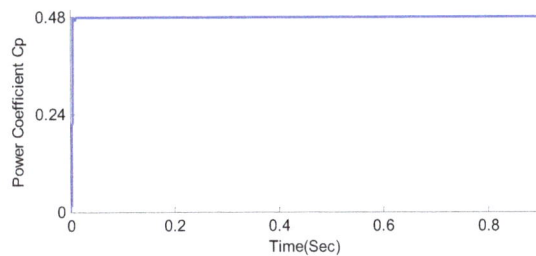

Fig. 10. *Power Coefficient Cp.*

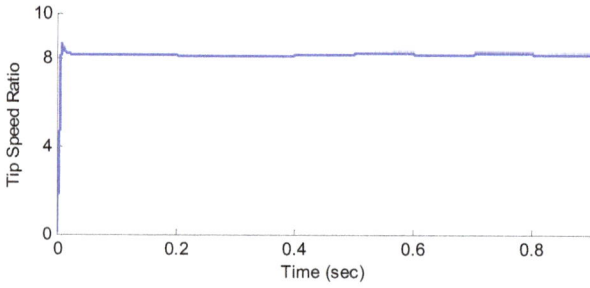

Fig. 11. Tip speed ratio.

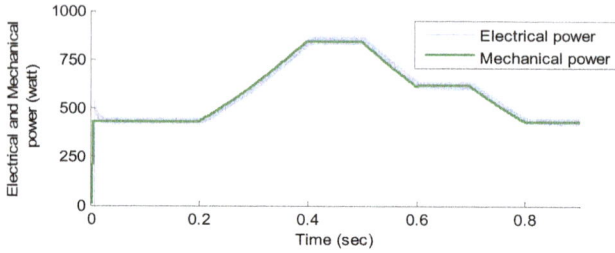

Fig. 12. Electrical and Mechanical Power of PMSG.

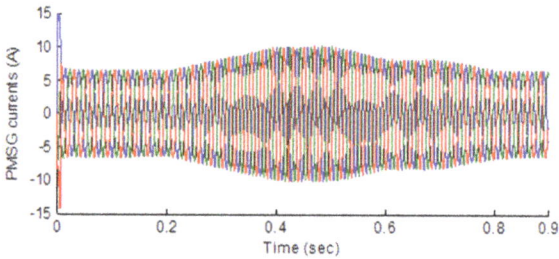

a)The generator currents

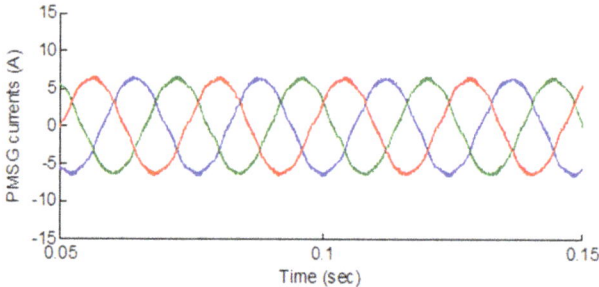

b)The generator current at Vw=8m/s

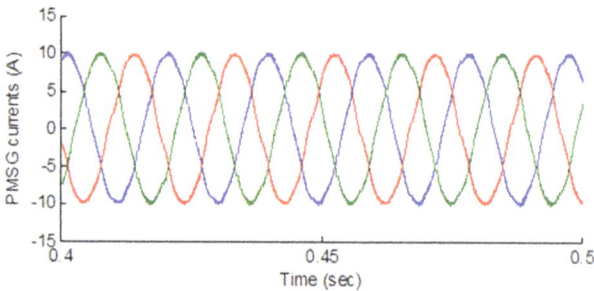

c)The generator current at V w=10 m/s

Fig. 13. The generator currents variation.

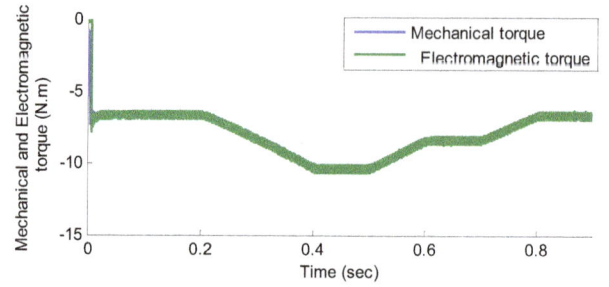

Fig. 14. Mechanical and Electromagnetic torque.

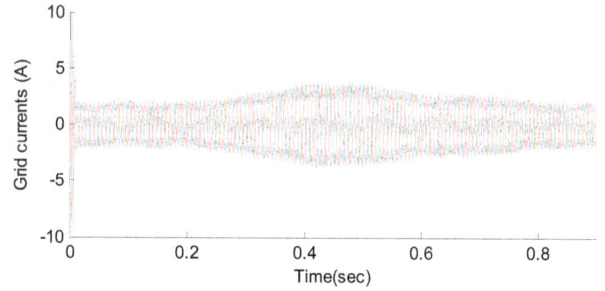

Fig. 15. Three-phase Grid Currents.

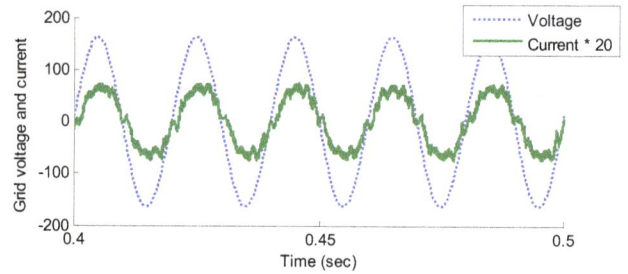

Fig. 16. Grid voltage and current.

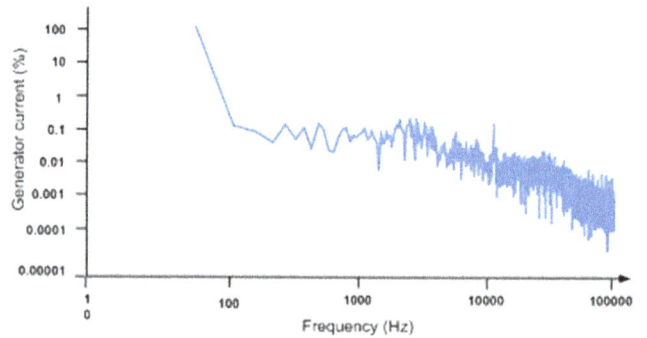

Fig. 17. FFT analysis for the Output currents of PMSG.

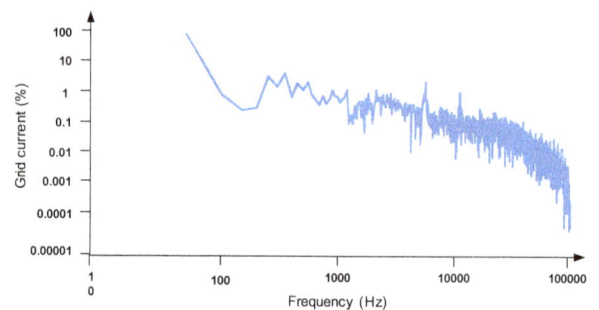

Fig. 18. FFT analysis for the Grid currents.

Appendix

Matlab Simulink parameters of the overall system shown in Fig.(8).

Table 1. System parameters.

Machine parameters	
No. pairs pole	P=4
Stator resistance	R_s=2.875Ω
Stator inductance	L_d=L_q=0.0085 H
Moment of inertia	J=0.00008 kg.m^2
Flux linkage	Ψ=0.175 wb
Grid parameters	
Phase voltage	V_a=220v
LC filter	
L= 15 mH C= 8 uF R_d= 47 Ω	

References

[1] J. M. Carrasco, L. G. Franquelo, J. T. Bialasiewicz, E. Galván, R. P. Guisado, M. A. Prats, J. I. León, and N. Moreno-Alfonso, "Power-electronic systems for the grid integration of renewable energy sources: A survey," Industrial Electronics, IEEE Transactions on, vol. 53, pp. 1002-1016, 2006.

[2] Z. Olaofe and K. Folly, "Energy storage technologies for small scale wind conversion system," in Power Electronics and Machines in Wind Applications (PEMWA), 2012 IEEE, 2012, pp. 1-5.

[3] T. Shanker and R. K. Singh, "Wind energy conversion system: A review," in Engineering and Systems (SCES), 2012 Students Conference on, 2012, pp. 1-6.

[4] S. Barakati, M. Kazerani, and X. Chen, "A new wind turbine generation system based on matrix converter," in Power Engineering Society General Meeting, 2005. IEEE, 2005, pp. 2083-2089.

[5] V. Agarwal, R. K. Aggarwal, P. Patidar, and C. Patki, "A novel scheme for rapid tracking of maximum power point in wind energy generation systems," Energy conversion, IEEE transactions on, vol. 25, pp. 228-236, 2010.

[6] Y. Izumi, A. Pratap, K. Uchida, A. Uehara, T. Senjyu, A. Yona, and T. Funabashi, "A control method for maximum power point tracking of a PMSG-based WECS using online parameter identification of wind turbine," in Power Electronics and Drive Systems (PEDS), 2011 IEEE Ninth International Conference on, 2011, pp. 1125-1130.

[7] Y. Errami, M. Maaroufi, and M. Ouassaid, "A MPPT vector control of electric network connected Wind Energy Conversion System employing PM Synchronous Generator," in Renewable and Sustainable Energy Conference (IRSEC), 2013 International, 2013, pp. 228-233.

[8] T.-F. Chan, "Permanent-magnet machines for distributed power generation: A review," in 2007 IEEE Power Engineering Society General Meeting, 2007, pp. 1-6.

[9] H. Polinder, F. F. Van der Pijl, G.-J. De Vilder, and P. J. Tavner, "Comparison of direct-drive and geared generator concepts for wind turbines," Energy conversion, IEEE transactions on, vol. 21, pp. 725-733, 2006.

[10] J. Elizondo, M. Macías, and O. Micheloud, "Matrix Converters Applied to Wind Energy Conversion Systems, Technologies and Investigation Trends," in Electronics, Robotics and Automotive Mechanics Conference, 2009. CERMA'09., 2009, pp. 435-439.

[11] E. Koutroulis and K. Kalaitzakis, "Design of a maximum power tracking system for wind-energy-conversion applications," Industrial Electronics, IEEE Transactions on, vol. 53, pp. 486-494, 2006.

[12] L. Zhang, B. Zhou, F. Cheng, and G. Zuo, "A novel maximum power point tracking control method suitable for a doubly salient electro-magnetic wind power generator system," in World Non-Grid-Connected Wind Power and Energy Conference, 2009. WNWEC 2009, 2009, pp. 1-6.

[13] S. M. Barakati, "Modeling and controller design of a wind energy conversion system including a matrix converter," University of Waterloo, 2008.

[14] S. M. Barakati, M. Kazerani, and J. D. Aplevich, "Maximum power tracking control for a wind turbine system including a matrix converter," Energy conversion, IEEE transactions on, vol. 24, pp. 705-713, 2009.

[15] M. Abdullah, A. Yatim, C. Tan, and R. Saidur, "A review of maximum power point tracking algorithms for wind energy systems," Renewable and Sustainable Energy Reviews, vol. 16, pp. 3220-3227, 2012.

[16] T. Nakamura, S. Morimoto, M. Sanada, and Y. Takeda, "Optimum control of IPMSG for wind generation system," in Power Conversion Conference, 2002. PCC-Osaka 2002. Proceedings of the, 2002, pp. 1435-1440.

[17] S. Morimoto, H. Nakayama, M. Sanada, and Y. Takeda, "Sensorless output maximization control for variable-speed wind generation system using IPMSG," in Industry Applications Conference, 2003. 38th IAS Annual Meeting. Conference Record of the, 2003, pp. 1464-1471.

[18] A. Mahdi, W. Tang, and Q. Wu, "Estimation of tip speed ratio using an adaptive perturbation and observation method for wind turbine generator systems," in Renewable Power Generation (RPG 2011), IET Conference on, 2011, pp. 1-6.

[19] J. S. Thongam and M. Ouhrouche, "MPPT control methods in wind energy conversion systems," Fundamental and Advanced Topics in Wind Power, pp. 339-360, 2011.

[20] H.-S. Shin, C. Xu, J.-M. Lee, J.-D. La, and Y.-S. Kim, "MPPT control technique for a PMSG wind generation system by the estimation of the wind speed," in Electrical Machines and Systems (ICEMS), 2012 15th International Conference on, 2012, pp. 1-6.

[21] D. Casadei, G. Grandi, G. Serra, and A. Tani, "Space vector control of matrix converters with unity input power factor and sinusoidal input/output waveforms," in Power Electronics and Applications, 1993., Fifth European Conference on, 1993, pp. 170-175.

[22] P. W. Wheeler, J. Rodriguez, J. C. Clare, L. Empringham, and A. Weinstein, "Matrix converters: a technology review," Industrial Electronics, IEEE Transactions on, vol. 49, pp. 276-288, 2002.

[23] A. Alesina and M. Venturini, "Solid-state power conversion: A Fourier analysis approach to generalized transformer synthesis," Circuits and Systems, IEEE Transactions on, vol. 28, pp. 319-330, 1981.

[24] A. Alesina and M. Venturini, "Analysis and design of optimum-amplitude nine-switch direct AC-AC converters," Power Electronics, IEEE Transactions on, vol. 4, pp. 101-112, 1989.

[25] J. Rodriguez, M. Rivera, J. W. Kolar, and P. W. Wheeler, "A review of control and modulation methods for matrix converters," IEEE Transactions on Industrial Electronics, vol. 59, pp. 58-70, 2012.

[26] J. Rodriguez, E. Silva*, F. Blaabjerg, P. Wheeler, J. Clare, and J. Pontt, "Matrix converter controlled with the direct transfer function approach: analysis, modelling and simulation," International journal of electronics, vol. 92, pp. 63-85, 2005.

[27] L. Zhang, C. Watthanasarn, and W. Shepherd, "Control of AC-AC matrix converters for unbalanced and/or distorted supply voltage," in Power Electronics Specialists Conference, 2001. PESC. 2001 IEEE 32nd Annual, 2001, pp. 1108-1113.

[28] M. M. Hussein, M. Orabi, M. E. Ahmed, and M. A. Sayed, "Simple sensorless control technique of permanent magnet synchronous generator wind turbine," in Power and Energy (PECon), 2010 IEEE International Conference on, 2010, pp. 512-517.

[29] L. V. Fausett, Applied numerical analysis using MATLAB: Pearson, 2008.

[30] W. H. Press, S. A. Teukolsky, W. T. Vetterling, and B. P. Flannery, Numerical recipes in C vol. 2: Citeseer, 1996.

[31] M. Y. Lee, P. Wheeler, and C. Klumpner, "Space-vector modulated multilevel matrix converter," Industrial Electronics, IEEE Transactions on, vol. 57, pp. 3385-3394, 2010.

[32] M. Jussila and H. Tuusa, "Comparison of direct and indirect matrix converters in induction motor drive," in IEEE Industrial Electronics, IECON 2006-32nd Annual Conference on, 2006, pp. 1621-1626.

[33] M. Rivera, J. Rodriguez, B. Wu, J. R. Espinoza, and C. A. Rojas, "Current control for an indirect matrix converter with filter resonance mitigation," Industrial Electronics, IEEE Transactions on, vol. 59, pp. 71-79, 2012.

25

MPPT Control Technique for Direct-Drive Five-Phase PMSG Wind Turbines with Wind Speed Estimation

Abdel-Raheem Youssef[1], Mahmoud A. Sayed[1], M. N. Abdel-Wahab[2], Gaber Shabib Salman[3]

[1]Dept. of Electrical Engineering, Faculty of Engineering, South Valley University, Qena, Egypt
[2]Dept. of Electrical Engineering, Faculty of Engineering, Suez Canal University, Ismailia, Egypt
[3]Dept. of Electrical Engineering, Faculty of Engineering, Aswan University, Aswan, Egypt

Email address:
abou_radwan@hotmail.com (Abdel-Raheem Y.), Mahmoud_sayed@ieee.org (M. A. Sayed),
mohamed_Nabil1973@yahoo.com (M. N. Abdel-Wahab), gabershabib@yahoo.com (G. S. Salman)

Abstract: This paper has presented comprehensive modeling of direct driven five-phase PMSG based grid-connected wind turbines along with the control schemes of the interfacing converters. Wind speed estimation has been achieved based on measured rotor speed. Five-phase to three-phase interface power converter based back-to-back common dc-link converter has been used to achieve the system objectives. The machine side converter (MSC) is used to track the maximum power point at different wind speed. The grid side converter (GSC) uses a vector current controller to inject pure active power to the grid. The effectiveness of proposed control scheme is validated through extensive simulation results by using MATALB/SIMULINK.

Keywords: MPPT, Five-Phase PMSG, MSC, GSC

1. Introduction

Wind power is today's most rapidly growing renewable energy source. A wind turbine operates either at a fixed or variable speed [1]. Most of wind turbine manufacturers are developing new different scale wind turbines based on variable-speed operation with pitch control using either a permanent magnet synchronous generator (PMSG) or a doubly fed induction generator (DFIG) [2]. Due to the intensified grid codes, a PMSG wind turbine with full VSC-based converters is becoming more favored by the wind power industry [3]–[7].The variable speed wind turbine with a multi-phase PMSG and full-scale/fully controllable voltage source converters (VSCs) is considered to be a promising, but not yet very popular, wind turbine concept [3]. The wind turbines based multi-phase PMSG configuration have many advantages such as gearless construction [4], elimination of a dc excitation system [5], full controllability of the system for maximum wind power extraction and grid interface, and ease in accomplishing fault-ride through and grid support [6].Therefore, the efficiency and reliability of a VSC-based PMSG wind turbine is assessed to be higher than that based DFIG [7].

Recently, multiphase machines have gained much interest due to their advantages over conventional three-phase machines. The use of multiphase permanent magnet synchronous generators PMSG to implement high power is an alternative to reduce the current rating of the converter power switches. Multiphase PMSG have many advantages such as reducing the amplitude and increasing the frequency of torque pulsations, higher reliability, and lowering the dc link current harmonics [8]-[11]. Therefore, multiphase PMSG are very suitable for the applications of high power, high reliability, and low dc bus voltage, such as renewable energy. These advantages have motivated the wind turbine manufacturers to use multiphase machines. For example, Spanish manufacturer Gamesa has developed a full-power 4.5 MW wind turbine with 6 parallel converters and 18-phase generator [13]. Some other topologies that use series connected generator-side converters have also been proposed to achieve medium voltage on the grid-side [14], [15].

Maximum power point tracking control in most of the conversion systems is implemented using wind speed data obtained from wind speed sensors [16]-[19]. However, accurate measurement of wind speed is not easy especially in case of large size wind turbines. Anemometer installed on the top of nacelle provides limited measurements of wind speed only at the hub height and cannot cover the whole span of large blades [20]. Moreover, due to the interaction between the rotor and the wind, anemometer, usually placed on nacelles, leads to inaccurate wind speed measurements in both upwind and downwind turbines. Therefore, Speed control of wind turbine based sensorless algorithms has gained many interests due to its accuracy and simplicity in tracking the maximum power point during wind speed variations [21, 22].

In this paper, maximum power point tracking for wind turbine based five-phase PMSG has been achieved by wind speed estimation technique. Estimation of the wind speed has been achieved measure rotor speed and the estimated load torque. A full scale power converter based five-phase has been used. The dc-link, connecting the back-to-back converters, allow fully decoupled control of the five-phase PMSG from the grid side. The MPPT has been achieved by controlling the PMSG speed at the generator side, whereas the grid side converter has been controlled to achieve unity

power factor at the grid side. The effectiveness of the proposed control technique in addition to the efficient operation of the wind turbine system has been verified using Matlab/Simulink.

2. Wind Energy Conversion System

Wind energy conversion system (WECS) converts kinetic energy of wind to mechanical energy by means of wind turbine rotor blades; then the generator converts the mechanical power to electrical power. The resulting electrical power is being fed to the electrical network through power electronic converters. In this paper, the WECS consists of a gearless wind turbine coupled to a five-phase PMSG, interfaced with the grid through back-to-back converters connected to each other through a common dc-link capacitor, as shown in Fig. 1.

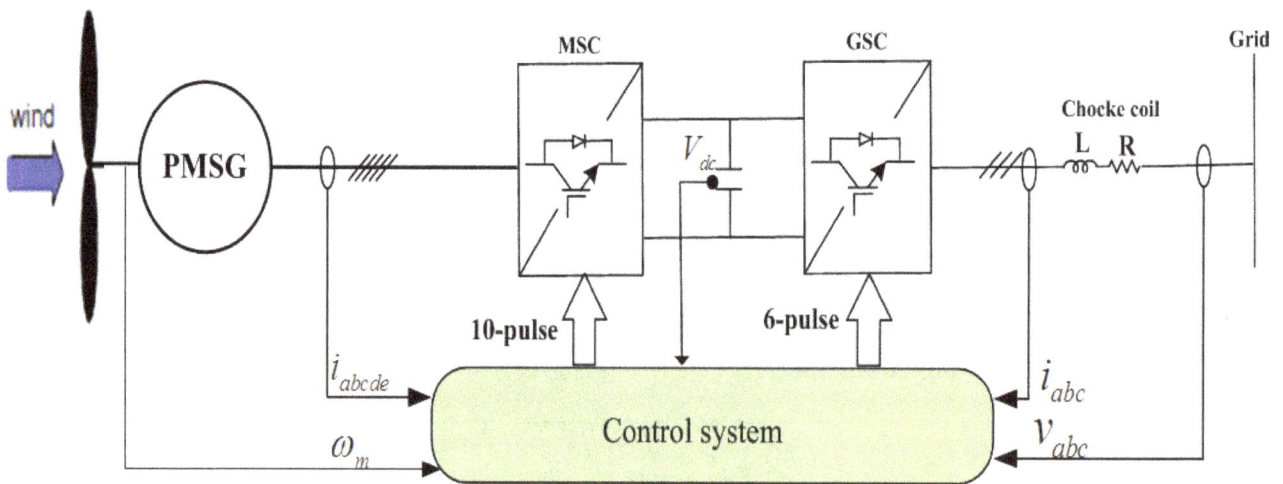

Fig. 1. Five phase PMSG based wind turbines.

2.1. Wind Turbine Model

The mechanical power captured from wind turbine can be formulated as follows:

$$P_m = \frac{1}{2}\rho A C_p(\lambda, \beta)v^3 \qquad (1)$$

Where P_m the mechanical output power (Watt), ρ is the air density (kg/m^3), A is the swept area (m^2), C_P is the power coefficient of the wind turbine, v is the wind speed (m/sec), λ is the tip speed ratio, β is pitch angle.

Consequently, the output energy is determined by power coefficient C_P, swept area, air density, tip speed ratio (λ) and pitch angle (β). If β is equal zero, the turbine power coefficient C_P and the tip speed ration λ can be formulated as follows:

$$C_P(\lambda) = \frac{60.04 - 4.69\lambda}{\lambda}\exp\left(\frac{-21 + 0.735\lambda}{\lambda}\right) + \frac{0.0068\lambda}{1 - 0.035\lambda} \qquad (2)$$

$$\lambda = \frac{\omega_m R}{v} \qquad (3)$$

Where ω_m is the rotor rotational speed (rad/sec), R is the

radius of blade (m).

The relation between c_p and λ when β equal zero degree is shown in Fig. 2. It can be noticed that the optimum value of c_p is about 0.48 for λ equal 8.1. Maximum power extraction from wind turbine can be achieved when the turbine operates at the optimum c_p. Therefore, it is necessary to control the rotor speed of the wind turbine at optimum c_p and λ during wind speed variation.

Based on the relations given in eq.(1) & (3), the optimum output power of the wind turbine can by formulated as follows:

$$P_{m\_opt} = \frac{1}{2}\rho A C_{p-opt}\left(\frac{\omega_{m-opt}R}{\lambda_{opt}}\right)^3 \qquad (4)$$

$$P_{m-opt} = K_{opt}(\omega_{m-opt})^3 \qquad (5)$$

Fig. 3 shows the relation between the mechanical powers generated by the turbine and the turbine rotor speed at different wind speeds. It is cleared that the maximum power point changes with the variation of wind speed and there is a unique maximum power point at each wind speed. The maximum power extraction can be achieved if the controller can properly follow the optimum curve with variation of wind speed, as shown in Fig. 3.

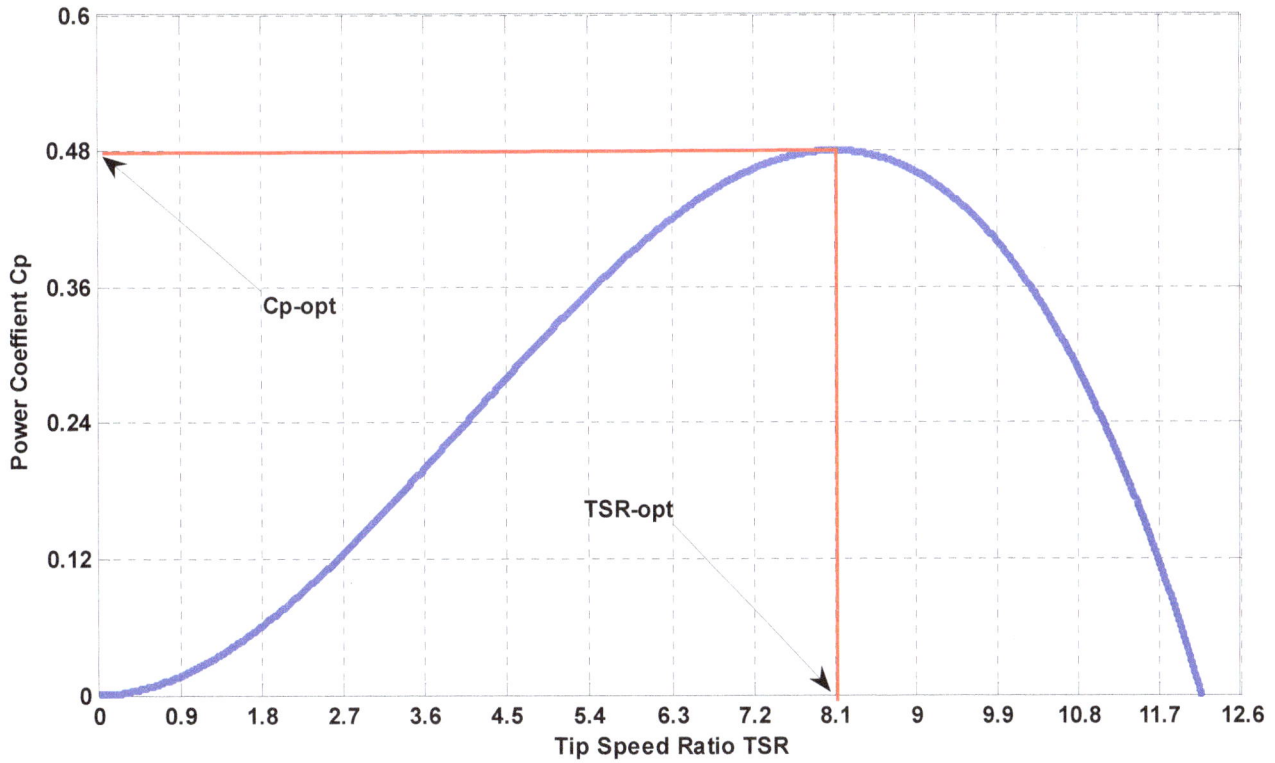

Fig. 2. The relation between power coefficient (Cp) and tip speed ratio (λ).

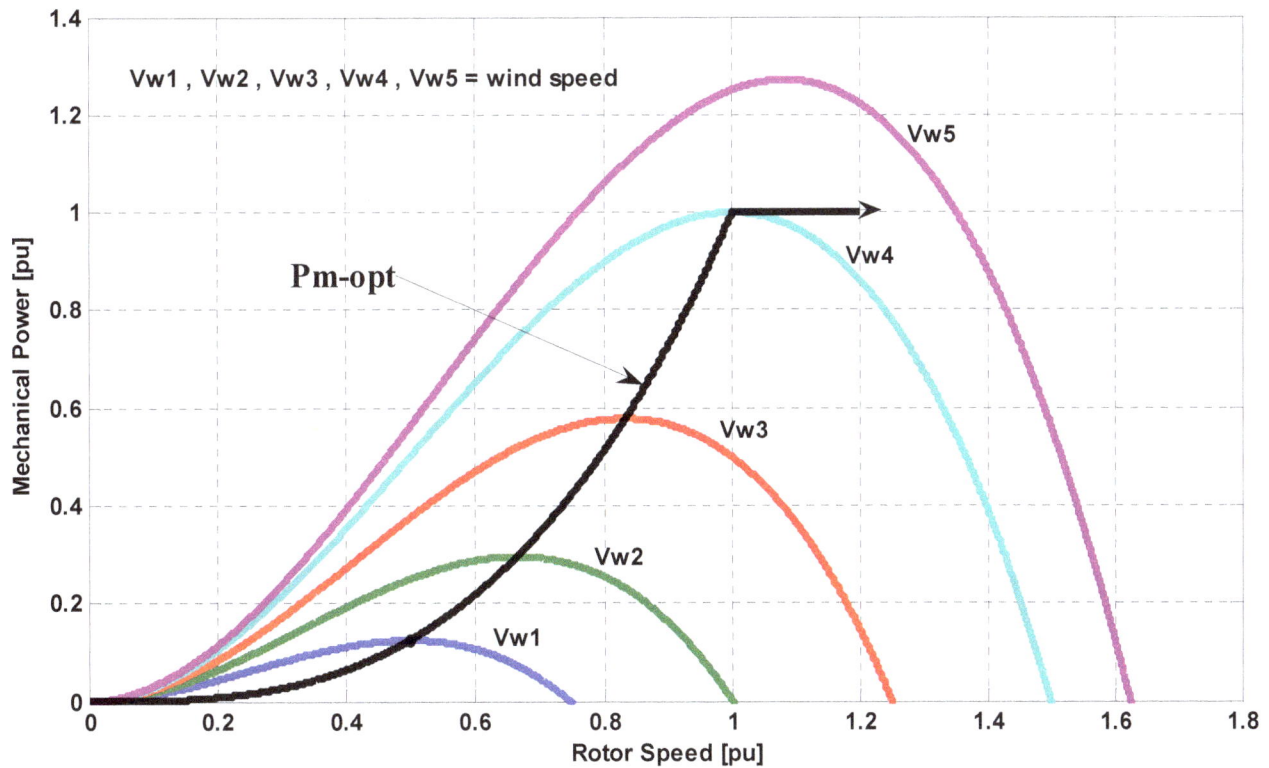

Fig. 3. The relation among generated mechanical powers and rotor speeds for different wind speeds.

2.2. Five-Phase PMSG Model

The voltage equations of five-phase permanent magnet synchronous generator expressed in the rotor reference frame using an extended park transformation (d_1,q_1 and d_2,q_2) axis can be described as follows:

$$V_{d1} = R_s i_{d1} + L\frac{di_{d1}}{dt} - L_q \omega_e i_{q1} \qquad (6)$$

$$V_{q1} = R_s i_{q1} + L\frac{di_{q1}}{dt} + L_d \omega_e i_{d1} + \psi \omega_e \qquad (7)$$

$$V_{d2} = R_s i_{d2} + L\frac{di_{d2}}{dt} \qquad (8)$$

$$V_{q2} = R_s i_{q2} + L\frac{di_{q2}}{dt} \qquad (9)$$

Where V_{d1}, V_{d2} and V_{q1}, V_{q2} represent the stator voltages in the (d, q) axis, i_{d1}, i_{d2} and i_{q1}, i_{q2} represent the currents in the (d,q) axis, R_s represent stator resistance, L represent armature inductance, L_d, L_q represents the (d, q) axis inductance, $\omega_e = p\omega_m$ (p is number of pole pairs, ω_m represent the turbine rotor angular speed and ψ is the permanent flux linkage).

The electrical torque of the five-phase PMSG can be formulated as:

$$T_e = \frac{5}{2} p \psi i_{q1} \qquad (10)$$

The mechanical equation of PMSG is given as follows:

$$T_m = T_e + f\omega_m + J\frac{d\omega_m}{dt} \qquad (11)$$

Where f is the friction coefficient, J the total moment of inertia and T_m is the mechanical torque produced by wind turbine, T_e is electromagnetic torque of PMSG.

3. Wind Speed Estimation Technique

Prior to explaining the wind speed estimation method, the nonlinear blade power coefficient curve needs to be approximated as third order polynomial [21] as follows;

$$C_p = a_0 + a_1\lambda + a_2\lambda^2 + a_3\lambda^3 \qquad (12)$$

Substituting (3) and (12) into (1) results in mechanical power as follows.

$$P_m = \frac{1}{2}\rho A\left(a_0 + a_1\frac{\omega_m R}{v} + a_2\frac{\omega_m^2 R^2}{v^2} + a_3\frac{\omega_m^3 R^3}{v^3}\right)v^3 \qquad (13)$$

Based on (13) wind speed can be formulated as follows

$$v^3 + \frac{a_1}{a_0}R\omega_m v^2 + \frac{a_2}{a_0}R^2\omega_m^2 v + \frac{a_3}{a_0}R^3\omega_m^3 - \frac{P_m}{0.5\rho A} = 0 \qquad (14)$$

Where
$P_m = T_m\omega_m$, $a_0 = 0.00715814$, $a_1 = -0.04454063$, $a_2 = 0.02899277$, $a_3 = -0.00202519$

The numerical solution for (14) generates three values for the wind speed. The second answer value is the more accurate empirical solution [23]. The mechanical power in (15) can be estimated using the detected rotor speed and the calculated torque as follows:

$$P_m = \omega_m(J\frac{d\omega_m}{dt} + f\omega_m + \frac{5}{2}p\psi i_{q1}) \qquad (15)$$

4. Control of Machine Side Converter (MSC)

Since the PMSG is a five-phase machine, the machine side converter has been built using five-leg of bidirectional IGBT switches, as shown in Fig. (4). The generator side converter is mainly used to control the wind turbine speed in order to extract maximum power P_{max}. In this case, the turbine should operate at maximum power coefficient $C_{p\,max}$. Therefore, it is necessary to keep the generator rotor speed ω_m at an optimum value of tip speed ratio $\lambda_{optimal}$. The PMSG rotor speed should be adjusted to follow the change of reference speed based on the change of wind speed and consequently adjust the turbine speed at wind variations. The MSC allows the generator to rotate at specified reference speed depending on wind speed variation. Fig. (4) shows the schematic diagram of the generator side converter control scheme.

In order to understand the speed control concept, the PMSG dynamic model should be studied [24]. The PMSG motion equation is given based on (11) as follows:

$$J\frac{d\omega_m}{dt} = T_m - T_e - f\omega_e \qquad (16)$$

The mechanical rotational speed of PMSG rotor is given by:

$$\omega_m = \frac{\omega_e}{p} = \omega_t G_r \qquad (17)$$

Where, ω_e electrical rotational speed of PMSG rotor (rad/s), ω_t turbine rotational speed and G_r gear ratio (if existed). For gearless PMSG based wind turbine $G_r = 1$. According to the characteristic of wind turbine at any value of wind speed, the rotational speed of the turbine rotor ω_m is regulated to the value $\omega_{m\,optimal}$ through generator side control hence:

$$\omega_{ref} = \omega_{m\,opt} = \frac{\lambda_{opt}v}{R} \qquad (18)$$

Therefore, the turbine power coefficient C_p is kept at its maximum value.

From (16), the speed control of generator can be achieved by the control of electromagnetic torque Te. From (10) the electromagnetic torque can be controlled directly by q-axis current component i_{q1}, therefore the speed can be controlled by controlling the q_1 axis current, as shown in Fig. (4). The reference q_1 axis current component can be formulated, based on (10), as follows.

$$i_{q1}^* = \frac{2}{5}\left(\frac{T_e^*}{p\psi}\right) \qquad (19)$$

(d_1, d_2 and q_2)-axis current components i_{d1}, i_{d2} and i_{q2} are set to zero to minimize the current and resistive copper losses for a given torque.

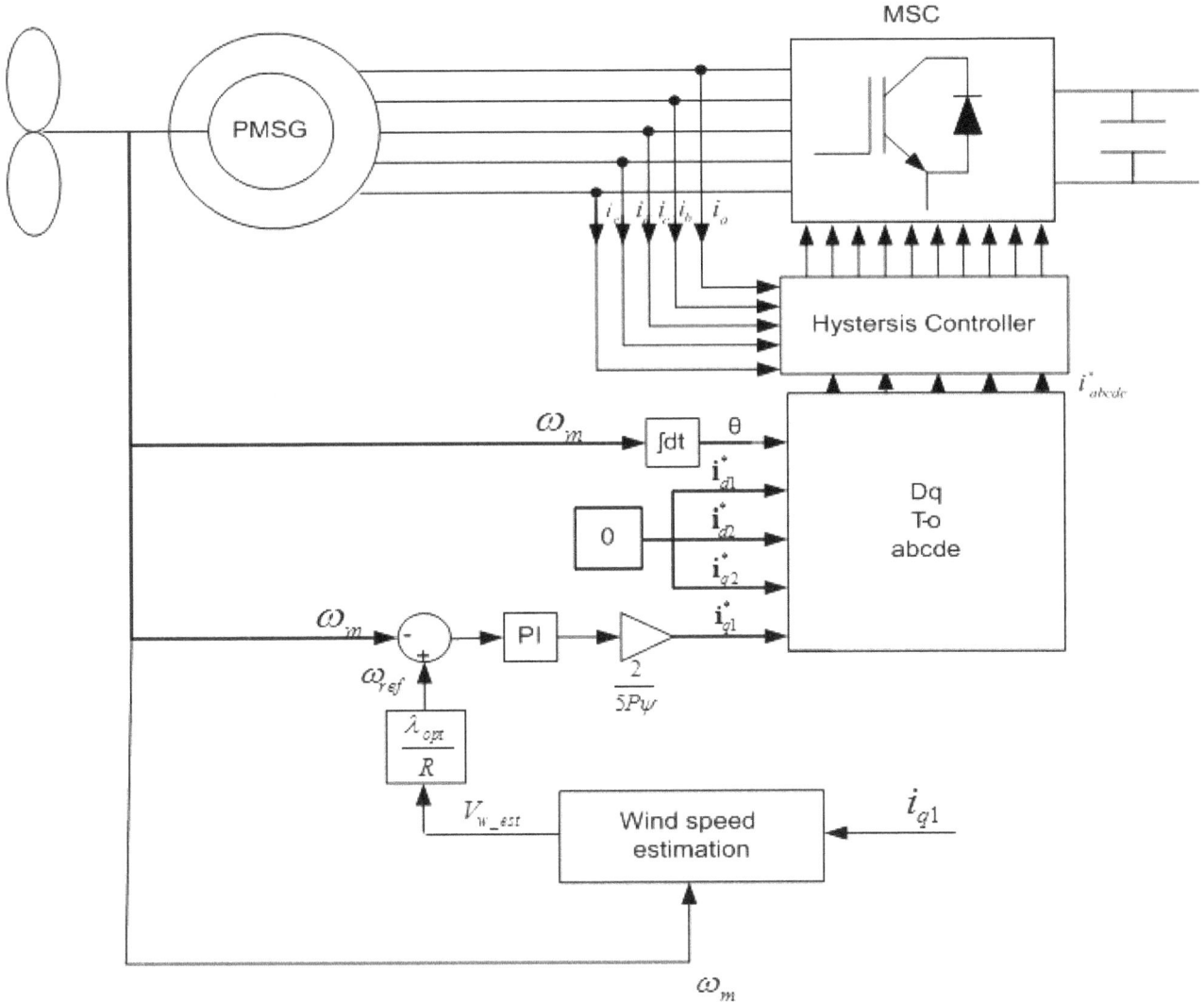

Fig. 4. *Generator Side Controller GSC.*

5. Grid-Side Converter Control

The objective of grid side converter control is to adjusts the DC link capacitor voltage at its reference value, and adjusts the active power and reactive power delivered to grid while wind changing. In grid side converter, a PI controller is used to stabilize the DC voltage reference value. The dynamic model of the grid connection, in reference frame rotating synchronously with the grid voltage, is given as follows [25]

$$V_{gd} = V_{id} - RI_{gd} - L\frac{d}{dt}I_{gd} + L\omega_g I_{gq} \qquad (20)$$

$$V_{gq} = V_{iq} - RI_{gq} - L\frac{d}{dt}I_{gq} - L\omega_g \qquad (21)$$

Where L and R are the grid inductance and resistance, respectively. V_{id} and V_{iq} are the d-q axis inverter voltage components. If the reference frame is oriented along the supply voltage, the grid vector voltage is:

$$V = V_{gd} + j0 \qquad (22)$$

Active and reactive power can be expressed as follows [25].

$$P_g = \frac{3}{2}V_{gd}I_{gd} \qquad (23)$$

$$Q_g = \frac{3}{2}V_{gd}I_{gq} \qquad (24)$$

It could be seen from above equations that we can control the active and reactive powers by respectively changing the d and q-current components. Also in order to transfer all the active power generated by the wind turbine the dc-link voltage must remain constant [26].

$$C\frac{dV_{dc}}{dt} = \frac{P_t}{V_{dc}} - \frac{P_g}{V_{dc}} \qquad (25)$$

Where subscript 'g' refers to the grid and 't' refers to the wind turbine.

Based on (25), if the two powers (the wind turbine power and the grid power) are equal there will be no change in the dc-link voltage. The grid side converter control scheme contains two cascaded loops. The inner loop controls the network currents and the outer loop controls the DC-link voltage. The inner loop regulates the power flow of the system by controlling the active and reactive power delivered to the

electrical grid. Further, unity power factor flow (zero reactive power exchange) could be easily obtained, unless the grid operators require different reactive power settings.

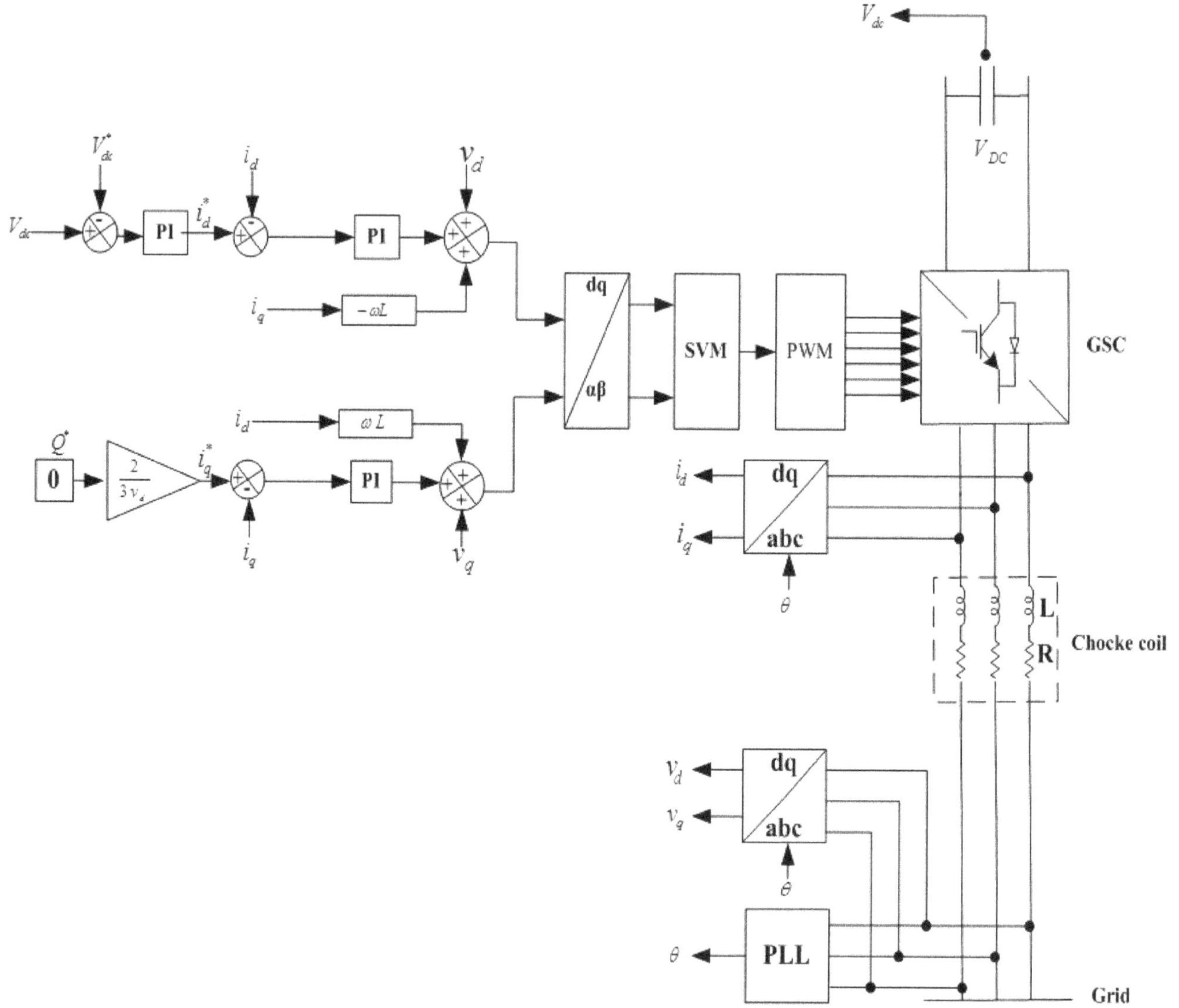

Fig. 5. *Grid side converter control.*

6. Simulation Results and Discussion

The model of wind turbine based five-phase PMSG in addition to the back-to-back- interface converters for grid connection have been carried out using Matlab/Simulink. The parameters of the system under study are given in appendix A. The proposed control scheme of the five-phase PMSG based variable speed WECS has been carried out using MATLAB/Simulink at different values of wind speed in order to investigate the wind speed estimation technique and the MPPT at the generator side in addition to the unity power factor control at the grid side and common dc-link capacitor voltage control of the interface converters.

6.1. Ramp Change Wind Speed

Fig.6 shows the actual and estimated wind speed, error in

wind speed, the reference and actual rotor speed, power coefficient, tip speed ratio, mechanical power, the mechanical and electromagnetic torque of the PMSG and the five-phase current of PMSG. According to wind turbine characteristic, the estimated and actual wind speed values coincide well and when wind speed varies the controller adjust PMSG rotor speed to follow the same value of ω_{m-opt}. It is clear that the actual and reference rotor speed agree well and the error between them is very small. Moreover, the power coefficient and tip speed ratio are almost constant following their optimal values for the whole simulation period. This in-turn prove that the MPPT has been achieved. The actual and estimated mechanical power agree well with the maximum power. The mechanical and electromagnetic torques of five-phase PMSG are varying according to the change in wind speed. Fig.7 shows the dc-link capacitor voltage, grid voltage and current, dq-axis grid current, power factor, and injected active and reactive power. It is the clear that sinusoidal grid voltage and current are in-phase for the whole simulation period to achieve unity power factor. The reference and actual dc-link voltage coincide well. The actual and reference q-axis current at grid side is always controlled to be zero to achieve unity power factor. Therefore, the injected reactive power is zero during the whole simulation time, whereas the injected active power has a changes according to the change in the wind speed.

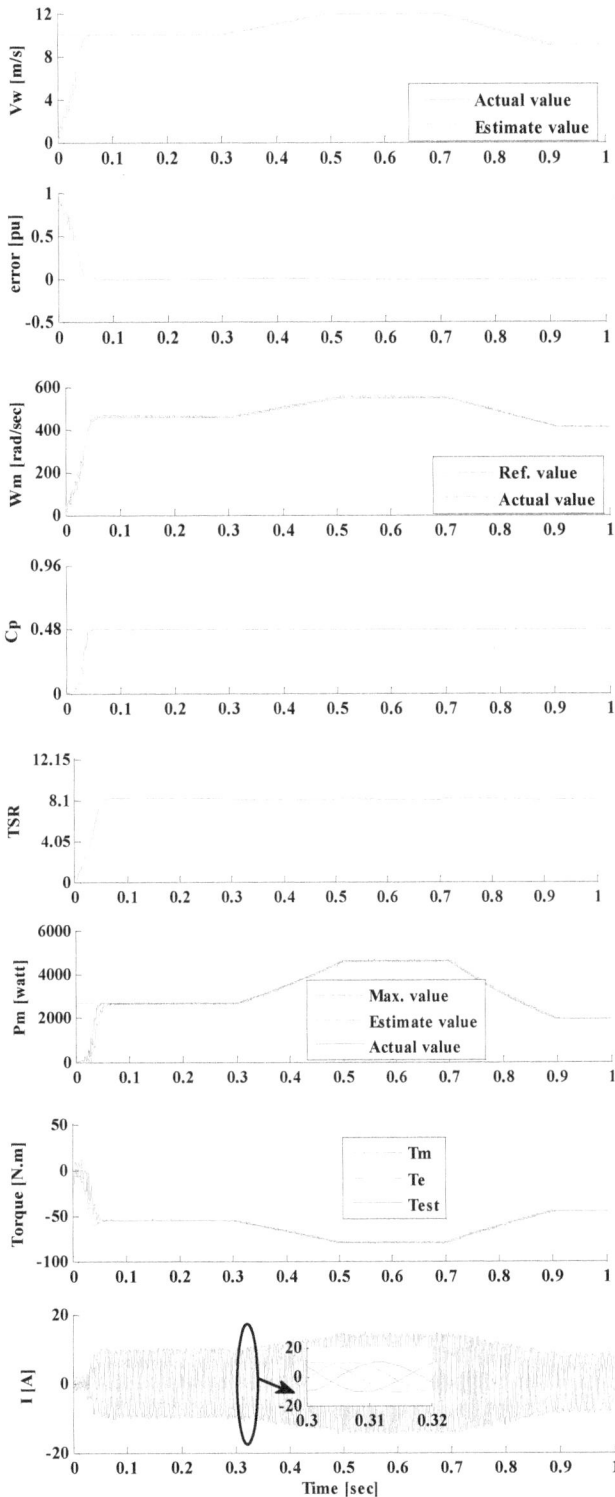

Fig. 6. Simulation results under machine side converter.

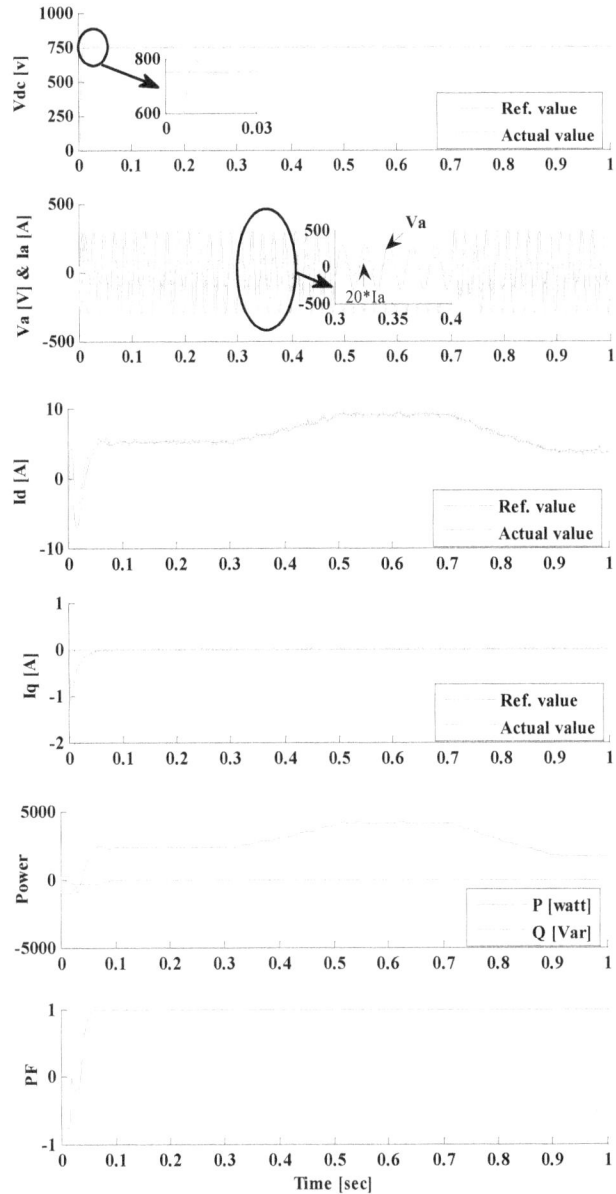

Fig. 7. Simulation results under grid side converter.

6.2. Random Wind Speed

The effectiveness of the proposed control techniques has been investigated with Random wind speed variation. Fig.8 shows the actual and estimated wind speed, error in wind speed, the reference and actual rotor speed, power coefficient, tip speed ratio, mechanical power, the mechanical and electromagnetic torque of the PMSG and the five-phase current of PMSG. It is clear that the difference between estimated and actual wind speed is very small, the turbine shaft speed is controlled to track its reference value. Achievement of MPPT is known, from the power coefficient, which is almost constant value at (0.48) and the change of the tip speed ratio that varies in a relatively small range around the optimal value of (8.1). The actual and estimated mechanical power agrees well with the maximum power. The mechanical and electromagnetic torques of PMSG coincides well. Fig. 9 shows the dc-link capacitor voltage, grid voltage and current, dq-axis grid current, power factor and injected active and reactive power. It is the clear that sinusoidal grid voltage and current are in-phase for the whole simulating period to achieve unity power factor. The reference and actual dc-link voltage coincide well. The actual and reference q-axis current at grid side is always controlled to be zero to achieve unity power factor. The injected reactive power is zero during the whole simulation time, whereas the injected active power has a step change according to the change in the wind speed.

Simulation results prove that the wind speed estimation algorithm has the ability to estimate the wind speed, the MSC has the ability to control the PMSG to extract the maximum power based on MPPT control technique, and the GSC has the ability to achieve unity power factor at the grid side.

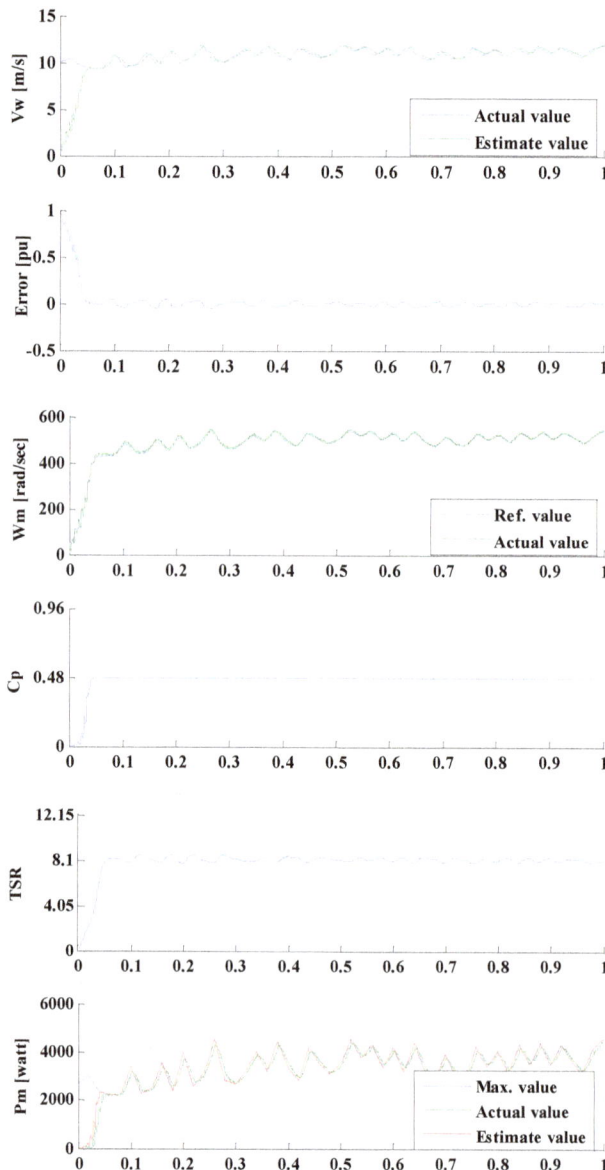

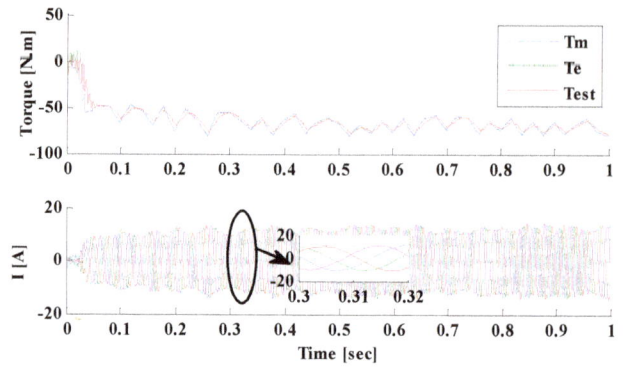

Fig. 8. Simulation results under machine side converter.

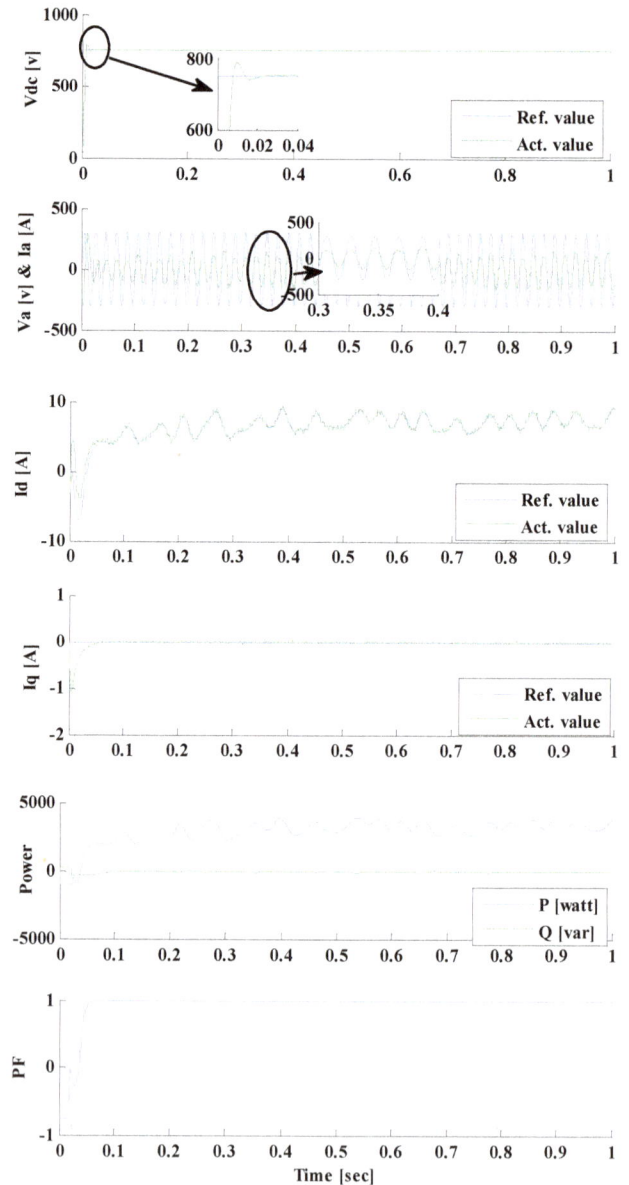

Fig. 9. Simulation results under grid side converter.

7. Conclusion

This paper has presented comprehensive modeling of direct driven five-phase PMSG based grid-connected wind turbines along with the control schemes of the interfacing converters. Wind speed estimation has been achieved based on measured rotor speed. Five-phase to three-phase interface power converter based back-to-back common dc-link converter has been used to achieve the system objectives. The generator side converter has been used to achieve maximum power operation point at each wind speed. The grid side converter has been used to inject sinusoidal current in-phase with the grid voltage in addition to controlling the common dc-link capacitor voltage. Vector current controller has been employed on the grid side VSI to obtain unity power factor. Simulation results prove that the proposed control scheme has a great capability to obtain unity power factor at the grid side and to achieve sensorless maximum power point tracking of wind turbine based five-phase PMSG during wind speed variation.

Appendix

Appendix 1. Specification of Wind Turbine

blade radius	$R = 1.8m$
Air density	$\rho = 1.225 \, kg/m^3$
Optimal tip speed ratio	$\lambda_{opti} = 8.1$
Maximum power Coefficient	$C_{p-max} = 0.48$

Appendix 2. Five-Phase PMSG Parameters

Pole pairs number	$n_p = 5$
Stator resistance	$R_s = 0.425\Omega$
Direct-axis inductance	$L_d = 0.00835H$
quadrature-axis inductance	$L_q = 0.00835H$
Moment of inertia	$J = 0.01197 kg.m^2$
Flux linkage	$\psi = 0.433wb$

Appendix 3. DC Bus and Gird Parameters

dc-link voltage	$V_{dc} = 750V$
Capacitor of the dc-link	$C = 2000\mu F$
Grid frequency	$F = 50hz$
Grid resistance	$R_g = 0.015\Omega$
Grid inductance	$L_g = 0.002H$

Appendix 4. Machine Side Control

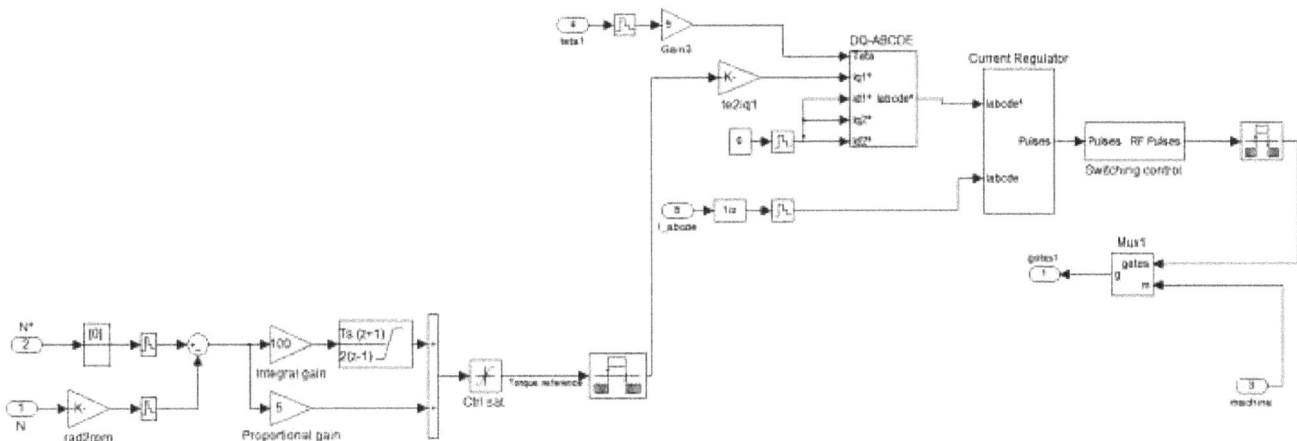

Hysteresis-current control of five-phase PMSG

Appendix 5. Grid Side Control

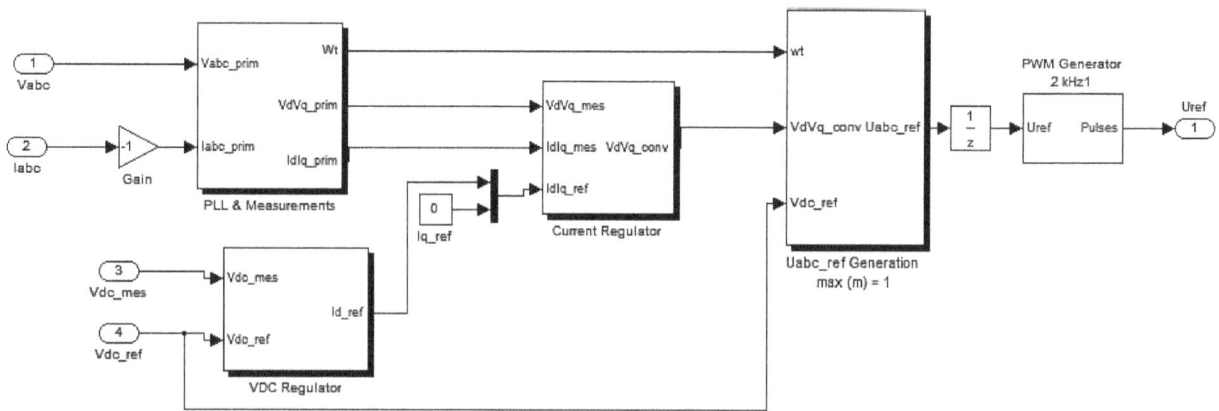

Vector control of GSC

References

[1] R. Zavadil, N. Miller, A. Ellis, and E. Muljadi, "Making connections: Wind generation challenges and progress," *IEEE Power Energy Mag.*, vol. 3, no. 6, pp. 26-37, Nov. 2005.

[2] Z. Chen, J. M. Guerrero, and F. Blaabjerg, "A review of the state of the art of power electronics for wind turbines," IEEE Trans. Power Electronic., vol. 24, no. 8, pp. 1859-1875, Aug. 2009.

[3] S. J. Ockel, "High energy production plus built in reliability - The new vensys 70/77 gearless wind turbines in the 1.5MW class," in Eur. Wind Energy Conf., Athens, Greece, 2006.

[4] Y. Chen, P. Pillary, and A. Khan, "PM wind generator topologies," IEEE Transactions Industry Applications vol. 41, no. 6, pp. 1619-1626, Nov./Dec. 2005.

[5] H. Polinder, S. W. H. de Haan, M. R. Dubois, and J. slootweg, "Basic operation principles and electrical conversion systems of wind turbines," in presented at the Nordic Workshop Power Ind. Electron., Trondheim, Norway, Jun.14-16,2004.

[6] G. Michalke, A. D. Hansen, and T. Hartkopf, "Control strategy of a variable speed wind turbine with multipole permanent magnet synchronous generator," in Eur.Wind Energy Conf. Exhib., Milan, Italy, May 7-10,2007.

[7] A. Graures, "Efficiency of three wind energy generator systems," IEEE Trans. Energy Convers., vol. 11, no. 3, pp. 650-657, Sep. 1996.

[8] Fei Yu, Xiaofeng Zhang, Huaishu Li, Zhihao Ye. "The space vector PWM control reserch of a multi-phase permanent magnet synchronous motor for electrical propulsion," Electrical Machines and Systems, 2003. ICEMS 2003 vol. 2, pp. 604-607, Nov. 2003.

[9] Ruhe Shi, H. A. Toliyat, "Vector Control of Five-phase Synchronous. Reluctance Motor with Space Pulse Width Modulation for Minimum. Switching Losses", Industry Applications Conference, 36th IAS Annual. Meeting. Vol. 3, pp. 2097-2103, 30 Sept.-4 Oct. 2001.

[10] Xue S,Wen X. H, "Simulation of A Novel Multi-phase SVPWM Strategy," in IEEE International Conf. on Power Electronics and Drive Systems (PEDS), 2005, pp. 756-760.

[11] Parsa L, H. A.Toliyat, "Multi-phase permanent motor drives," in Industry Applications Conf., 38th IAS Annual Meeting.Vol.1, 12-16 Oct. 2003, pp. 401-408.

[12] Z. Xiang-Jun,Y. Yongbing, Z. Hongato, L. Ying, F. Luguang, and Y. Xu, "Modelling and control of a multi-phase permanent magnet synchronous generator and efficient hybrid 3L-converters for large direct-drive wind turbines," IET Electric Power Applications, vol. 6, no. 6, pp. 322-331, 2012.

[13] B. Andresen and J. Brik, "A high power density converter system for the Gamesa G10x4.5MW wind turbine," in Proc.European Conf. on Power Electronics and Applications EPE, 2007.

[14] M. J. Duran, S. Kouro, B. Wu, E. Levi, F. Barrero, and S. Alepuz, "Six-phase PMSG wind energy conversion system based on medium voltage multilevel converter," in Proc. Europen Conf. on power Electronics and Applications EPE, 2011.

[15] H. S Che, W. P. Hew, N. A. Rahim, E. Levi, M. Jones, M. J Duran "A six-phase PMSG wind energy induction generator system with series connected DC-Links," IEEE Power Electronics for Distrubated Generation Systems PEDG, pp. 26-33, 2012.

[16] W. M. Lin, C. M. Hong, and F. S. Cheng, "Fuzzy neural network output maximization control for sensorless wind energy conversion system," Energy, vol. 35, no. 2, pp. 592-601, Feb. 2010.

[17] R. Chedid, F. Mrad, and M. Basma, "Intellignet Control of a class of wind energy conversion systems," IEEE.Trans. Energy Conv., vol. 14, no. 4, pp. 1597-1604, 1999.

[18] J. S. Thongam, P. Bouchard, H. Ezzaidi and M. Ouhrouche, "Artifical Neural Network Based Maximum Power Point Tracking Control for Varaible Speed Wind Energy Conversion Systems," in proc. of IEEE MSC2009, July 8-10,2009.

[19] A. Youssef, M. A. Sayed, M. N. Abdelwhab and F. A. Khalifa ""Control Scheme of Five-Phase PMSG Based Wind Turbine"," in MEPCON'2014, cairo, Dec.2014.

[20] K. E. Johnson and L. Y. Pao, "A tutorial on the dynamics and control of wind turbines and wind farms," in Proc. American Control Conf. (ACC09), june 10-12,2009, pp. 2076-2089.

[21] C. X. Hye-Su Shin, J.-M. Lee, J.-D. La and Y.-S. Kim, ""MPPT control technique for a PMSG wind generation system by the estimation of the wind speed," in Electrical Machines and Systems (ICEMS), 2012 15th International Conference, Sapporo, 21-24 Oct. 2012.

[22] Alaa Eldien M. M. Hassan, Mahmoud A. Sayed, Essam E. M. Mohamed. "Three-phase Matrix Converter Applied to Wind Energy Conversion System for Wind Speed Estimation". International Journal of Sustainable and Green Energy. Vol. 4, No. 3, 2015, pp. 117-124.

[23] L. V. Fausett, Applied numerical analysis using MATLAB: Pearson, 2008.

[24] W. Qiao, L. Qu, and R. G. Harely "Control of IPM Synchronous Generator for Maximum Wind Power Generation Considering Magnetic Saturation," IEEE Trans. Industry application, vol. 45, no. 3, May/June 2009.

[25] Chinchilia. M, Arnaltes S, Burgos J. Control of permanent-magnet generator applied to variable-speed wind-energy systems connected to the grid. IEEE Trans Energy Convers 2006;21(1).

[26] Muyeen SM, Takahashi R, Murata T, Tamura J. A variable speed wind turbine control strategy to meet wind farm grid code requirements. IEEE Trans Power Syst 2010;25(1).

26

Hydrogen Production by Water Electrochemical Photolysis Using PV-Module

Sergii Bespalko[*]**, Anton Kachymov, Kostiantyn Koberidze, Oleksandr Bespalko**

Department of Energy Technologies, Cherkasy State Technological University, Cherkasy, Ukraine

Email address:

sergiibespalko@gmail.com (S. Bespalko)

Abstract: The experimental research on hydrogen production by water electrochemical splitting is presented in the article. In the study low temperature electrolytic unit with 26th% KOH liquid solution and small-scale photovoltaic module (PV-module) were used to convert solar energy into molecular hydrogen. Speeds and volumes of average monthly hydrogen production are defined for Kyiv insolation using experimental facilities. The method applied can be proposed to estimate hydrogen amount generated when combining the conventional electrolysis process and photovoltaic module for compensating the long term fluctuations of solar photovoltaics.

Keywords: Hydrogen, Electrolysis, Photovoltaic Module, Solar Energy

1. Introduction

Stabilizing future atmospheric CO_2-levels at less than a doubling of pre-industrial levels will be a difficult task because it requires a continuous flow of new carbon-free power 2-3 times greater than today's energy supply to sustain economic development for a global population approaching 10 billion people by the middle of 21st century [1].

The sun and wind are the two largest sustainable sources of carbon-free power. However, to realize their potential, they must overcome a key hurdle – a challenge of their intermittent nature. Unlike other forms of renewable energy such as hydropower and geothermal energy, the energy generated by wind and photovoltaics fluctuates. This fluctuation poses a sizable challenge to their power grid integration and a widespread adoption as the mainstream power sources [1, 2].

There are several potential answers to the intermittency challenge and one of the more viable solutions is a credible form of electricity storage [2].

Power storage can improve the efficiency and reliability of the electric utility system by reducing the spinning reserve requirements to meet peak power demands. This makes better use of efficient base load generation and allows greater use of intermittent renewable energy technologies. Energy storage technologies include utility battery storage, flywheel storage, superconducting magnetic energy storage, compressed air energy storage, pumped hydropower, and super capacitors.

Additionally, hydrogen may be used as an energy storage medium [3, 4].

Concerning pumped hydropower and compressed air energy storage systems, hydrogen storage has somewhat higher investment costs and a lower efficiency. Simultaneously, it has significantly higher energy density and hence, significantly higher energy capacity. This, combined with fuel cell technology, makes hydrogen storage most appropriate for the compensation of long-term fluctuations [2].

To have a highly effective and efficient renewable-hydrogen system, hydrogen should be used at the chosen time. When renewable resources are available, e.g. the sun is shining, and electricity is needed, the electric current should be used immediately. To meet even higher electricity demands, energy can be supplied directly from renewable resources as well as from hydrogen stores. As demands decrease, the extra electricity from renewables can be converted and stored as hydrogen.

Additionally, hydrogen provides a connecting point between renewable electricity production, transportation, and portable energy needs. In transportation applications, hydrogen provides a way to convert renewable resources to fuel for vehicles. In portable energy, hydrogen with fuel cells can be used as an important power source for mobile electronic devices, offering key advantages over conventional batteries. It will increase operating times, it will reduce the weight, and it can be recharged easily. At the same time, hydrogen can store energy for a long period without any power dissipation.

This entire portfolio of options makes renewable hydrogen systems more effective in providing flexible and reliable energy in the most necessary forms [5].

2. Fundamental Principal

The water electrochemical photolysis is a method exploiting photovoltaic modules for generating low-grade electric energy used for hydrogen production by conventional water electrolysis [6, 7].

Electrolysis is a process that occurs when direct current passes through the electrolytic system composed of an anode, a cathode and electrolyte. The resulting reaction is as follow [8, 9]:

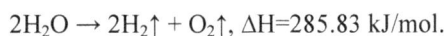

$$2H_2O \rightarrow 2H_2\uparrow + O_2\uparrow, \Delta H = 285.83 \text{ kJ/mol}.$$

Hydrogen production by the conventional water electrolysis obeys the Faraday's law of electrolysis [10]:

$$m = K \times q,$$

with m – separated substance mass, K – electrochemical equivalent, q – electrical charge passed through the electrolyte.

In turn, the electrical charge is defined as follow [10]:

$$q = I \times \tau,$$

with I – electrical current, τ – operating time of electrolyzer.

The electrochemical equivalent K of a chemical element is the mass transported by one coulomb of electricity, e.g. the electrochemical equivalent for hydrogen is $1.045 \cdot 10^{-8}$ kg/C [8].

Hydrogen production by conventional electrolysis process has the following advantages over other hydrogen producing methods [6, 11]:

- the produced hydrogen is about 99% pure,
- the electrolytic cell is simple, continuous, automatic, and without gear motion,
- the most widespread chemical substance notably water is used in electrolysis,
- and finally, there is possibility of using renewable energy sources for hydrogen production.

Objective of the research is to study the average monthly hydrogen productions by water electrochemical photolysis using PV-module for Kiev insolation.

3. Experimental Method

Principle circuit of the experimental facility that was used for hydrogen production by the conventional water electrolysis is presented in fig. 1.

The photovoltaic module generated electric energy and the electric current went to the electrolyzer by cords. The electrolyzer was a tank with two carbon electrodes immersed in the electrolyte. Transparent plastic tubes were located over carbon electrodes for capturing hydrogen and oxygen bubbles. These tubes had graded scales with a division value equal to 0.2 ml.

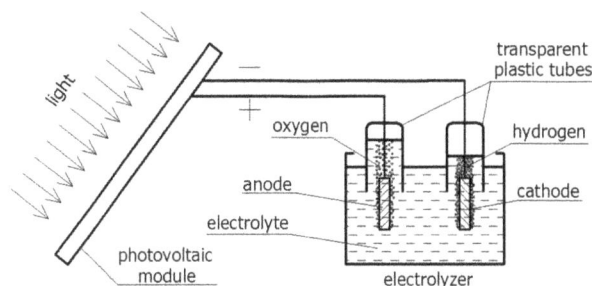

Figure 1. *Principle circuit of the experimental facility. [12].*

When light acted on the sloping surface of the photovoltaic module, the electrolysis process began. After that molecular hydrogen was created on the cathodic surface and oxygen was created on the anodic surface. Hydrogen and oxygen bubbles rose to hollow tubes displacing the electrolyte. Hydrogen production speed was determined by captured gas quantity per unit time.

In the experiment 26th% KOH liquid solution was used as the electrolyte for electrolysis process.

Characteristics of the used PV-module and electrolyzer are presented in the table 1 and table 2 respectively.

Table 1. *Characteristics of the photovoltaic module [12].*

Type of photovoltaic module	KV-10W/12V
Type of silicon	mono
Overall PV-module dimensions, mm	527×233×34
Maximum power, W	10±3%
Efficiency, %	10
Voltage at maximum power, V	16.5
Current at maximum power, A	0.7
Open-circuit voltage, V	20
Short-circuit current, A	0.84
Active surface area, m²	≈0.1

Table 2. *Characteristics of the electrolyzer [12].*

Pressure	standard
Temperature, °C	75
Type of electrolyte	KOH liquid solution
Electrolyte concentration, %	26
Electrolyte volume, liters	1
Electrode material	carbon
Electrode surface area, mm²	1120
Maximal current density, A/cm²	0,025
Distance between electrodes, mm	16

Photo of the used experimental facility is illustrated in fig. 2.

4. Implementation and Results

When determining the PV-module average monthly current-voltage curves, the standard procedure was used. However instead of a constant light source, the light source with adjustable radiation intensity was utilized. It enabled to influence the PV-module sloping surface by controlling radiation intensities, which were equal to the Kiev average monthly solar intensities (table 3).

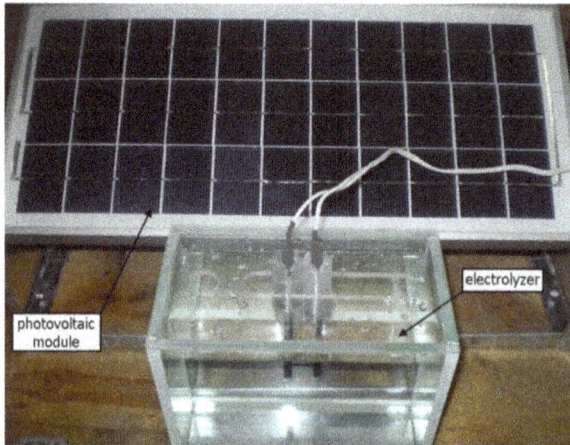

Figure 2. Photo of the experimental facility. [12].

As a result, the average monthly current-voltage curves were determined for Kiev insolation (fig. 3). Thus, in this diagram the larger average monthly solar intensity (June) satisfies the greater amount of electrical current.

Table 3. *Kiev insolation data.*

Month	Average monthly solar intensities that act on sloping surface of PV-module, W/m²
January	77.8
February	106.4
March	153.8
April	170.7
May	197.5
June	213.1
July	206.4
August	198.7
September	183.1
October	137.1
November	59.9
December	52.3

The slope angle of the photovoltaic module was equal to 50° (latitude angle of Kiev).

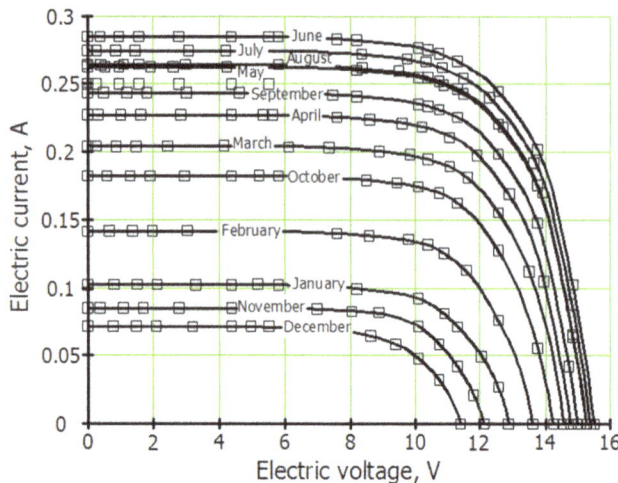

Figure 3. *Average monthly current-voltage curves of the photovoltaic module for Kiev insolation.*

For determination of the hydrogen producing operating points in the electrolysis process, the experimental current-voltage curve of electrolyzer with 26th% KOH liquid solution was tested (fig. 4). Here, with low electrolysis voltage less than 1.23 V, hydrogen production does not occur. Practical zero value of the electrical current confirms non-hydrogen generation. Increasing the voltage to more than 1.23 V generates hydrogen exponentially.

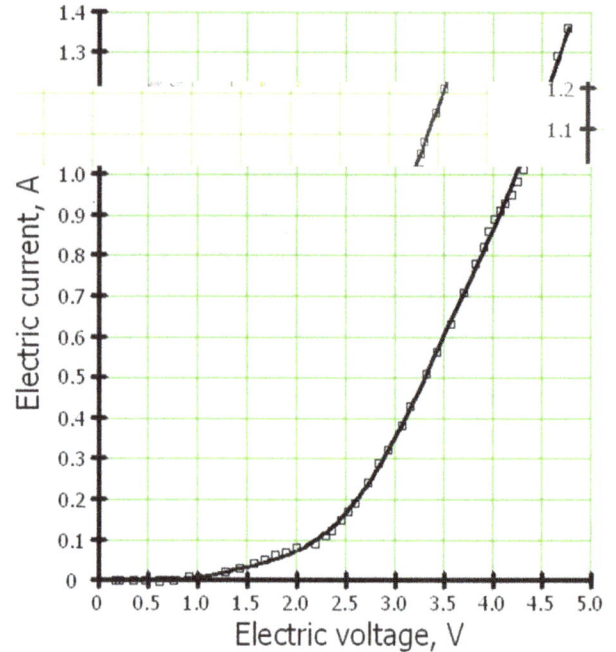

Figure 4. *Electrolysis current-voltage curve (electrolyte is 26th% KOH liquid solution).*

The intersections of the electrolysis current-voltage curve and the average monthly current-voltage curves of the photovoltaic module determined the operating points of hydrogen production for each month (fig. 5). In this diagram, operating electric currents are equal to the short-circuit currents of the photovoltaic module. Thus, the PV-module operation with an electrolyzer corresponds to the short-circuit conditions.

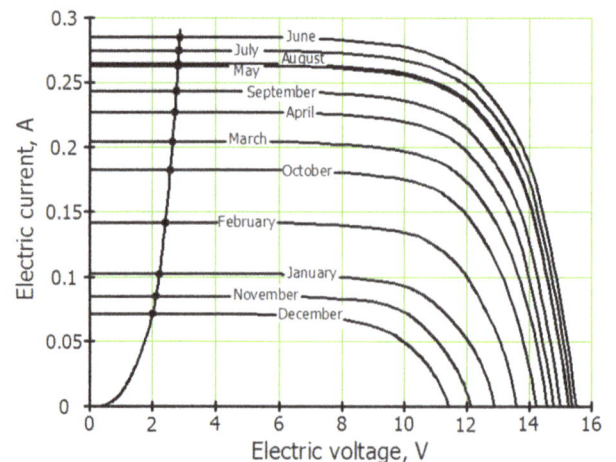

Figure 5. *Operating points of hydrogen production by the experimental facility for Kiev insolation.*

A photo of the hydrogen producing process is shown in fig. 6.

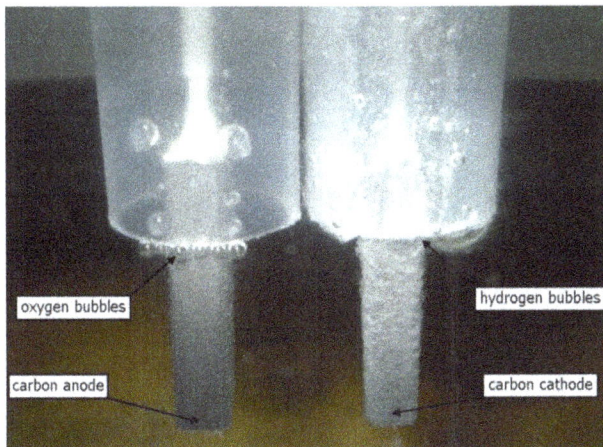

Figure 6. Process of hydrogen production. [12]

Thus, in the issue of performed experiments, the speeds of hydrogen production by the electrolysis process were determined for each month (fig. 7). As expected in June, the speed of hydrogen production is maximal (about 0.035 *milliliters/second*) and minimal speed is in December (about 0.009 *milliliters/second*).

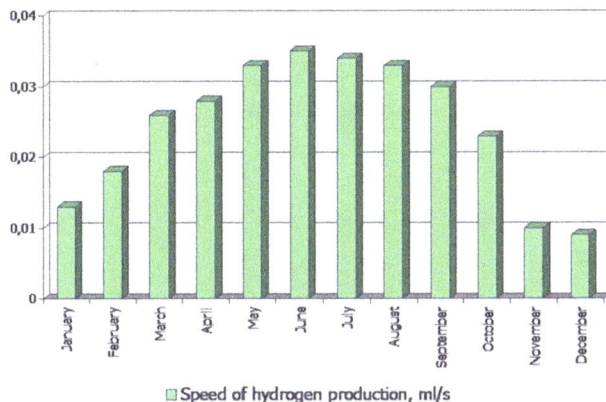

Figure 7. Speeds of hydrogen production by the experimental facility for Kiev isolation.

Volumes of produced hydrogen by the experimental facility for Kiev insolation are shown in fig. 8.

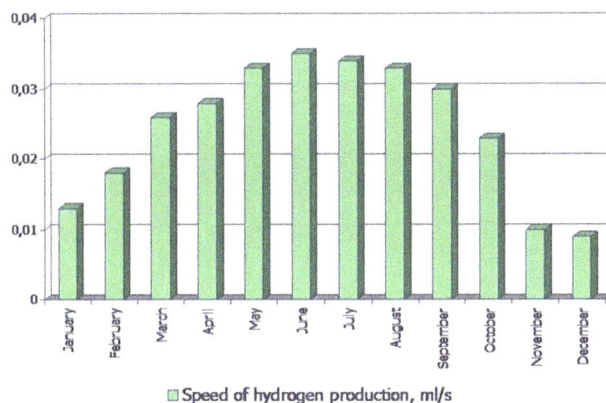

Figure 8. Volumes of hydrogen production by the experimental facility for Kiev insolation.

Thus, in June and July volumes of produced hydrogen undoubtedly are the highest (about 0.092 m^3/month).

The average energy required for producing a normal cubic meter of hydrogen and 0.5 cubic meter of oxygen by the experimental facility is about 5.7 kW-hours.

Overall efficiency of the solar energy conversion into molecular hydrogen is about 5.2%. In the first place this very low efficiency is due to low efficiency of used PV-module (about 10%).

5. Conclusion

Average monthly hydrogen generation can be estimated using a method proposed in the paper. The method is based on utilization of the Faraday's law of electrolysis and operating electric currents defined for each month through intersection of the electrolysis current-voltage curve and the average monthly current-voltage curves of the PV-module.

The method proposed can be applied for estimation of speeds and volumes of hydrogen production combining the conventional water electrolysis and photovoltaics, e.g. when compensation of long term fluctuation of solar PV is needed.

The energy required for producing a cubic meter of hydrogen and 0.5 cubic meter of oxygen by the experimental facility is about 5.7 kW-hours.

Currently this method of hydrogen production, notably using PV-modules has very low efficiency, e.g. the overall efficiency of solar energy conversion into molecular hydrogen is about 5.2% only. First of all this very low efficiency is due to low efficiency of the used PV-module. Therefore for increasing the overall efficiency of the solar energy conversion into molecular hydrogen it is necessary to use more effective PV-modules and also improve efficiency of the electrolysis process.

References

[1]   Berry, G. (2004). Present and future electricity storage for intermittent renewables, Workshop proceedings "The 10-50 Solution: Technologies and Policies for a Low-Carbon Future", The Pew Center on Global Climate Change and the National Commission on Energy Policy, Washington, pp. 217-220.

[2]   Pieper, C., Rubel, H. (2010). Electricity Storage: Making Large-Scale Adoption of Wind and Solar Energies a Reality. The Boston Consulting Group, Inc.

[3]   BPL Global, A Smart Plan for Electric Utilities, viewed 17 December 2014 <http://www.bplglobal.net/eng/knowledge-center/download.aspx?id=130>.

[4]   Carnegie, R., Gotham, D., Nderitu, D., Preckel, P. (2013). Utility Scale Energy Storage Systems: Benefits, Applications, and Technologies. State Utility Forecasting Group.

[5]   Fuel Cell and Hydrogen Energy Association (FCHEA), Renewable Hydrogen Production Using Electrolysis, viewed 17 December 2014, <http://ftp.fchea.org/core/import/PDFs/factsheets/Renewable%20Hydrogen%20Production%20Using%20Electrolysis_NEW.pdf>.

[6] Dincer, I., Joshi, A. (2013). Solar Based Hydrogen Production Systems, SpringerBriefs in Energy.

[7] Pyle, W., Healy, J., Cortez, R. (1994). Solar Hydrogen Production by Electrolysis. Home Power, #39, pp. 32-38.

[8] Häussinger, P., Lohmüller, R., Watson, A. (2004). Hydrogen, Ullmann's Encyclopedia of Industrial Chemistry, Wiley-VCH, NewYork and Weinheim.

[9] Santos, D., Sequeira, C. (2013). Hydrogen production by alkaline water electrolysis. Quim. Nova, Vol. 36, No. 8, 1176-1193.

[10] Atkins, P., Julio de Paula. Physical Chemistry. W.H. Freeman and Company. New York, 2006.

[11] Steward, D., Saur, G., Penev, M., Ramsden, T. (2009). Lifecycle Cost Analysis of Hydrogen Versus Other Technologies for Electrical Energy Storage: Technical Report. National Renewable Energy Laboratory / TP-560-46719.

[12] Bespalko, S., Kachimov, A., Koberidze, K. (2013). Experimental Facility for Investigation of the Hydrogen Generation by Water Electrochemical Splitting Using PV-Module, Proceedings of the 12th International Research Conference "Physical Processes and Fields of Technical and Biological Objects", pp. 3-5.

27

Removal of Phenol and Parachlorophenol from Synthetic Wastewater Using Prepared Activated Carbon from Agricultural Wastes

Muzher Mahdi Ibrahem AL-Doury[1], Suha Sameen Ali[2]

[1]Petroleum and Minerals Engineering College, Tikrit University, Tikrit, Iraq
[2]Chemical Engineering Dept, College of Engineering, Tikrit University, Tikrit, Iraq

Email address:
sar31205@yahoo.com (M. M. I. AL-Doury), engineer.suha@yahoo.com (S. S. Ali)

Abstract: The aim of the present study is to remove Phenol and Parachlorophenol from synthetic wastewater using prepared activated carbon from agricultural wastes (rice husk, (RH) date stones, (DS) and palm fronds, (PF)) utilizing chemical and physical activation methods. Three principal operating parameters have been considered which are activation temperature, activation time, and impregnation ratio. The results showed that the activated carbon prepared from rice husk by chemical method at a temperature of $700\,^{\circ}C$, activation time of 3 hours, and impregnation ratio of 3: 1 gives highest removal of phenol and Parachlorophenol. The results also showed that the activated carbon prepared by physical method from rice husk gives higher removal efficiency for phenol and Parachlorophenol than that of activated carbon prepared from date stones and palm fronds. The results also showed that the rate of removal of Phenol and Parachlorophenol increases with the increase of activation temperature, activation time, and impregnation ratio. The highest removal of phenol and Parachlorophenol are (84.38% and 94.65%) respectively. Application of the most prominent adsorption models shows acceptable agreement with Langmuir and Temkin models. Maximum adsorption capacity for the prepared activated carbon is found to be 39 mg/g and 38.82 mg/g for Phenol, 44.64 mg/g and 44.94 mg/g for Parachlorophenol respectively.

Keywords: Activated Carbon, Adsorption, Phenol, Parachlorophenol, Rice Husk

1. Introduction

Activated carbon is commonly defined as a carbonaceous material showing a well-developed surface area and porous texture. As a consequence, activated carbon has been widely used as adsorbent and in catalysis or separation processes (Kalderis et al., 2008; Tamai et al., 2009). The characteristics of activated carbon depend on the physical and chemical properties of the precursor as well as on the activation method (Demiral et al., 2008). Activated carbon can be produced from any carbonaceous solid precursor which may be either natural or synthetic. The choice of precursor is largely dependent on its availability, cost, and purity. Due to environmental considerations, agricultural wastes are considered to be a very important precursor because they are cheap, renewable, safe, and available at large quantities; in addition they have high carbon and low ash content (Kalderis et al., 2008).

Rice husk, date stones, and palms fronds are agricultural wastes, have been reported as a good adsorbents for many heavy metals and hydrocarbon compounds, lignocellulosic composition promotes the preparation of activated carbon from these precursors (Bouchelta et al., 2008).

There are two processes for the preparation of activated carbon: physical activation and chemical activation. Physical activation involves carbonization of a carbonaceous materials followed by activation of the resulting char in the presence of activating agents such as CO_2 or steam. In chemical activation, a raw material is impregnated with an activating reagent such as $ZnCl_2$, H_3PO_4, KOH, etc., and the impregnated material is heated in an inert atmosphere. There will be a reaction between the precursor and the activating agent leads to a development of porosity. Chemical activation is preferred over physical activation owing to the higher yield, simplicity, lower temperature, and shorter time needed for activation, and good development of the porous structure (Guo and Rockstraw, 2007).

Phenolic compounds are classified to be extremely toxic for human beings and for all aquatic life. One of the most hazardous polluting phenolic compounds to the environment is phenol, which can exert negative effects on different biological processes and their presence even at low concentrations can cause unpleasant taste and odor of drinking water and can be an obstacle to the use of wastewater (Dabrowski et. al., 2005). Phenol was designated as priority pollutants by the USEPA, which takes the 11th place in the list of 129 chemicals; Environmental Protection Agency (EPA) has set a limit of 0.1 mg/L of phenol in wastewater (Salame and Bandosz. 2003). Other important polluting phenolic compound is Parachlorophenol. It can enter the human body through all routes and reacts easily in the blood to convert hemoglobin to methamoglobin, thereby preventing oxygen uptake even at low concentration.

Removing these pollutants or decreasing their concentration levels in the wastewater to the allowable or permitted levels is a target that needs to be achieved and reached by several environmental agencies and governments. Several ways have been developed to remove phenolic compounds from wastewaters, including electrochemical oxidation, chemical coagulation, solvent extraction, membrane separation, photo catalytic degradation, and adsorption. Among these methods adsorption is still the most popular and widely used technique for phenols removal, because of its simple design, easy operation, it can remove both organic as well inorganic constituents even at very low concentration, no sludge formation, and the adsorbent can be regenerated and reused again (Nevskaia et. al., 2004). Moreover the process is economic because it requires low capital cost and there are abundant low cost materials available which can be used as adsorbents (Halouli and Drawish., 1995). It is a mass transfer process that involves contact of a solid (adsorbent) with a fluid contacting the target solute (adsorbate). The efficiency of the adsorption process is mainly determined by the characteristic of the adsorbent such as high surface area, high adsorption capacity, microporous structure, and special surface reactivity. To the present day, activated carbon is the most widely used adsorbent because it has good capacity for adsorption of hydrocarbon compounds and heavy metals from wastewater.

Mohammed, N. A., Aseel, A. H., and Firas S. A. (2013) used Iraqi rice husk (IRH) to remove phenol from wastewater under various operating conditions. Results show that the higher removal efficiency was 89.73% and this efficiency is decreased with the increase of initial concentration, flow rate, and pH while it is increased with the increase of inlet concentration, bed height, and feed temperature.

Khu Le Van' and Thu Thuy Luong Thi (2014) had prepared four activated carbon (AC) samples from rice husk under different activation temperatures. The specific surface area of AC sample reached 2681 $m^2 g^{-1}$ under activation temperature of 800 °C. The AC samples were then tested as an electrode

material.

Huaxing Xu et.al.(2014) prepared nanoporous activated carbon from the waste rice husks (RHs) by precarbonizing RHs and activating with KOH. The nanoporous carbon has the average pore size of 2.2 nm and high specific surface area of 2523.4 $m^2 g^{-1}$.

Mervette, El Batouti, Abdel-Moneim M. Ahmed(2014) conducted batch adsorption of Ni(II) onto activated carbon prepared from rice husk and stated that the adsorption process depends on the initial concentrations, adsorbent dose, contact time, and pH.

Mohammad, Y.S.(2014)examined the applicability of rice husk activated carbon in an adsorption column for the treatment of phenolic refinery wastewater under various bed depth and flow rate. They found that the adsorption capacity is 28mg/g. The performance of the adsorption column is affected by bed depth and flow rate.

Chitaranjan Dalaia, Ramakar Jhab, and Desaic a V.R. (2015) used rice husk based Activated Carbon (RHAC) to remove Iron and Manganese from groundwater. They stated that Iron and Manganese are 100% removed.

In spite of this; it suffers from a number of disadvantages. Activated carbon is expensive, and the higher the quality is the greater the cost. Consequently, there has been a growing interest in developing and implementing various potential adsorbents for the removal of hydrocarbon compounds and heavy metals from wastewater.

2. Experimental Work

Figure (1) represents a schematic diagram of the activated carbon preparation unit. It consists of the following parts:

1. Tubular reactor is made of stainless steel 310. Its height is 76cm and its internal diameter is 6.35cm, sample of raw material is placed in.
2. Electrical coil, to heat the reactor which contains the sample.
3. Nitrogen and carbon dioxide cylinders, to supply gas for the reactor during the preparation of activated carbon.
4. Flow meter, to measure the gas flow rate.
5. The reactor contains in the upper end an exhaust to release the produced gases during the process.
6. Temperature recorder (Digital Recorder), used to measure and control the temperature of reactor.

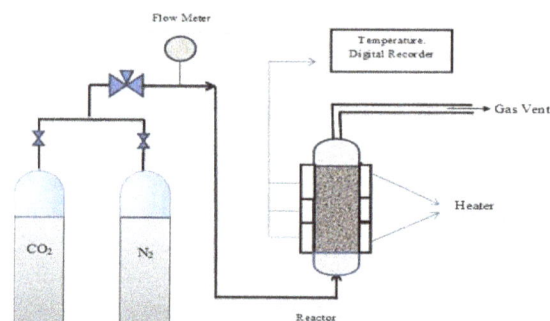

Fig. (1). Schematic diagram of the activated carbon preparation unit.

3. Material

Precursor: Rice husk, date stones, and palms fronds are used as the precursors in the preparation of activated carbon. These are first washed with water to get rid of impurities, dried at 110°C for 24 hour, and sieved. Only the fraction of particle sizes between 1 and 3 mm is selected for the preparation.

Activators: Zinc chloride (purchased from Didactic company) of 99.9% purity is used as a chemical reagents.

Adsorbate: Phenol (Ph) and Paraclorophenol (PCP) of purities higher than 99% are used as an adsorbate in this study.

Chemicals: Other chemical used are hydrochloric acid, sodium chloride, and acetone.

3.1. Preparation of Activated Carbon

3.1.1. Physical Activation Method
This method involves the following steps:
1. Raw material is washed with deionized water and subsequently dried at 105°C for 24 hour to remove moisture content.
2. The dried raw material is cut and sieved to a particle size of 1- 3mm.
3. Carbonization; raw material of step 2 is heated at a rate of 10°C/minute till the temperature reached 700°C under nitrogen of 99.9% purity flows at a rate of 150 ml/minute.
4. Activation; carbonized material is exposed to an oxidizing atmospheres, carbon dioxide at a temperature of 700°C for 3hours.
5. The activated product was then cooled to room temperature under nitrogen flow.
6. The samples are stored in a closed flask for adsorption applications.
7. Rice husk is washed with deionized water and subsequently dried at 105°C for 24 hour to remove moisture content.
8. The dried rice husk is cut and sieved to a particle size of 1- 3mm.

3.1.2. Chemical Activation Method
This method involves the following steps:
1. Chemical activation of the dried raw material was then done using activating agents $ZnCl_2$ using weight ratio of (1:1, 2:1, 3:1) and then deionized water is then added to dissolve ZnCl2 pellets.
2. After chemical activation, sample is dried at 110°C for 15 hour.
3. Carbonization; rice husk of step 4 is heated at a rate of 10°C/minute till the temperature reached 500°C, 600°C, and 700°C under nitrogen of 99.9% purity flows at a rate of 150 ml/minute.
4. The activation step is done using the same reactor as in carbonization step. Once the final activation temperature reached, the gas flow is switched from nitrogen to CO_2 at a flow rate of 150 ml/minute for 1, 2,

and 3hours.
5. The activated product is then cooled to room temperature under nitrogen flow.
6. The activated product is washed with deionized water and hydrochloric acid (0.1M) until the pH of the washed solution reached 7 and subsequently dried at 105°C.
7. The sample is stored in closed flask for adsorption applications.

3.2. Batch Experiments

A series of batch experiments are carried out to determine the adsorption isotherms of Ph and PCP solutions onto activated carbon (AC), 0.2gram, at pH of 7. Activated carbon is placed into 250 ml flask containing phenol or parachlorophenol of 100ppm concentration. The mixture is stirred; the pH values are controlled by adding 0.1 N NaOH or 0.1 N HCl. The adsorption experiments are conducted at room temperature for 280 minute to achieve equilibrium. Then, the solutions are filtrated, and their concentration are determined by UV-Visible Spectrophotometer (Shimadzu UV-530) at λ_{max} =269nm and λ_{max} =280nm for Ph and PCP respectively. The adsorption capacity q_e (mg/g) of each solute onto AC was calculated using Equation (1)

$$q_e = \frac{V(C_o - C_e)}{W} \qquad (1)$$

Where;
C_o and C_e are the initial and equilibrium concentration of adsorbate respectively (mg/L),
V is the volume of solution (L), and
W is the weight of AC (gram).

4. Results and Discussion

In the present work, samples of activated carbon are produced from different Iraqi waste materials (rice husk,(RH) date stones,(DS) and palm fronds(PF)) using two activation methods which are physical activation and chemical activation method. The prepared activated carbon samples are used to conduct many batch adsorption experiments for phenol and parachlorophenol of 100 mg/L concentration.

4.1. Results of Physical Activation Experiments

In physical activation method, 700°C is used as an activation temperature while the activation time is 3 hours. The results of batch adsorption experiments for phenol and parachlorophenol of 100mg/L concentration on raw materials (RH, DS, and PF) and on the prepared activated carbon samples by physical activation method (RH1, DS1, and PF1) are shown on Figures (2-5).

Figs.(2)&(3) indicated that phenol and parachlorophenol adsorption on raw materials (RH, DS, and PF) quickly reaches equilibrium (within 1 hour) with low removal for Ph and PCP. This is due to low surface area of the raw materials giving low adsorption sites. Adsorption process depends on many parameters including the type and nature of adsorbent and adsorbate (Al-Sultani and Al-Seroury, 2012). However, parachlorophenol removal is higher than phenol due to its higher affinity (Ahmaruzzaman and Sharma, 2005). Moreover, rice husk, gives higher phenol and parachlorophenol removal (8.64% & 17.95% respectively) as compared with that of PF (4.38% & 10.65) and DS (5.69% & 8.36%). This is due to the main components of rice husk are carbon and silica; it has the potential to be used as an adsorbent. Thus, the ionic structure of silica provides a capability of adsorbing phenolic compounds which are polar molecules (Kumar, 1987).

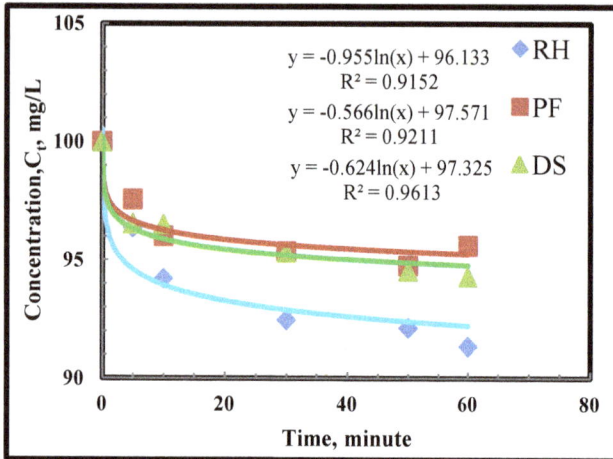

Fig. (2). *Phenol adsorption on raw materials (RH, DS, and PF).*

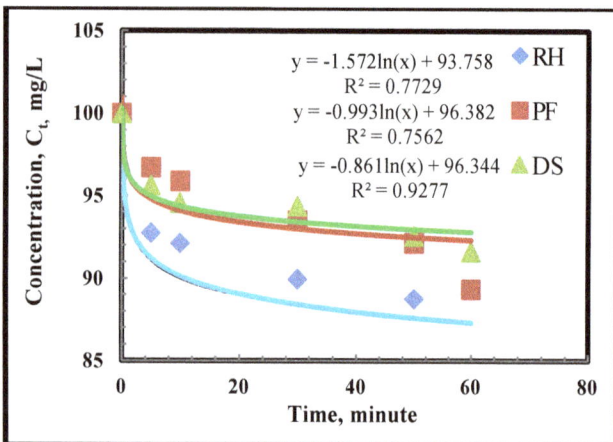

Fig. (3). *Parachlorophenol adsorption on raw materials (RH, DS, and PF).*

Figs.(4) & (5) represent the batch adsorption for phenol and parachlorophenol on the produced activated carbon samples (RH1, DS1, & PF1). These Figures indicated that (i) equilibrium time (about 4 hours) and phenol and parachlorophenol removal are higher than the corresponding of raw materials, (ii) the removal of parachlorophenol (64.42%, 51.78%, & 46.72% for RH1, PF1, &DS1 respectively) are higher than the corresponding values (58.4%, 40.21%, & 36.78) of phenol removal, and (iii) RH1 gives higher removal for both pollutants as compared with PF1 which is in turn gives higher removal as compared with DS1. These results are due to: physical activation enables

moisture loss as well as lignin decomposition of raw materials (Lapuerta et al., 2004). The treatment of raw material by physical activation also reduces the content of hemicellulose, lignin, and cellulose crystals which leads to an increase of the surface area compared to raw material (Daffalla et. al., 2010).

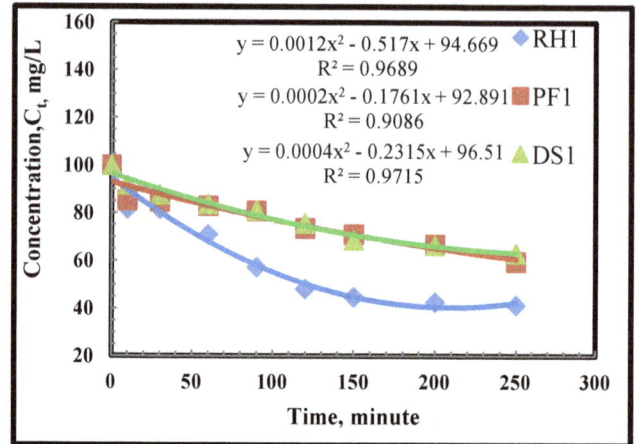

Fig. (4). *Phenol adsorption on RH1, PF1, DS1.*

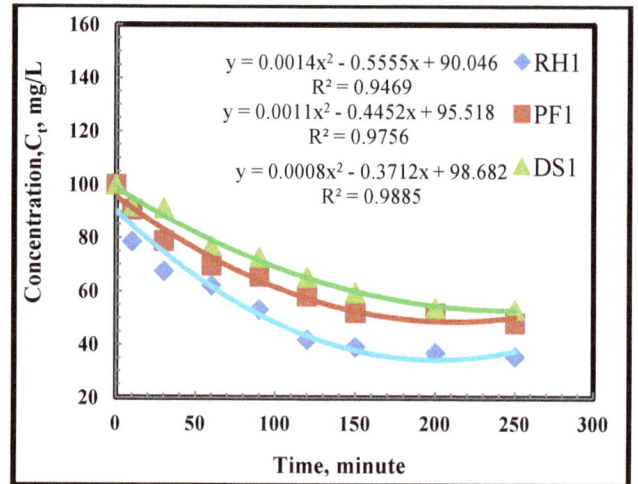

Fig. (5). *Parachlorophenol adsorption on RH1, PF1, and DS1.*

4.2. Results of Chemical Activation Experiments

Since rice husk gives the best adsorption performance for phenol and parachlorophenol when it is activated by physical activation method, it is used to perform chemical activation method under various operating conditions which are: activation temperature of 500°C, 600°C, and, 700°C activation time of 1, 2 and, 3 hour: impregnation ratio $ZnCl_2$/Rice husk of 1:1, 2:1, and 3:1. Table (1) represents the operating conditions and the name given for each sample of the prepared activated carbon.

Phenol and parachlorophenol batch adsorption experiment is conducted for each sample of the prepared activated carbon in order to test the effect of each operating parameter on the adsorption performance. The adsorption isotherms are shown graphically on Figs.(6-23).

Table (1). *Operation conditions for chemical activation.*

Run No.	Activation, Temperature,T,°C	Impregnation, Ratio, IR	Activationtime,AT, hour	Produced, Activated, Carbon name
1	500	3:1	1	RHC1
2	500	3:1	2	RHC2
3	500	3:1	3	RHC3
4	500	2:1	3	RHC4
5	500	1:1	3	RHC5
6	600	3:1	1	RHC6
7	600	3:1	2	RHC7
8	600	3:1	3	RHC8
9	600	2:1	3	RHC9
10	600	1:1	3	RHC10
11	700	3:1	1	RHC11
12	700	3:1	2	RHC12
13	700	3:1	3	RHC13
14	700	2:1	3	RHC14
15	700	1:1	3	RHC15

4.3. Effect of Activation Temperature

Figures (6-11) represent adsorption data for phenol and parachlorophenol on various activated carbon samples under different operating conditions.

It is clear that the removal of phenol and parachlorophenol is increased with the increase of activation temperature with the highest removal for phenol (84.38%) and for parachlorophenol (94.06%) obtained for RH13 which is activated under 700 °C, 3:1 IR, and 3 hours activation time.

This is due to the formation of well-developed mesopores structures on the adsorbent during activation. The activation temperature is very essential parameter for pore structure of activated carbon, which determines the adsorption capacity (Hu et al., 1995). The volatiles from the samples continue to evolve with increasing carbonization temperature. The devolatilization process further develops the pore structure and creates new porosities and thus, increasing surface area (Kalderis et al., 2008). Increases in phenol and parachlorophenol removal with activation temperature indicate the increase in suitability of pore size to accommodate the phenol and parachlorophenol. Similar result was reported by (Theydan and Ahmed, 2012). According to (Guo and Lua, 2001), increasing temperature within the range of 500-900°C will increase the released volatiles. The decrease of yield is maximum between 200°C and 800°C due to rapid carbonization in this region. It is also unsuitable to prepare activated carbon when the carbonization temperature is higher than 800°C since the successive decrease in volatile matter is minimum above this range. This is usually accompanied with an increase of fixed carbon and ash content which may be attributed to the removal of volatile matter during carbonization process. It is also noticed that the adsorbed parachlorophenol is higher than phenol due to its lower solubility in aqueous solution and because of it is more non-polar, lesser affinity (Cooney,1999). In addition, the difference in adsorption behavior of phenol and parachlorophenol might be due to the different affinities of the two phenolic species for the reactive

functional groups in the ARH. Similar result was reported by (Streat et al., 1994), (Jung et al., 2001), (Termoul et al., 2006) and (Al-Roubaiaay,2011), where the adsorption capacity for parachlorophenol is greater than that for phenol.

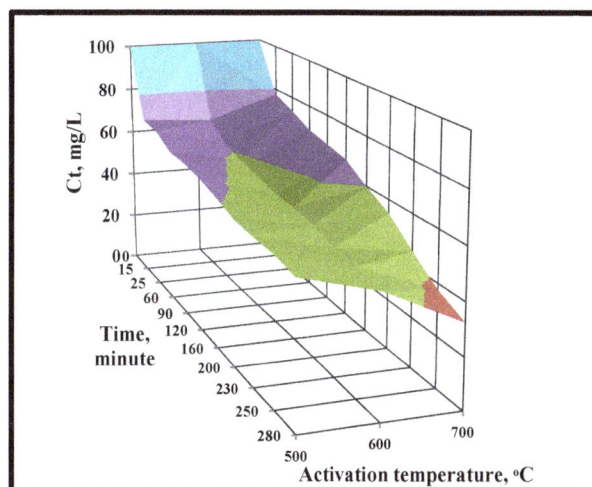

Fig. (6). *Phenol concentration versus time for various activation temperature(T), IR= 3:1, and AT=1 hour.*

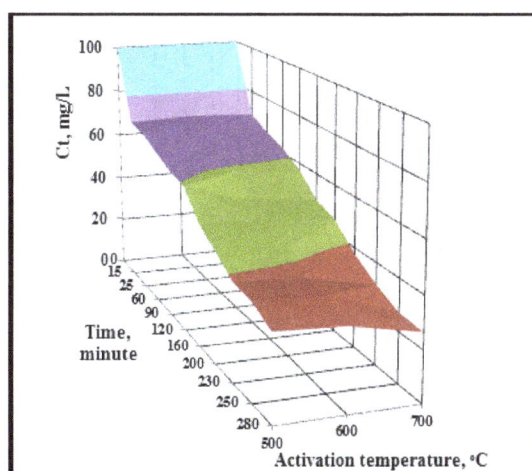

Fig. (7). *Parachlorophenol concentration versus time for various activation temperature(T), IR= 3:1, and AT=1 hour.*

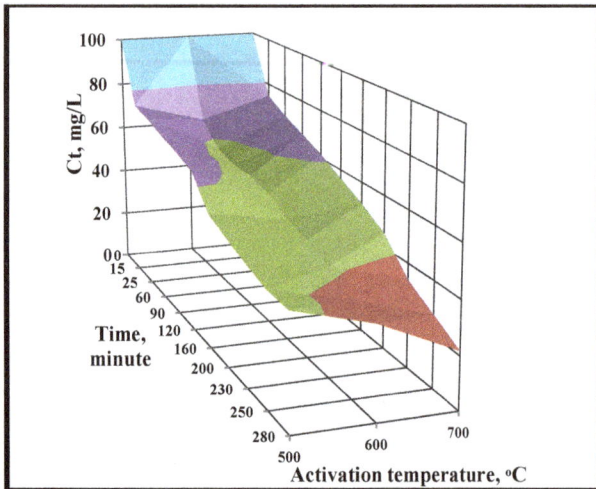

Fig. (8). *Phenol concentration versus time for various activation temperature(T), IR= 3:1, and AT=2 hour.*

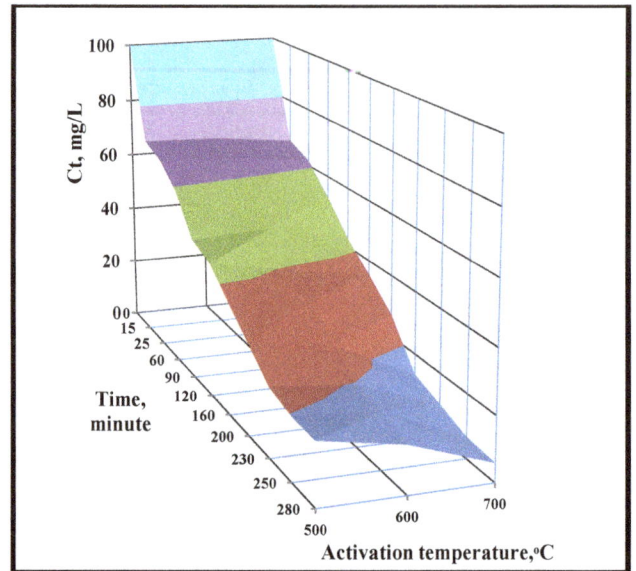

Fig. (9). *Parachlorophenol concentration versus time for various activation temperature(T), IR= 3:1, and AT=2 hour.*

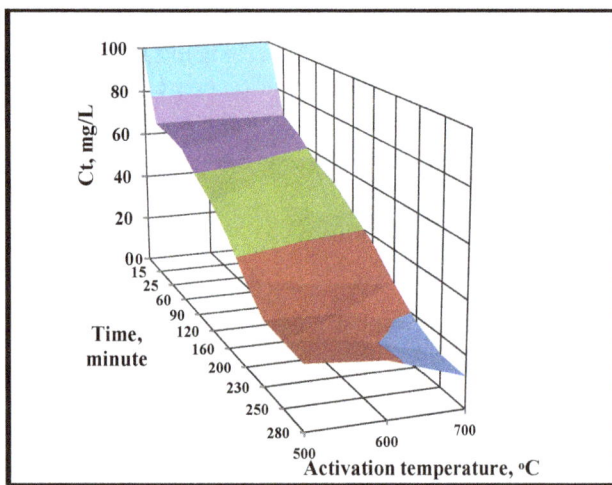

Fig. (10). *Phenol concentration versus time for various activation temperature(T), IR= 3:1, and AT=3 hour.*

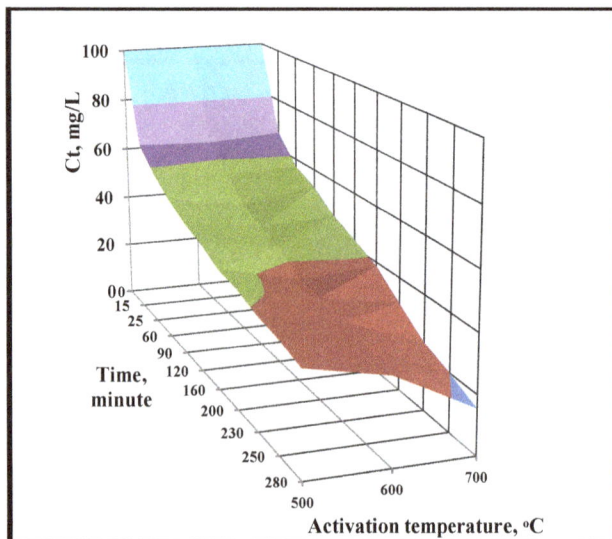

Fig. (11). *Parachlorophenol concentration versus time for various activation temperature(T), IR= 3:1, and AT=3 hour.*

4.4. Effect of Activation Time

The activation time has a significant effect on the porous network development in the activated carbon. Therefore, in order to obtain activated carbon having desired properties, precursor must be treated at a suitable carbonization time. Fig.s (12-17) represent adsorption data for phenol and parachlorophenol on various activated carbon samples under different activation time, 1, 2, & 3 hour. It is clear that the removal of phenol and parachlorophenol is increased with the increase of activation time. The highest removal for phenol (84.38%) and parachlorophenol (94.06%) obtained for RH13 which is activated under 700°C, 3:1 IR, and 3 hours activation time. This is because increasing activation time will give more chance for hot flowing gases to do its work in increasing the pores and surface area (Hameed et al., 2009). i.e. higher time will increase the amount of volatile matter during carbonization process and higher removal giving more pores and higher surface area. Also at high activation times, the change (quality and quantity) in functional groups may subsequently affect the adsorption of phenol and parachlorophenol. Generally at high activation times the increase in the aromatic content of AC functional groups has been well recorded (Baçaoui, Yaacoubi et al. 2001; Wang, S. et al. 2005). These results agree with the result published by (Chatterjee and Kumar, 2012; Yang and Lua, 2003).

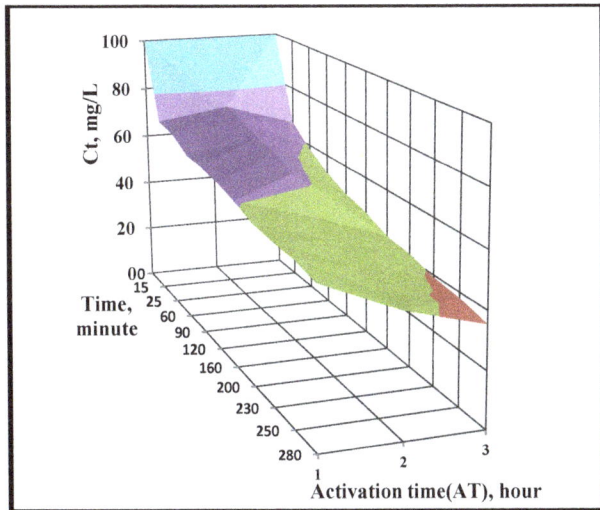

Fig. (12). *Phenol concentration versus time for various activation time(AT), IR= 3:1, and T=500 ᵒC.*

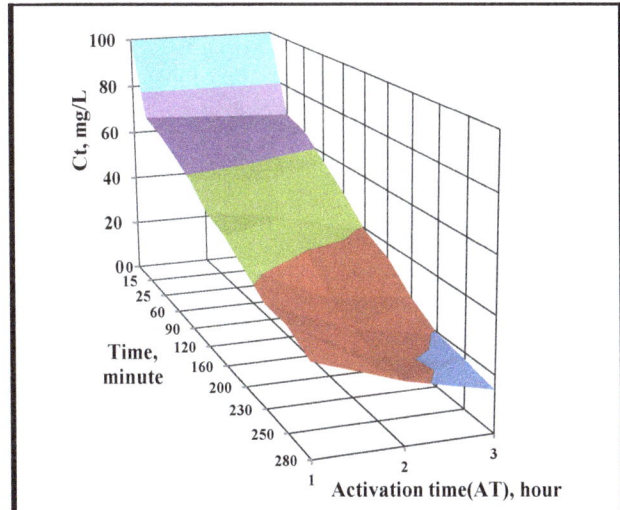

Fig. (15). *Parachlorophenol concentration versus time for various activation time(AT), IR= 3:1, and T=600 ᵒC.*

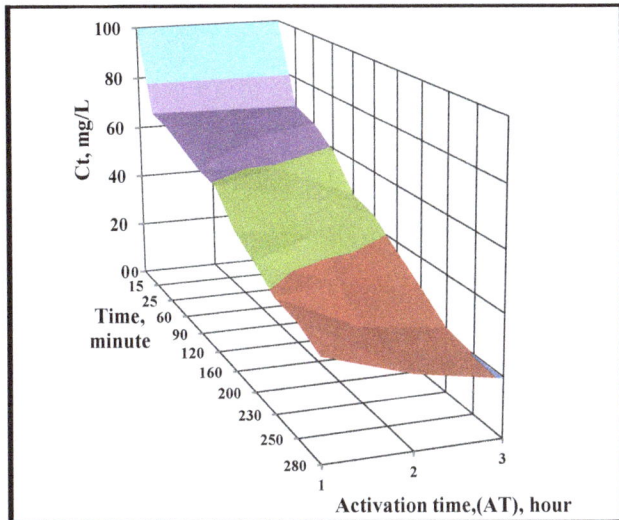

Fig. (13). *Parachlorophenol concentration versus time for various activation time(AT), IR= 3:1, and T=500 ᵒC.*

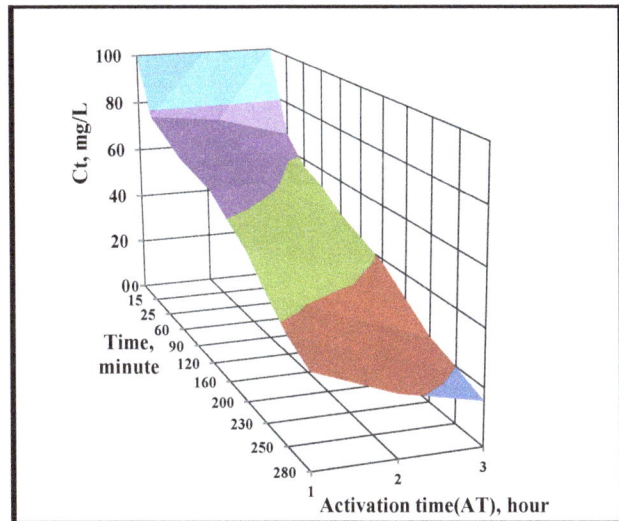

Fig. (16). *Phenol concentration versus time for various activation time(AT), IR= 3:1, and T=700 ᵒC.*

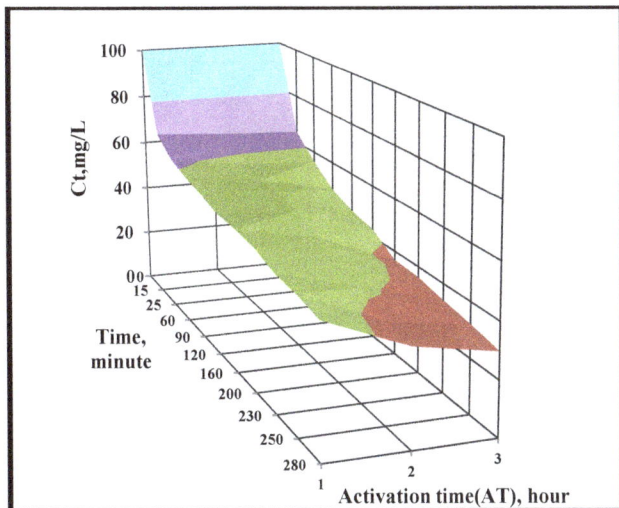

Fig. (14). *Phenol concentration versus time for various activation time(AT), IR= 3:1, and T=600 ᵒC.*

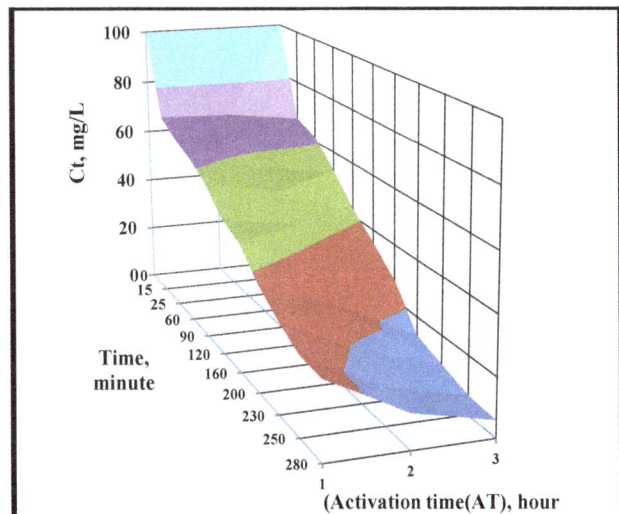

Fig. (17). *Parachlorophenol concentration versus time for various activation time(AT), IR= 3:1, and T=700 ᵒC.*

4.5. Effect of Impregnation Ratio,IR, of ZnCl₂/Rice Husk

Figures (18-23) represent adsorption data for phenol and parachlorophenol on various activated carbon samples under different Impregnation ratio (IR) 1:1, 2:1, and 3:1. The highest removal for phenol (84.38%) and parachlorophenol (94.06%) obtained for RH13 which is activated under 700°C, 3:1 IR, and 3 hours activation time. It is clear that the removal of phenol and parachlorophenol is increased with the increase of IR. This is because rice husk which is treated with ZnCl₂ led to a removal of the cellulosic component of the adsorbent. It is well known that activation with zinc chloride prevents the accumulation of tar on the carbon surface and provides further decompositions and thus, develops the microporosity when using cellulosic and lignocellulosic precursors in the manufacture of activated carbon. More activator perhaps leads to excessive dehydration and destruction of mesopores turning them into larger pores which reduces the adsorption efficiency (Kim et al., 2001). Similar results are reported by (Al- Roubaiaay2011 and Kalderis.2008).

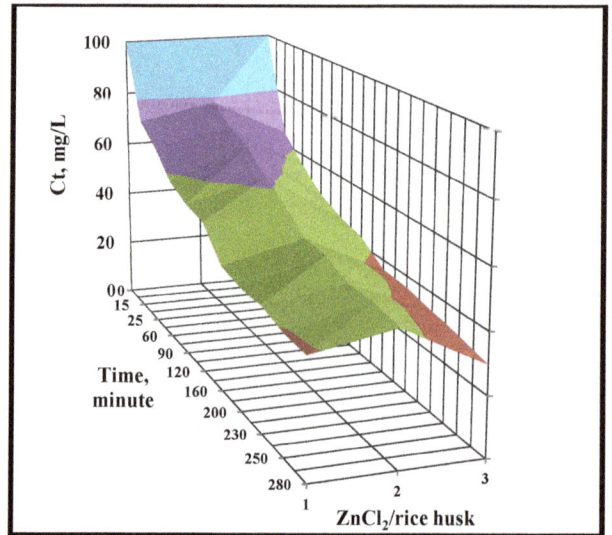

Fig. (20). Phenol concentration versus time for various Impregnation ratio, IR, (AT)=3hour, T=600 °C.

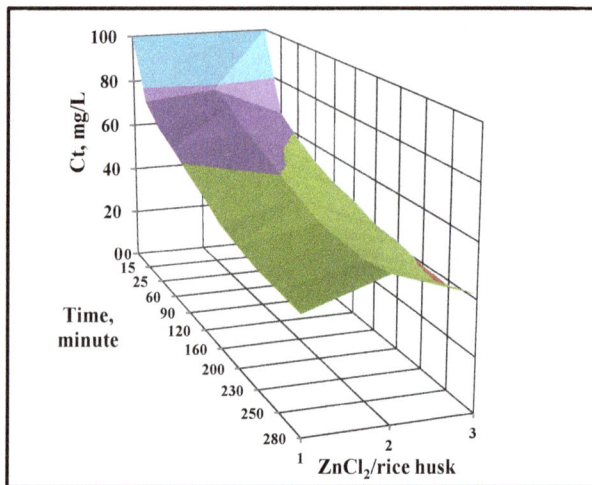

Fig. (18). Phenol concentration versus time for various Impregnation ratio, IR, (AT)=3hour, T=500 °C.

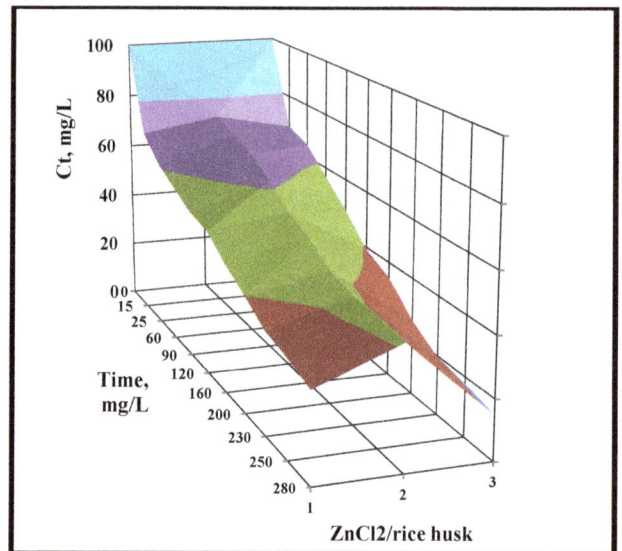

Fig. (21). Parachlorophenol concentration versus time for various Impregnation ratio, IR, (AT)=3hour, T=600 °C.

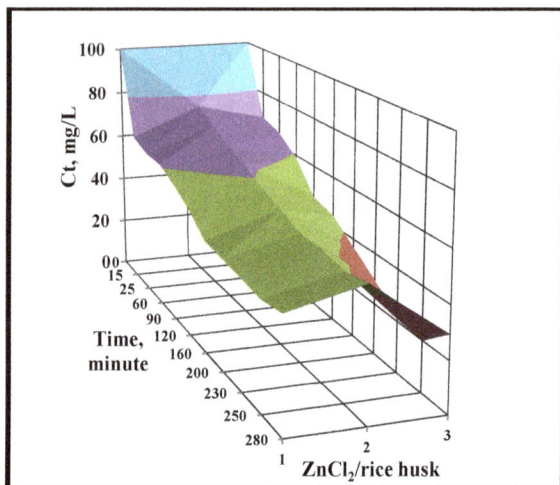

Fig. (19). Parachlorophenol concentration versus time for various Impregnation ratio, IR, (AT)=3hour, T=500 °C.

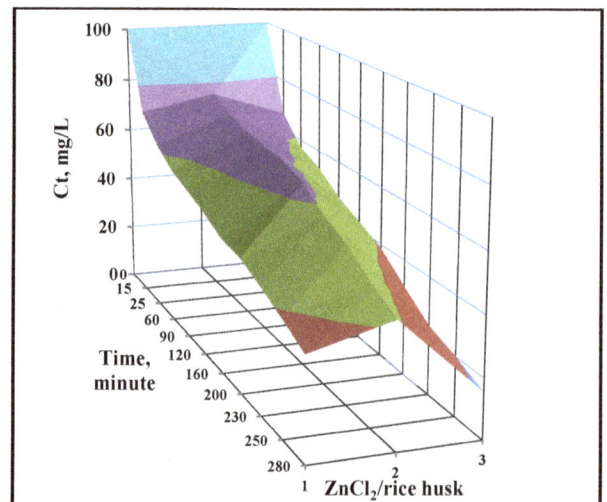

Fig. (22). Phenol concentration versus time for various Impregnation ratio, IR, (AT)=3hour, T=700 °C.

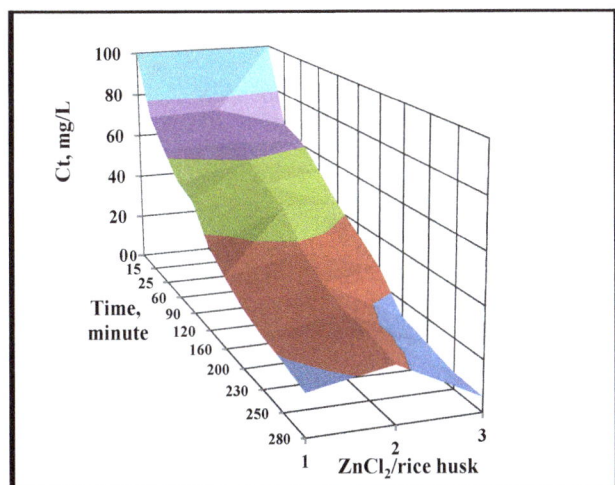

Fig. (23). Paraclorophenol concentration versus time for various Impregnation ratio, IR, (AT)=3hour, T=700 °C.

4.6. Applicability of Adsorption Isotherms

Adsorption data are analyzed using four different models: Freundlich, Langmuir, Temkin, and BET. The ranges of the coefficients of determinations (R^2) for these models are listed in Table (2). From this table, it seems that Temkin model gives the highest range of R^2 for phenol and parachlorophenol. This means that this model fairly fits experimental data. Langmuir and Freundlich models have close values of R^2 ranges which are lower than that of Temkin model while BET model gives the lowest R^2 ranges.

The high fitness of the Langmuir model for the adsorption process of activated rice husk indicates the monolayer concealment of phenol and parachlorophenol on the outer surface of activated rice husk, in which the adsorption occurs uniformly on the active part of the surface (Abdel Wahab et al., 2005). The adsorption capacity for Phenol and Parachlorophenol using Langmuir model are 39 and 44.64 mg/g respectively. The corresponding values for Temkin model are 38.82 and 44.94 mg/g.

Table (2). Ranges of R^2 of the used adsorption models.

Model	R^2 range for:	
	phenol	parachlorophenol
Freundlich	0.9090-0.9817	0.9074-0.9899
Langmuir	0.9032-0.9876	0.9035-0.9887
Temkin	0.9624-0.9967	0.9269-0.9977
BET	0.7642-0.9638	0.7461-0.9101

5. Conclusions

Chemical activation method is found to be superior than physical activation method for all raw materials used (RH, PF and DS) and the best raw material is found to be RH which gives maximum removal of 84.38% and 94.06% for phenol and parachlorophenol respectively under activation temperature of 700°C, activation time 3 hours, and impregnation ratio 3:1.

The removal of Phenol and Parachlorophenol is increased with the increase of activation temperature, activation time, and impregnation ratio and the experimental data is found to agree Temkin model better than other adsorption models. Maximum adsorption capacity for RH calculated by Langmuir model is 39 and 44. 64 mg/g for Phenol and Parachlorophenol respectively.

Nomenclature

C_t Concentration of solute in solution at any time, (mg/l)
C_0 Initial concentration of adsorbate (mg/l)
C_e Concentration of solute in solution at equilibrium,(mg/l)
q_e Adsorption capacity, (mg/g)
R^2 Correlation coefficient
t Time, (minute)
V Volume of solution, (L)
W Weight of adsorbent, (gm)

References

[1] Abdel Wahab, O., Nemr, A. E., Sikhaily, A.E., and Khaled, A., "Use of Rice Husk for Adsorption of Direct Dyes from Aqueous Solution: Direct F. Scarlet", Egyptian Journal of Aquatic Research, Vol. 31, No.1, pp. (1110 - 0354), 2005.

[2] Ahmaruzzaman, M., Sharma, D.K., "Adsorption of phenols from Wastewater", Journal of Colloid and Interface Science, Vol. 287, PP. (14-24), 2005.

[3] Al-Roubaiaay, N. N., "Using Dual Media Filters to Remove Phenol and Parachlorophenol Compounds by Commercial and Manufactured Date Stones Based Activated Carbon", M.Sc. Thesis, College of Engineering, Tikrit University, 2011.

[4] Al-Sultani, K. F. and Al-Seroury, F.A., "Characterization the Removal of Phenol from Aqueous Solution in Fluidized Bed Column by Rice Husk Adsorbent", Research Journal of Recent Sciences, Vol. 1, PP. (145-151), 2012.

[5] Baçaoui, A., Yaacoubi, A., "Optimization of conditions for the preparation of activated carbons from olive-waste cakes." Carbon 39(3): PP. (425-432), 2001.

[6] Bouchelta, C., Medjran, M. S., Bertrand, O. and Bellat, J., "Preparation and characterization of activated carbon from date stones by physical activation with steam", J. Anal. Appl. Pyrolysis, 28, PP. (70-77), 2008.

[7] Chatterjee, S., Kumar, A., "Application of Response Surface Methodology for Methylene Blue Dye Removal from Aqueous Solution Using Low Cost Adsorbent", Chemical Engineering Journal, PP. (289-299), 2012.

[8] Chitaranjan Dalaia, Ramakar Jhab , and Desaica V.R.,"Rice Husk and Sugarcane Baggase Based Activated Carbon for Iron and Manganese Removal", International Conference on Water Resources, Costal and Ocean Engineering (ICWRCOE 2015)

[9] Cooney, David, O., "Adsorption design for wastewater Treatment", Lewis Publisher, Washington, D.C. USA, 1999.

[10] Dabrowski, A., Podkoscielny, P., Hubicik, Z. and Barczak, M., "Adsorption of Phenolic Compounds by Activated Carbon - A Critical Review", Chemosphere, 58, PP. (1049- 1070), 2005.

[11] Daffalla, S. B., Mukhtar, H. and Shaharun, M. S., "Characterization of Adsorbent Developed from Rice Husk, Effect of Surface Functional Group on Phenol Adsorption", Journal of Applied Sciences, 2010.

[12] Demiral, H., Demiral, I., Tumsek, F., and Karabacakoglu, B., "Pore Strucuture of Activayted Carbon Prepared from Hazelnut Bagasse by Chemical Activation", Surface Interface Analysis, 40, PP. (616-619), 2008.

[13] Guo, Y. and Rockstraw, D. A., "Physicochemical Properties of Carbons Prepared From Pecan Shell by Phosphoric Acid Activation", Bioresour. Technol., 98, PP. (1513-1521), 2007.

[14] Guo, J. and Lua, C.A., "Kinetic Study on Pyrolytic Process of Oil-Palm Solid Waste Using Two-Step Consecutive Reaction Model", Biomass and Bioenergy, 20, PP. (223–233), 2001.

[15] Halouli, K. A. and Drawish, N. M., "Effect of Ph and Inorganic Salts on the Adsorption Of Phenol From Aqueous Systems on Activated Decolorizing Charcoal", Sep. Sci. Technol., 30, PP. (3313-3324), 1995.

[16] Hameed, B. H. and Salman, J. M., "Adsorption Isotherm and Kinetic Modeling of 2, 4-D Pesticide on Activated Carbon Derived from Date Stones", Journal of Hazardous Materials, 163(1), PP. (121-126), 2009.

[17] Hu, Z. and Vansant, E. F., "A New Composite Adsorbent Produced by Chemical Activation of Elutrilithe with Zinc Chloride", J. Colloid Interface Sci. 176(2), PP. (422–431), 1995.

[18] Huaxing Xu, Biao Gao, Hao Cao, Xueyang Chen, Ling Yu, Kai Wu, Lan Sun, Xiang Peng, and Jijiang Fu," Nanoporous Activated Carbon Derived from Rice Husk for High Performance Supercapacitor", Journal of Nanomaterials, Vol. 2014 (2014).

[19] Jung, M. W., Ahn, K. H., Lee, Y., Kim, K. P., Rhee, J. S., Park, J. T. and Paeng, K. J., "Adsorption Characteristics of Phenol and Chlorophenol on Granular Activated Carbons (GAC) ,"Microchemical Journal, 70, PP. (123-131), 2001.

[20] Kalderis, D., Koutoulakis, D., Paraskeva, P., Diamadopoulos, E., Otal, E., Olivares, Valle, O. J. and Fernandez-Pereira, C., " Adsorption of Polluting Substances on Activated Carbons Prepared from Rice Husk and Sugarcane Bagasse", Chemical Engineering Journal, 144, PP. (42–50), 2008.

[21] Kim, J. W., John, M. H., Kim, D. S. and Kwon, Y. S., "Production of Granular Activated Carbon from Waste Walnut Shell and its Adsorption Characteristic for Cu+2 Ion", J. Hazard. Mater, 85, PP. (301-315), 2001.

[22] Khu Le Van, ,Thu Thuy Luong Thi, "Activated carbon derived from rice husk by NaOH activation and its application in supercapacitor", Progress in Natural Science: Material International, Vol. 24, Issue 3, June 2014.

[23] Kumar, S. Upadhyay, S. N., Upadhya, Y. D., "Removal of Phenols by Adsorption on Fly Ash", J. Chem. Technol. Biotechnol, 37, PP. (281 287), 1987.

[24] Lapuerta, M., Hernández, J. J. and Rodríguez, J., "Kinetics of Devolatilization of Forestry Wastes from Thermo gravimetric Analysis", Biomass Bioenergy, Vol. 27, No. 4, PP. (385 – 391), 2004.

[25] Mervette. El Batouti, Abdel-Moneim M. Ahmed," Adsorption Kinetics Of Nickel (II) Onto Activated Carbon Prepared From Natural Adsorbent Rice Husk", International Journal of Technology Enhancements and Emerging Engineering research, Vol. 2, ISSUE 5 145.(2014)

[26] Mohammed N. A., Aseel A. H., and Firas S. A.,(2013)" Phenol Removal from Wastewater Using rice Husk", Diyala journal for pure sciences, Vol. 9 No.4, October 2013.

[27] Mohammad, Y.S. , Shaibu-Imodagbe, E.M. , Igboro, S.B. , Giwa, A., and Okuofu, C.A. ," Adsorption of Phenol from Refinery Wastewater Using Rice Husk Activated Carbon", Iranica Journal of Energy & Environment, 5 (4).(2014)

[28] Nevskaia, D.M., Castillejos-Lopez, E., Munoz, V. and Guerrero-Ruiz, A., "Adsorption of Aromatic Compounds from Water by Treated Carbon Materials", Environ. Sci. Technol., 38(21), PP. (5786-5796), 2004.

[29] Salame, I. I. and Bandosz, T. J., "Role of Surface Chemistry in Adsorption of Phenol on Activated Carbons", Journal of Colloid and Interface Science, 264, PP. (307–312), 2003.

[30] Streat, M., Patrick, J. W. and Camporo-Perez, M. J., "Sorption of Phenol and Parachlorophenol from Water Using Conventional and Novel Activated", Water Research, Vol.29, No. 2, PP. (467-472), 1994.

[31] Tamai, H., Nobuaki, U., and Yasuda, H., (2009), "Preparation of Pd supported mesoporous activated carbons and their catalytic activity", Mater. Chem. Physi. , 114, PP. (10-13), 2009.

[32] Termoul, M., Bestani, B., Benderdouche, N., Belhakem, M. and Naffrechoux, E., "Removal of Phenol and 4-Chlorophenol by Chemically Activated Olive Stones", ADSORPTION Science Technology, 24, PP. (375), 2006.

[33] Theydan, K. S. and Ahmed, J. M., "Optimization of Preparation Conditions For Activated Carbons from Date Stones Using Response Surface Methodology", Powder Technology, 224, PP. (101–108), 2012.

[34] Wang, S., Z. H. Zhu, et al. (2005). "The physical and surface chemical characteristics of activated carbons and the adsorption of methylene blue from wastewater." Journal of Colloid and Interface Science 284(2): 440-446.

[35] Yang, T. and Lua, A.C., "Characteristics of Activated Carbons Prepared from Pistachio-nut Shells by Physical Activation", Journal of Colloid and Interface Science, 267(2), PP (408-417), 2003.

Permissions

List of Contributors

Adenike Boyo and Olasunkanmi Kesinro
Department of physics, Lagos State University, Ojo, Lagos state, Nigeria.

Henry Boyo
Department of Physics, University of Lagos, Lagos State, Nigeria.

Jean Baptiste Nduwayezu, Theoneste Ishimwe, Ananie Niyibizi and Alexis Munyentwali
Institute of Scientific and Technological Research (IRST), P.O. Box 227 Butare, Rwanda.

P. A. Nwofe and P. E. Agbo
Division of Materials Science and Renewable Energy, Department of Industrial Physics, Ebonyi State University, Abakaliki, Nigeria.

Joseph Martin Petersen
Washington State University, School of Electrical Engineering and Computer Science, Richland, Washington, USA
Pacific Northwest National Laboratory Energy Policy & Economics, Richland, Washington. USA

Ana Godson R. E. E. and Udofia Bassey G.
Department of Environmental Health Sciences, Faculty of Public Health, College of Medicine, University of Ibadan, Ibadan.

Adenike Boyo and Olasunkanmi Kesinro
Department of physics, Lagos State University, Ojo, Lagos state, Nigeria.

Henry Boyo
Department of Physics, University of Lagos, Lagos State, Nigeria.

Lei Wen
Department of Economics & Management, North China Electric Power University, Baoding, Hebei, China
The Academy of Baoding Low-Carbon Development, Baoding, Hebei, China.

Ye Cao
Department of Economics & Management, North China Electric Power University, Baoding, Hebei, China.

Abdulhameed Danjuma Mambo
Department of Building, Federal University of Technology, PMB 65, Minna, Nigeria.

Mahroo Eftekhari
School of Civil and Building Engineering, Loughborough University, LE11 3TU, UK.

Thomas Steffen
Department of Aeronautic and Automotive Engineering, Loughborough University, LE11 3TU, UK.

Ramadan Abdiwe and Markus Haider
Institute for Energy Systems and Thermodynamics, Vienna University of Technology, Wien, Austria.

Ana Godson R. E. E. and Sokan Adeaga Adewale Allen
Department of Environmental Health Sciences, Faculty of Public Health, College of Medicine, University of Ibadan, Ibadan, Nigeria.

Patrick Boampong-Ohemeng
Graduate School, Ghana Technology University College/Coventry University, Kumasi, Ghana.

Simonov Kusi-Sarpong
School of Management Science and Engineering, Dalian University of Technology, Dalian, PR of China.

Adam Sandow Saani
Procurement Department, Tamale College of Education, Tamale, Ghana.

Martin Agyemang
School of Business Management, Dalian University of Technology, Dalian, PR of China.

Matarazzo-Neuberger Waverli Maia, Alves Luiz Roberto and Bernardes Marco Aurélio
Sustainability Center at the Methodist University of Sao Paulo, Methodist University, Sao Bernardo do Campo, Brazil.

Bala Gambo Jahun
Department of Agricultural and Bioresource Engineering Abubakar Tafawa Balewa University, Bauchi, Nigeria.

Fati Adamu Astapawa
Department of Agricultural and Environmental Engineering, Modibbo Adama University of Technology, Yola, Nigeria.

Balogun Shuaibu Alani
Department of Mechanical Engineering, Federal Polytechnic, Bauchi, Nigeria.

Nwogu Chukwunonso
Department of Mechanical Engineering, Michael Okpara University of Agriculture Umudike, Umuahia, Nigeria.

Marwan Mosad Ghanem, Omar Mohamed Al Wassal, Abdelrahman Ahmed Kotb and Mohamed Ayman El-Shahhat
STEM Egypt High School for Boys, 6th of October City, Egypt.

Kevin R. Anderson, Maryam Shafahi and Arthur Artounian
Mechanical Engineering Department, Solar Thermal Alternative Renewable Energy Lab, College Engineering, California State Polytechnic University, Pomona, CA, USA.

Adam Chrisman
SUNEARTH Inc., Fontana, CA, USA.

Sameer Saadoon Al-Juboori and Ali Hlal Mutlag
Electronic and Control Engineering Dept., Kirkuk Technical College, Kirkuk, Iraq.

Ehsan Fadhil Abbas Al-Showany
Refrigerating and Conditioning Engineering Dept., Kirkuk Technical College, Kirkuk, Iraq.

Shuvankar Podder
Department of Electrical and Electronic Engineering, Bangladesh University of Engineering and Technology, Dhaka, Bangladesh.

Md. Minarul Islam
Department of Electrical and Electronic Engineering,Shahjalal University of Science and Technology, Sylhet, Bangladesh.

Lipeng Zhang
Civil Engineering Department, Technical University of Denmark, Anker Engelunds Vej Building 118, Kgs. Lyngby, Denmark
Danfoss A/S, District Energy Division, Application Center, Nordborgvej 81, Nordbrg, Denmark.

Oddgeir Gudmundsson
Danfoss A/S, District Energy Division, Application Center, Nordborgvej 81, Nordbrg, Denmark.

Hongwei Li and Svend Svendsen
Civil Engineering Department, Technical University of Denmark, Anker Engelunds Vej Building 118, Kgs. Lyngby, Denmark.

Usman Yahaya
Forestry Research Institute of Nigeria, Trial Afforestation Research Station, Kaduna, Nigeria.

Umar Yahaya Abdullahi and Denwe Samuel Dangmwan
Department of Biological Sciences, Nigerian Defence Academy, Kaduna, Nigeria.

Muhammad Muktar Namadi
Department of Chemistry, Nigerian Defence Academy, Kaduna, Nigeria.

Song Jae Lee
Electronics Engineering Department, Chungnam National University, Daejeon, Korea.

Asha Jyothi U. H. and Mamatha B.
Department of Resource Management, Smt. V.H.D. Central Institute of Home Science, Bangalore University, Bangalore, India.

H. S. Surendra
Department of Statistics, University of Agricultural Sciences, GKVK, Bangalore, India.

Protik Kumar Das and Mir Ahasan Habib
Dept of Mechanical Engineering, Rajshahi University of Engineering & Technology (RUET), Rajshahi, Bangladesh.

Mohammed Mynuddin
Dept of EEE, Atish Dipankar University of Science and Technology, Dhaka, Bangladesh.

Alaa Eldien M. M. Hassan, Mahmoud A. Sayed and Essam E. M. Mohamed
Department Electrical Engineering, South Valley University, Qena, Egypt.

Abdel-Raheem Youssef andMahmoud A. Sayed
Dept. of Electrical Engineering, Faculty of Engineering, South Valley University, Qena, Egypt.

M. N. Abdel-Wahab
Dept. of Electrical Engineering, Faculty of Engineering, Suez Canal University, Ismailia, Egypt.

Gaber Shabib Salman
Dept. of Electrical Engineering, Faculty of Engineering, Aswan University, Aswan, Egypt.

Sergii Bespalko, Anton Kachymov, Kostiantyn Koberidze and Oleksandr Bespalko
Department of Energy Technologies, Cherkasy State Technological University, Cherkasy, Ukraine.

Muzher Mahdi Ibrahem AL-Doury
Petroleum and Minerals Engineering College, Tikrit University, Tikrit, Iraq.

Suha Sameen Ali
Chemical Engineering Dept, College of Engineering, Tikrit University, Tikrit, Iraq.

Index